SUBJECTS OF THE SUN

ELEMENTS   *A series edited
by Stacy Alaimo and Nicole Starosielski*

SUBJECTS OF THE SUN

SOLAR ENERGY IN THE SHADOWS
OF RACIAL CAPITALISM

MYLES LENNON

DUKE UNIVERSITY PRESS   Durham and London   2025

Project Editor: Michael Trudeau
Typeset in Chaparral Pro and Knockout by BW&A Books, Inc.

Library of Congress Cataloging-in-Publication Data
Names: Lennon, Myles, 1984– author.
Title: Subjects of the sun : solar energy in the shadows of
racial capitalism / Myles Lennon.
Other titles: Elements (Duke University Press)
Description: Durham : Duke University Press, 2025. | Series: Elements
Includes bibliographical references and index.
Identifiers: LCCN 2024044746 (print)
LCCN 2024044747 (ebook)
ISBN 9781478031789 (paperback)
ISBN 9781478028567 (hardcover)
ISBN 9781478060765 (ebook)
Subjects: LCSH: Solar energy—Social aspects—United States. |
Clean energy industries—United States. | Environmental justice—
United States. | Environmentalism in mass media. | Energy policy—
Social aspects—United States.
Classification: LCC TJ809.95 .L46 2025 (print) | LCC TJ809.95 (ebook)
DDC 304.20973—dc23/eng/20250224
LC record available at https://lccn.loc.gov/2024044746
LC ebook record available at https://lccn.loc.gov/2024044747

Cover art: Courtesy Shutterstock/ck Photo.

This book is dedicated to my family—Movah, Chaz, and Luke (*sleeypang!*)—and the ancestral homelands of the Lenape people (known to many of us as New York City).

CONTENTS

At its core, this is a book about different people's responses to climate change. But it is focused less on planet Earth and more on Google Earth, less on power plants and more on PowerPoints, less on melting ice sheets and more on making spreadsheets. This emphasis on everyday "screenwork" might seem misaligned with the grave matter of environmental collapse.[1] But the greatest existential threat to life on earth is quite often experienced through life in the cloud. While we don't need digital mediation to grasp the ever-present consequences of a warming world—especially in communities of color—online visuals and screenwork enable a wide range of people to understand and engage with *the empirical fact* of anthropogenic carbon emissions as well as the massive energy infrastructure that drives those emissions. Indeed, many of us in the Global North encounter the ubiquitous electricity generation system that powers our lives not through direct experience but through datafication.

Yet the civic discourse on energy transitions often focuses on policies, financial instruments, and political will, overlooking the everyday ways in which activists, technocrats, and laypeople actually go about transforming energy infrastructure through virtual and visual media. This elision matters because in a "society of the spectacle," such media often distort, obfuscate, and maintain the extractive order of racial capitalism, leaving us more alienated from the very world that clean energy is supposed to heal.[2] *Subjects of the Sun* thus looks critically at the quotidian visual interfaces of community organizers and white-collar technologists, Black anticapitalists and white neoliberals, environmental justice nonprofits and cleantech corporations—an array of differently positioned actors united by their commitment to clean energy futures. In doing so, this book illuminates the underacknowledged disjuncture between utopic visions for solar power and the banal digital routines through which so many of us try to make those visions real.

Admittedly, I feel some guilt writing a book about renewable energy

—a vaunted climate solution—that focuses critically (some might say obsessively) on minute matters like the aesthetics of an Instagram post. The present political moment only accentuates the ostensible folly of this exercise. I write this on the precipice of a second Trump administration—a ruling regime that will almost certainly dismantle environmental regulations, weaken the renewable energy industry, and discard environmental justice protections. While most of the fieldwork for this book took place during the first Trump administration, the heightened stakes this time around point to the need for multiscalar political action, not myopic media analysis. As a concerned climate activist and scholar, I can surely find more productive things to do than nitpick over how a solar corporation uses Google Earth. Except . . . I'm not so sure. After years of working as a sustainable energy policy advocate and conducting fieldwork for this book, I have found that the digital platforms that my colleagues, interlocutors, and I use to transform environments that we otherwise have limited relations with are perhaps as symptomatic of our climate challenges as a corrosive fossil fuel refinery in a poor neighborhood, or the frightening weather events that increasingly harm marginalized communities. Tending to this alienation requires a close look at the screen-based world that the energy transition operates in. At first glance, then, this book's ethnographic documentation of solar's virtual and visual imprint might seem disconnected from the urgent material imperatives of renewable energy (e.g., reducing life-threatening emissions that disparately impact communities of color). But it is precisely this disconnect that I hope to destabilize by theorizing the paradoxically two-dimensional nature of a profoundly physical transformation in electricity generation.

While *Subjects of the Sun* does not address the incoming administration's plans to impede this transformation, these plans are all the more reason for disentangling the energy transition from an insidious mediascape that aided the rise of Trump and his extractivist policies. Put differently, both progressive digital activism and the politics of viral misinformation depend on screenwork, and while these contiguous dependencies don't mean we should wholly unplug our efforts, they underscore the need for an energy transition that is not fully enmeshed in a virtual decarbonization paradigm in which power is concentrated in our laptops.

This argument is rooted in an understanding that energy infrastructure is not just a material force but also an ideological force. As Timothy Mitchell has documented, the materiality of fossil fuels—their spatiality, movability, and extractability—directly shapes political possibilities and social imaginaries.[3] *Subjects of the Sun* expands on this analysis, illuminating how solar electricity materializes through an interconnected physical landscape and digital mediascape in ways that simultaneously upend and uphold the political order of fossil fuels. I contend that reckoning with this unsettling simultaneity can usher in a more radical energy transition centered on reconfiguring our relations with the spaces that solar is supposed to safeguard—the environmental senses and sensibilities that screenwork neglects.

This book is therefore quite critical of hegemonic environmental stewardship and its hardworking, committed practitioners. At times it might even read like a smug exposé. But as someone who spent years in New York City's environmental policy ecosystem, I have zero interest in such ungenerous gotcha tactics. Instead, the book's targeted ethnographic analysis aims to uncover how the most mundane facets of digital life enable intelligent, competent, empathetic people to reproduce the structures of power that they wish to uproot. For my interlocutors have a sophisticated understanding of the world and often effectively enact change (they are not, for instance, the clueless technocratic dupes that James Ferguson encountered in Lesotho).[4] But they remain ensnared in the underexplored contradictions of a late liberal mode of environmental stewardship that is arguably no less productive of environmental degradation than fossil fuels.[5] *Subjects of the Sun* elucidates this paradoxical praxis with an eye toward an energy transition that is not so much immune to contradiction as attuned to that which digitalized carbon reduction obfuscates: the multifarious forms of life on the ground in even the most developed urban spaces. As I show in this book, moving from screenwork to the soil can help realize the transformative potential of energy generated directly from the luminescent orb in the sky that defines life as we know it.

How odd it is to tour someone's living space without their consent. To see their kitchen when their apartment door is ajar, to view their vestibule and corridors as they leave their building for work. I couldn't shake this sense of impropriety, feeling like an unwelcome interloper as I walked in a group of spectators led by a professionally clad white man through the Marcus Garvey Apartments (MGA). MGA is a low-income housing complex in the poor Black community of Brownsville, Brooklyn. MGA loosely resembles the community where I grew up a few miles away in Queens—a space that would never attract a sightseeing tour. But MGA differs from all of New York City's housing complexes in one crucial way: it's home to one of the first solar-powered microgrids in the United States.

A microgrid is a small-scale power station that provides locally generated electricity (as opposed to centrally generated electricity) to a geographically bounded network of electricity users, enabling communities to be energy self-sufficient in the event of grid malfunction. Brownsville is perhaps the last place you'd expect to find such innovative infrastructure, as the neighborhood is marked by all the familiar indicators of structural violence—high incarceration rates, pervasive unemployment, poor public health statistics, and so on.[1] *Solar-powered* microgrids, in particular, are complex and costly, requiring not only considerable financial investment but also technocratic aptitude, well-resourced institutions, and a stable built environment (e.g., intact rooftops)—assets that impoverished communities of color like Brownsville lack.

How, then, had MGA achieved something that even the most environmentally conscious wealthy white communities only dream of—not simply transitioning to renewable energy but also adopting cutting-edge climate-resilient infrastructure? And how could other poor communities of color follow their lead? These questions brought me and several of my activist comrades to MGA; we wanted to see, with our own two eyes, a beacon of an equitable sustainable future in the middle of the 'hood. We all worked as either employees or volunteers for Environmental Quality for All, or EQUAL for short—an environmental justice (EJ) nonprofit in a low-income community of color resembling Brownsville.[2] When we spoke with the white-run property management firm who owns MGA to arrange our visit, I didn't consider that the people who actually lived there would have no clue who we were. I was too preoccupied with the abstract concept of a Black solar-powered utopia to give much thought to what it might feel like actually being on the ground in a space that many of us mythologized.

Before our visit, MGA felt less like a real place and more like a fantasy uploaded to Instagram. For it was there, on the world's preeminent image-sharing platform, that the concept of a microgrid in the middle of the 'hood first felt palpable to me—even if I could only "visually 'touch'" a pixelated version of it from afar.[3] I had noted, in particular, the images in figure I.1, which sit side by side with no blank space separating them in a post on a real estate company's Instagram page.[4] With text accompanying these and other images, the post celebrated MGA's then-recent renovations—including the microgrid installation. The post caught my attention because it juxtaposes solar with a Black woman presenting an image of poor Black children smiling in school uniforms. It thus situates solar in a raced and classed context that challenges the conventional wisdom regarding renewable energy. More than simply accessible, solar here is a part of the community's broader rejuvenation; the respectable children and the woman's podium signal investments in Brownsville's future—a future powered by the solar panels to their left. The images suggest, then, that the kids of the ghetto *can* be at the forefront of a renewable energy revolution, that a neglected Black community *can* generate the energy it needs without investor-owned utilities, illustrating the elusive justice-oriented energy transition (hereafter *just transition*) that my comrades and I had long dreamed about.

But it was tough accessing this Instagram imaginary in a group of spectators led by a white man dressed for work. I kept to the back of the group, and before we entered the electrical room, a couple of middle-aged men politely stopped me to ask what we were doing. "We're here to see the microgrid!" I explained, only to be met with blank stares.

"What's that?"

I pointed to the rooftops. "Oh, you know, the solar panels."

"Hold up, we don't got solar panels here."

I couldn't believe what I heard. Had the management company really

FIGURE I.1. Images from an Instagram post on the Marcus Garvey Apartments (MGA) in Brooklyn. GDSNY, "Marcus Garvey Village," Instagram, June 10, 2017, https://www.instagram.com/p/BVKwT1lhLpD.

withheld information about their sustainable infrastructure from their residents? In a revealing move, one of the company's office managers who had joined us looked suspiciously at the men and curtly explained that we were on a tour and that they weren't allowed to come to the electrical room with us, positioning the mere sight of the microgrid as a privilege denied to the very residents it was supposed to benefit. We descended to the underground space as the two men who had never seen it looked on at us, this group of strangers, with bewildered frustration.

I began to suspect that the Instagram post wasn't telling the whole story, that solar wasn't the raced and classed intervention I had seen on my phone. Perhaps, then, the post was a disingenuous simulacrum of inclusivity—propaganda that intentionally concealed the exclusionary reality on the ground.

But I would soon learn that social media's nefarious dimensions could not account for the disconnect between the ebullient Black faces in the post and the bewildered Black faces in real life. Instead, I argue in this book that New York City–based activists, experts, and laypersons alike often mistakenly conceptualize solar infrastructure as a force of equity, democracy, and social mobility because most of us have only ever known and interacted with such infrastructure through online platforms and screenwork. Indeed, differently positioned New Yorkers primarily relate to sustainable energy technologies through not hands-on labor or in-person interactions but instead a two-dimensional solar "visual economy" of which the Instagram post is just one part.[5] As I show in this book, though, this visual economy cannot be reduced to the ethers of "the cloud"; it also emerges from solar technology's *material properties*—its spatiality, modularity, and shimmering surfaces, for instance. I contend that these material properties work in tandem with cloud-based platforms and screenwork to produce a facile solar imaginary that figure I.1 only partly illustrates. Consider, for instance, *the diffuse spatiality* of solar energy—the fact that it can be generated in everyday spaces such as a destitute housing complex like MGA. This spatiality positions solar as an equalizing force in the images in figure I.1, as solar's decentralized topography centers marginalized subjects in many online renderings of high-tech sustainability.

But when we peer beneath this screenwork's shiny facade, we can find a solarized world that is not as equitable as that Instagram post suggests: a solarized world produced through the structures of racial capitalism that uphold the fossil fuel status quo. I got only a glimpse of this insidious dynamic when I locked eyes with the two bewildered Black men who were denied a view of a microgrid that I—myself a Black man—was improbably given access to. As I would learn, this seemingly minor act of marginalization was but a miniscule snapshot of the opaque ways in which solar reproduces the structural hierarchies that the Instagram post suggests it upsets.

Subjects of the Sun contends that we can begin to realize the radical potential of solar simulated in figure I.1 when we understand how such visuals obfuscate the raced and classed structures that simultaneously produce solar and hide it from MGA residents (in ways our tour would further reveal, as discussed later in this introduction). I wager that this sober recognition can address the contradictions of an ostensibly equalizing green energy infrastructure contingent on the toxic ravages of racial capitalism. This project first requires an understanding of how contemporary subjects often relate to not just solar but energy more broadly through visual media—a topic I now explore before returning to Brownsville.

ELECTRICAL SPECTACLE

The fossil fuel industry doesn't want you to see how they produce energy. Their propaganda prominently highlights modern essentials like planes, medical technologies, and heated homes but not the toxic infrastructure that enables those essentials: oil rigs, pipelines, power plants. In contrast, the solar industry saturates their media with images of sun-drenched solar panels, calling on you, the energy consumer, to actively envision how your energy *could* be produced. This is partly why the veiling of MGA's microgrid was so jarring; solar is usually highlighted for *all* to see—even if this revelation often happens on a screen.

Consider the two images in figures I.2 and I.3—both focused on New York City. The first of these images appears in a promotional video for the American Petroleum Institute (API) while an authoritative narrator states, "There's energy everywhere. But sometimes it can be hard to see."[6] The image attempts to address this visual conundrum by branding NYC's buildings and bustle as "energy," bringing some visible form to the ubiquitous force of life that often eludes our sight. After presenting the city skyline, API visualizes energy through images of grassy mountains, gas stoves, oceans, suburbs, space travel, home appliances, and cell phones—seemingly anything other than the extractive infrastructures through which the fossil industry produces energy.[7] In contrast, the second image—a post from a Brooklyn-based solar company's Instagram page—frames the city's built environment by drawing direct attention to a solar array.[8] In this context, energy is rendered visible through the spectacle of infrastructure.

Thus, as solar transforms the political economy of energy production, it also shifts the *visual* economy of energy consumption, generating an optics of electricity infrastructure that counters the relative invisibility of fossil fuels. Indeed, the dirty refineries and extractive zones that

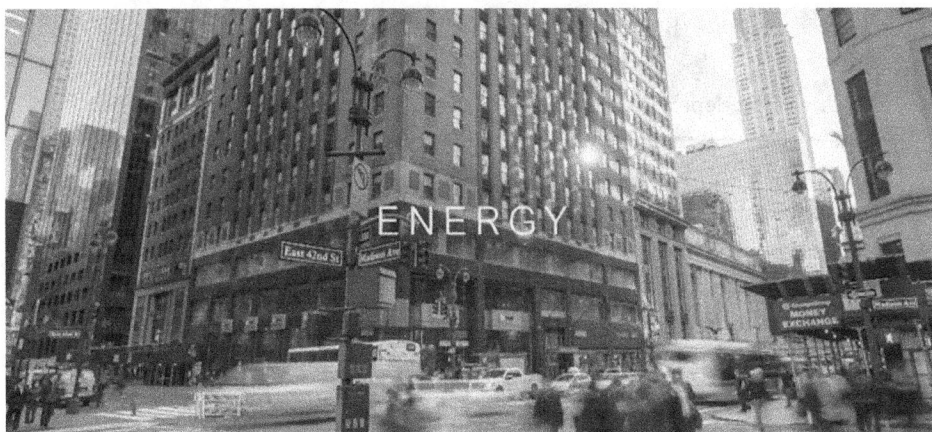

FIGURE I.2. Energy and New York City's built environment. Still from American Petroleum Institute, "Intro—2022 State of American Energy."

FIGURE I.3. Energy and New York City's built environment. Brooklyn Solar-Works, #solarpower in the Big 🍎, Instagram, November 18, 2022, https://www .instagram.com/p/ClHkdaqOCHz.

many of us depend on for survival are often hidden from our everyday view, such that images of solar invite us to envision a sustainable version of that which we usually can't even see. More generally, visuals of "clean energy" infrastructure iconize that *invisible* force that powers modern society—electricity—giving a visible form to streams of charged electrons that the human eye can't detect.

Crucially, this aesthetic is a medium for reimagining the raced, classed, and gendered order of racial capitalism, enabling the sort of utopic dreaming through which I romanticized Brownsville's microgrid. Put differently, solar imagery often visualizes infrastructures of electrical power as a means of redressing entrenched structures of intersectional power.

Consider the two images in figures I.4 and I.5, both taken from the Ins-

SUPPORT BLACK CO-OPS
SOLAR UPTOWN NOW SERVICES

FIGURE I.4. Blue-collar solar labor on a Black worker co-op's Instagram page. Solar Uptown Now Services, "Support Black Co-Ops," Instagram, accessed November 25, 2022, https://www.instagram.com/solar.uptown.now.services.

tagram page of a Black, worker-owned solar installation cooperative in NYC.[9] By visually linking solar with Black worker co-ops, the first image suggests that renewable energy can empower the Black proletariat to own the means of production. In this way, it offers a measured contrast to the world of fossil energy, which is dominated by white executives and power brokers.[10] The second image employs a geometric aesthetic reminiscent of constructivist Soviet propaganda to visualize rooftop solar as a catalyst for not only forging "stronger communities" but also empowering blue-collar women workers. Given that the co-op who posted this image does not have any women worker-owners, the image simulates an imagined horizon of gender-inclusive diversity—a spectacle of worker

FIGURE I.5. Blue-collar solar labor. IWW Environmental Unionism Caucus, "Green Jobs and Intergenerational Justice: Trump's Climate Order Undermines Both," March 30, 2017, https://ecology.iww.org/node/2169.

solidarity with women, who comprise a sliver of the blue-collar work-force.[11] When viewed together, both images render solar as a medium for redressing the economic inequality, anti-Blackness, and patriarchy of racial capitalism, suggesting that renewable energy can transform both our infrastructures of electrical power *and* entrenched structures of intersectional power.

But as my MGA tour suggested, looks can be deceiving. These images obfuscate what my interlocutors' bewildered Black faces hinted at: oppressive systems can inhere in even the greenest technologies.

IN AESTHETICIZING a Black worker cooperative, community power, feminist labor politics, and a microgrid in the 'hood, the Instagram images I've discussed gesture to a sentiment that many of their viewers more explicitly endorse: solar energy must be central in the struggle against racial capitalism.[12] On the surface, this sentiment makes sense. After all, solar can free poor communities of color from the chokehold of petro industries, empowering these communities to own the energy they depend on.[13] It can challenge the extractive paradigm at the heart of both fossil capital and its antecedent, the transatlantic slave trade, centering energy generation on sustaining life, not bottom lines.[14] And it can enable many of us to engage with energy infrastructures that often sit quietly in the backdrop of our consciousness, addressing our alienation from the slow violence of electricity production.

But when we look beneath the shiny facade of solar imagery, this "clean power" instead appears enmeshed in the machinations of racial capitalism. Solar technology is made from toxic materials mined from the earth; manufactured with the physical labor of precarious people of color; transported transnationally with fossil-fueled ships; installed by exploited, low-wage workers; and could potentially become a significant e-waste problem in marginalized communities throughout the world.[15] From child cobalt miners in the Democratic Republic of the Congo to poisoned manufacturing communities in China to imprisoned Black and brown workers in the United States, the terrain of solar production is molded in those long-standing "sacrifice zones" that constitute racial capitalism.[16] More generally, solar and other renewable energy technologies can be deployed to increase the efficiency of the existing capitalist structure, generating new markets and optimizing existing ones. And as I began to suspect during the MGA tour, even when corporations deploy solar in the 'hood, it is not always evident that it will empower local residents—a hunch that my research would only corroborate.

Subjects of the Sun addresses this tension between, on the one hand, a transformative vision for solar rooted in racial and economic justice— what the Instagram images aestheticize, what I had hoped to encounter in Brownsville—and, on the other, an emergent structure of "sustainable" capital accumulation that is no less extractive than its fossil forebears. Specifically, this book explores how work for a "just transition" often unintentionally reproduces the intersectional hierarchies and modes of exploitation it is supposed to supplant. I center my analysis on the image-oriented, graphic, and screen-based work of sustainable energy technocrats at for-profit solar corporations and of climate justice activists at EJ organizations like EQUAL in my hometown, NYC. I conducted ethnographic research there from 2015 to 2018 as a grad student after working for almost a decade on sustainable energy policy as a young professional with EJ activists and labor unions throughout the city.

I contend that the images, graphics, and screen-based platforms that New Yorkers use to envision the just transition overlook the symbiotic relationship between racial capitalism and modern energy in even its greenest forms. In exploring this elision, my analysis focuses less on my interlocutors' moral shortcomings and more on an agent without a clear political consciousness: solar technology itself. I consider how this technology takes form in a dense urban landscape to suggest that solar is not simply a material infrastructure to generate electrical power from the sun. It is also an affective infrastructure whose physical form can shape our senses and sensibilities, orienting social imaginaries toward uncharted political horizons—as evinced by my Brownsville romanticism. In this vein, solar technology has reconfigured aspirations for racial and economic justice in communities grappling with intersecting climate, housing, and employment precarities, as the Instagram images partly demonstrate. I argue, though, that this affective power paradoxically solidifies the hegemony of racial capitalism, as it inspires aspirations for a solarized political economy that appears outside of that which constitutes it—an ecosocialist vision of Black and brown communities fighting climate change through technologies that are embedded in the very economic order that that vision disavows.

In dissecting this vision, my analysis highlights the centrality of sight and spectacle to dominant forms of contemporary energy governance, exploring images that are coterminous with those on the previous pages. As electricity is at once imperceptible to everyday sight and foundational to a modern world centered on screens and surveillance, it necessitates forms of visual representation that render it visible and thereby governable.[17] This ocularcentric imperative has particularly insidious effects when it comes to solar. Specifically, I argue that what I call *late liberal screenwork* interacts with solar's material properties in ways that conceal the extractive logics of racial capitalism—perpetuating those very logics in the process. Put differently, visualizations of solar often suggest that it can move us beyond the ontological divisions through which racial capitalism operates. As Cedric Robinson has shown, capitalism is rooted in a racialized framework of "differential value" in which capital accumulates through social structures of racial difference.[18] I want to suggest that solar's physical forms (its shine, its rooftop placement, its customizability) work in conjunction with certain visual forms (PowerPoints, smart boards, social media) with the effect of obfuscating this differential value.

I uncover this obfuscatory work by exploring several material properties of solar infrastructure, namely, its decentralized terrain, its shiny appearance, its modularity, its electrical quantifiability, and the corporeality of its requisite labor. I argue that these material properties work closely with images, digital platforms, and quantitative graphics to shape the progressive vision animating my preconception of Brownsville—a vision in which renewable energy can eradicate the constitutive tensions of racial capitalism (white versus other, society versus nature, corporations versus workers). But solar is no less likely to extract value from these dualistic differences than other, dirtier forms of industrialized energy. It relies on extractive zones, exploitative markets, elite expertise, and an anthropocentric order, even when it appears to flatten these structures.[19] My point, then, is not that solar power is the same as petro power but rather that it is no less enmeshed in differential value—a point this book expands on at length. As a corrective to an energy politics that obfuscates differential value through images and graphics, *Subjects of the Sun* calls for a just transition that centers the senses and sensibilities neglected by screenwork: one's haptic care for their local environment; the full-bodied feel of infrastructural labor; the sublime affect of the nonhuman world.

Before exploring this sensorial horizon, though, I now further contextualize the relationship between racial capitalism and energy infrastructure. As I show, screenwork deceptively suggests that solar can overcome this entrenched relationship.

ENERGY'S EXTRACTIVE RELATIONS

A range of scholarship contends that solar has the power to either sustainably maintain the capitalist status quo in the face of climate crisis or disrupt the dominant political economy, as if we have a choice between extending fossil capital's inequitable growth paradigm or democratizing the productive output of that paradigm.[20] I want to suggest that these sorts of analyses narrowly understand solar as a technological force that can impact carbon emissions, energy prices, job creation, and liberal governance, as opposed to a relational force that shapes how people engage with each other and the more-than-human world of which we are a part. *Subjects of the Sun* instead adopts the latter perspective to illuminate how solar at once upends and upholds the relations of differential value that constitute racial capitalism, destabilizing clean energy teleologies.

This analysis requires that I first clarify what I mean by "relations of differential value" and how these relations are inextricable from energy infrastructure. The English word *relations* is at once a conceptual tool for linking disparate phenomena, a way of characterizing interdependencies and contingencies among a range of persons and agents, the name for the fundamental "building blocks of society," and a synonym for positive interpersonal connections and kinship ties.[21] Difference is immanent in all relations, as the word can characterize either the similarities, shared attributes, or structural binds of distinct entities, on the one hand, or the points of convergence among (ostensibly) dissimilar things, on the other.[22] At its core, racial capitalism is a structure that configures relations and thereby organizes difference in ways that orient ontological hierarchies, social formations, and interpersonal connections around the pursuit of surplus value. For example, racial capitalism organizes humans' relations with nonhumans around extraction, calcifying the perceived difference between the two into an inflexible, exploitative hierarchy.[23] It also generates new axes of relational difference, reconceptualizing group identities and social formations through racial and ethnic schemas that enable some humans to exploit the labor power of others.[24] *Differential value* names the profits and privileges that these exploitative configurations of difference yield.[25]

Energy infrastructure has long been a central medium for materializing difference and producing differential value under racial capitalism.[26] Broadly speaking, infrastructure is a means of giving material form to social relations, enabling the linkages, formations, and connections that constitute society.[27] Racial capitalism ushered in *industrial-scale* infrastructures: material forms that facilitate relations among labor, environments, "natural resources," and capital to generate surplus value through economies of scale. If we understand energy to be not electricity, heat, or fuels but instead a capacity immanent in those things—that is, the ability to transform matter—then the first industrial-scale energy infrastructure was not the transcontinental railroad, the electricity grid, or the interstate highway but instead an earlier approach to scaling up this transformational capacity: the transatlantic slave trade.[28] Modern energy, then, emerged from a plantation system powered by Indigenous genocide and Black life—a capacity to transform matter in ways that reorganized human relations around (anti-Black) differential value. These relational dynamics persisted even as fossil fuel combustion replaced slavery as the dominant energy infrastructure. From coal mines to railroads to refineries, the fossil-fueled industrial order continued to designate Black and brown lives and land as extractable and disposable, maintaining the plantation economy's hierarchies through labor exploitation, environmental injustices, and, by extension, the everyday relations through which Black and brown people experience quotidian indignities.[29]

A just transition to renewables must therefore transform the relations that constitute energy infrastructure, generating new ways for people to connect to one another and to the more-than-human world, unbeholden to the paradigm of differential value. This book contends that the images, graphics, and digital platforms through which many New Yorkers relate to solar infrastructure misleadingly simulate such a transgression, configuring political imaginaries that often don't meaningfully reconfigure racial capitalist relations. As I show, the mystifying potency of these media is a testament to the ways they interact with the infrastructural hardware and physical configurations of electricity technologies. That said, these material-discursive entanglements are not particularly novel. For starters, energy infrastructure transmits information and is thus a form of media in its own right. Furthermore, numerous scholars have explored how the transmission of information dialectically co-constitutes infrastructural hardware.[30] But in zeroing in on the renewable energy transition, *Subjects of the Sun* shows how this dialectical configuration alienates differently positioned subjects from the infrastructures they hope to steward by paradoxically choreographing a world in which they overcome that alienation.

I therefore argue that a just transition must enable us to foster relations with energy that are not centered on the screenwork through which we visualize energy. Toward this end, the book's final chapter considers the many emotional, sensorial, and bodily ways we relate to energy that are not preoccupied with digital media, pointing to affective resonances that can frustrate the extractive, hierarchical relations that produce differential value. From stewarding the earth with solar-powered tools to on-the-ground organizing for solar cooperatives to cultivating a consciousness of the sun, humans can work with solar energy without relying solely on cloud-based platforms, countering the hierarchical logics of racial capitalism in the process.

To underscore the imperative of this sensorial reorientation, I now return to the screen-based simulations of the Brownsville microgrid—visual artifacts that illuminate the broader spectacle of solar that this book explores in depth.

THE BROWNSVILLE MICROGRID

As a Black New Yorker from a community marked by environmental injustice, I was somewhat taken aback when I saw the Brownsville microgrid on Instagram. For in aestheticizing the upward mobility of a marginalized Black community vis-à-vis solar commodities, the Instagram images in figure I.1 affiliate Blackness—the paradigmatic form of racial difference—with the hegemonic project of high-tech sustainability, suggesting that eco-infrastructure can redress the structural schisms of racial capitalism.[31]

I want to suggest, therefore, that the MGA post offers a helpful starting point for understanding how screenwork and cloud-based platforms work in service of what Elizabeth Povinelli calls *late liberalism*. Povinelli conceptualizes late liberalism as the market-oriented governance of racial difference pioneered by states to counter—and often co-opt—the decolonial, antiracist, and feminist movements that challenged their legitimacy, while also managing the social and economic fallout from the retreat from Keynesianism.[32] Povinelli argues that late liberalism simulates an image of multicultural unity and reform—an *appearance* of cultural recognition staged by the state—to detract from more radical forms of structural redress.[33] But the visual imprint of late liberalism is not limited to this sort of highly choreographed image from above; as I show in this book, the market-based governance of racial difference also occurs through everyday screenwork like the Brownsville Instagram post.[34]

Specifically, this book documents *late liberal screenwork*: cloud-based platforms and digital content that render commerce and commodities as antidotes to raced and classed inequality—often unwittingly undermining social movement work. This concept highlights how my interlocutors' mundane ways of relating to the world vis-à-vis screens (e.g., scrolling on their phones) double as a mode of experiencing remarkable structural transformations (e.g., the installation of a climate-resilient microgrid in a poor Black community), paradoxically endowing the everyday texture of life under racial capitalism with the feeling of moving beyond differential value. Of course, when I first saw the images in figure I.1, I wasn't explicitly thinking in these terms because the post that broadcast them had no didactic antiracist message whatsoever. Yet given the metonymic relationship of Blackness and otherness, this and other similar visual renderings of Black people invariably evoke race despite their dearth of commentary on racial difference.[35] To do so through a celebratory image of solar infrastructure is to comment on the inclusive potential of the clean energy economy, irrespective of what the image curators intended to convey. Next, I ethnographically situate these images to elucidate the misleading sense of possibility that they produce, revealing the ways my interlocutors and I myopically relate to energy infrastructure through the abstractions of cloud-based platforms.

THE VIEW FROM NOWHERE

Months before I saw the images in figure I.1, Mark, a white engineer and PhD candidate from California, typed something into his phone and showed me the screen (figure I.6).

There, before my eyes, was evidence of something that otherwise struck me as impossible: a solar microgrid in Brownsville! I had met Mark only minutes before—we were at an academic conference in Barcelona—and on learning what I researched, he began gushing about the (then-uncompleted) microgrid, baffled that I didn't know about it. "It's gonna have a lithium battery—the first in New York! You sure you've never heard of this?" I was rather certain, though, that he was mistaken—until he showed me his phone. In a "hyperlinked society," cell phone browsers are disruptive in this way, uprooting habituated skepticism in the course of factual disputes.[36] This sort of screen-based intervention has an almost scripted rhythm. First, one summons the facts via Google, then one quickly flashes the facts to nonbelievers, and then these nonbelievers instantly capitulate, ceding authority after only a cursory beat of visual verification. While stubborn skeptics will insist on reading the facts closely, others—like me—will humbly defer to an evocation of Google, knowing that the certainty they harbored only a moment prior is likely unjustified. In the social field of everyday conversation, the browser car-

FIGURE I.6. A screenshot of the headline of the microgrid article that Mark showed me. Julian Spector, "Tackling New York's Rising Power Demand with an Affordable Housing Microgrid," *Greentech Media*, December 16, 2016, https://www.greentechmedia.com/articles/read/new-york-grid-microgrid-bqdm-con-ed-peak.

ries an epistemological weight that instantly unsettled my assumptions, sketching a horizon of possibility in which infrastructural transformation led by a marginalized community—*not the state*—presages otherwise unlikely social change. Thus, the tinted text on Mark's phone evoked a late liberal world where racial progress emerges not through political struggle but through commodified technologies, countering my calcified impression of Brownsville as structurally oppressed. Indeed, everyday screenwork advances late liberalism when it stimulates imaginaries at odds with a radical political consciousness—whether it is intended to or not.

Of course, I wasn't so naive to think that the microgrid had enabled Brownsville to overcome its subaltern positioning. Furthermore, my broader skepticism about this microgrid remained. I wondered, suspiciously, why a white scholar with no connection to NYC knew more than I about the infrastructure of a Black community in the city where I was born, raised, and conducting research. Why was a privileged outsider so knowledgeable about a project in the middle of a 'hood that he had no relationship with?

Putting this question aside, I was excited to hear that Brownsville was improbably developing a solar microgrid, for several reasons. First, solar technology—whether connected to a microgrid or not—reduces dependence on fossil fuels and electricity bills. This is particularly impactful in low-income communities of color where residents spend a higher percentage of their income on electricity and are disproportionately vulnerable to fossil-fueled climate change.[37] Microgrids that are powered by solar technology amplify these positive impacts. As Mark pointed out, solar microgrids are connected to energy storage batteries so their users can access low-carbon electricity when their solar technology can't draw directly from the sun (notably after sunset or on cloudy days). In this vein, every microgrid—whether powered by solar or anything else—dramatically reduces its users' dependence on the central grid by enabling them to draw a significant share of their electricity from local energy generation. This localization coupled with battery storage ensures that solar microgrid users can consistently access electricity even when a blackout occurs or when the grid is unreliable. For this reason, solar microgrids symbolize resilience—celebrated for not just mitigating emissions but also empowering communities that are vulnerable to infrastructural disruption due to climate change. Additionally, the installation of solar microgrids demands good-paying blue-collar labor, and it can therefore generate quality job opportunities for communities struggling with underemployment.

But there is another, often overlooked potentiality of microgrids in communities like Brownsville: they can empower people who have traditionally been excluded from the spaces where electricity is controlled to take a leading role in the power sector that their lives depend on. Microgrids could therefore transform marginalized people from passive consumers to active agents in the political economy of energy generation, emboldening them to foreground their racial and economic concerns in broader energy transition efforts. This political potentiality is rooted in how microgrids reconfigure material power. When a community is given the power to generate part of its own electricity, the power to sell excess electricity that it generates to the central grid, the power to exercise some economic autonomy from unaccountable investor-owned utilities, and the power to survive with minimal disruption in the face of climatic volatility, it can leverage this power to reimagine itself as a self-sufficient collective. By decentralizing electricity generation, microgrids can engender a sense of possibility—a *feeling* that everyday people can upend the corporate system that controls industrialized modernity. When decentralized electricity affords us new power, this power could transform our politics; infrastructural transition could instigate structural change.

Could. But not necessarily. Microgrids' decentralized power could just as easily maintain the status quo of electricity governance—a world in which marginalized people have little knowledge of energy generation. Electricity's differential value stays intact when privileged experts like Mark have insider information on power production while everyone else is left in the dark. So on my flight back home, I considered the oddity of learning about Brownsville's microgrid from a white man in Europe, wondering if my late liberal response to his browser was warranted.

MY CURIOSITY and uncertainty about Brownsville's microgrid only grew when I took part in an ostensibly run-of-the-mill conference call several months after I returned to NYC. The call was with a solar technical assistance company and EQUAL—the aforementioned EJ organization. I was volunteering for EQUAL's campaign to bring solar to a working-class community of color that resembled Brownsville. A senior representative from the technical assistance company mentioned the Brownsville microgrid during our discussion, lauding its transformative capabilities. Once again, I was struck by the novelty of hearing a technology expert—a privileged white man with no relationship to Brownsville—gush over the technological innovations of a community with a reputation that couldn't be further from that of Silicon Valley. As he spoke about the microgrid's unprecedented lithium battery, I wondered if he was, on any level, thinking about Black people or poverty or police brutality or negligent landlords or any of the other fraught things often affiliated with Brownsville—or if, alternatively, he viewed the community as the home of an innovative infrastructure and nothing more. I didn't realize then how solar works with late liberal screenwork to animate white-collar care for destitute spaces—a central theme of *Subjects of the Sun*.

Regardless, I was glad he mentioned the microgrid to EQUAL's activists, as it inspired a curiosity that had gone missing from their work. How, they wondered, had folks in Brownsville managed to bring such an expensive beacon of high-tech sustainability to the 'hood? For the previous three years, many of EQUAL's community organizing efforts had focused on this very goal, but a local microgrid never felt like it was more than a pipe dream. While EQUAL had long talked the talk of an "inclusive green economy" that fosters worker ownership and sovereignty from investor-owned utilities, my comrades had grown hopelessly aware of the practical limitations of this vision when it came to developing a local microgrid—a prohibitively expensive and structurally challenging aspiration.

This hopelessness, though, was rooted not only in the particular difficulties of microgrid development but also in the broader everyday routines of EJ organizing in marginalized spaces: holding your breath while slipping flyers into mailboxes in piss-scented lobbies; traversing the gray blur of treeless boulevards in EQUAL's bright yellow T-shirt trying to find someone—*anyone*—who cares enough about truck pollution to talk to; standing in the run-down community center with puke-colored walls while you offer the same welcome spiel to an unchanging group of opinionated local retirees every other week, seemingly in perpetuity; feeling despondent when you see Leroy, the local Ballantine-sipping panhandler, kneeling on his milk crate before the newly gentrified bodega that's stocked with oat milk and white yuppies who avert their eyes when he asks for a dollar; sitting at an infantilizing cubby of a desk trying to find a moment of quiet in the stale indoor air and harsh overhead lights of EQUAL's office before it's inundated with the chatty participants of a community meeting. In these and countless other habituated bearings, EQUAL's staff moved through their community with a sense of stasis; familiar streets, buildings, elevators, scaffolding, vestibules, pul-

pits, hallways, and meeting rooms were heavy with the mind-numbing-
ness of organizing work that often yielded little in the way of structural
change. "New day, same shit," mused Dante, an EQUAL organizer, as we
marched through familiar streets distributing flyers on solar, speaking
with a complacent levelheadedness that felt essential to working there.
When you're ostensibly stuck in these spaces for years on end, trying,
futilely, to unearth the violent social order that made them what they
are, they become marked with the unmovable stench of raced and classed
power, this dispiriting sense that *this is just what it's like to be here*. So the
microgrid idea—as much as we loved it—often felt incompatible with
our everyday experiences in such a Black and brown community. While
EQUAL staff never hesitated to excitedly share a generic vision for a local
solar microgrid—proclaiming, without specificity, *We can develop one in
our own backyard!*—this vision's dearth of detail betrayed the seldom-
spoken sentiment that *stuff like that don't happen here*—an indifference
that hung in their heavy breaths between the words they spoke on streets
that they had long grown tired of.

Subjects of the Sun explores this disjuncture between the quotidian disappointment of EJ activism, on the one hand, and a bold vision for an
inclusive clean energy economy, on the other. I attribute this disjuncture
to the uneven affective texture of racial capitalism, moving beyond ideological polemics to illuminate an ambivalent environmentalism focused
on the contradictory task of forging an inclusive solar-powered utopia in
the dystopic shadows of free market hegemony—the complacent mien
that my comrades wore while canvassing their dilapidated terrain. Yet
this ambivalence seemed to evaporate when they learned that a community comparable to theirs had improbably realized the pipe dream of
a local microgrid. Brownsville became a much-needed jolt to our sense
of possibility, manifesting in our persistent cry after the call: *But how
did they do it?!* Put simply, MGA's microgrid, like any late liberal icon,
appeared to defy the differentiating logics of racial capitalism. *Subjects of
the Sun* attempts to account for this defiance, theorizing how the spectacle of solar animates imaginaries of moving beyond differential value—
aspirations that would call on us to witness the microgrid with our own
eyes.

AFTER OUR conference call, Selena, one of EQUAL's leaders, googled the
Brownsville microgrid and pulled up the aforementioned Instagram post.
There she saw not only the photos in figure I.1 but also two photos of the

sidewalk in front of MGA on a gray winter day. The post positioned these unremarkable landscape visuals right below the pictures of solar panels and Black children with no white space separating them, enfolding a high-tech "climate solution" into the mundaneness of the children's built environment. It was precisely this mundaneness that caught Selena's attention—the fact that a remarkable technology appeared as an *unremarkable* part of the community. "It would be so cool if we could do that here," she said. "You're just going about your day like normal, except it's powered by a resilient solar microgrid." *Your day like normal.* These words marked an interest in the innocuous, lacking the visionary zeal that initially animated EQUAL's curiosity regarding Brownsville.

Crucially, this innocuousness was well aligned with EQUAL's visual aesthetic. The organization's small office was adorned with selfies of local residents in yellow EQUAL T-shirts posing throughout their community— this banal mosaic of half-smiling Black and brown people gesturing to nothing in particular except the EQUAL logo on their chests, seemingly just living their normal lives as community activists. The leadership of EQUAL constantly instructed everyone involved in their work to take these nondescript selfies when doing even the most quotidian activity related to the organization, like sitting at a community fair's information booth or attending a community meeting on workforce opportunities. The organization's visual aesthetic was therefore resonant with the microgrid Instagram post, centered as they both were on normal Black people uniformed in T-shirts happily posing in their everyday spaces while gesturing obliquely to inclusive environmental improvement. So as Selena scanned the Instagram post with a soft contentment—as if its quaint *look* of inclusive energy management had touched her in a bromidic way—I considered how this visual aesthetic had redirected her personal energy from the more radical EJ politics she had initially brought to the organization and toward the more generic appearance of marginalized communities improving their situation through eco-friendly technologies. How, in other words, had the visuals that she frequently peppered her PowerPoints with—images all over EQUAL's office—transformed her political commitments such that they now included the curation and consumption of photos?

I never doubted the sincerity of these commitments; sharp witted and no-nonsense, Selena worked tirelessly for transformative EJ legislation. Yet when she'd follow her colleagues' instructions and assiduously choreograph a group selfie at a protest or put together a PowerPoint that was light on policy details and saturated with such selfies, she seemed slightly possessed by late liberal screenwork, as though the simulation of empowered people of color was a goal in and of itself. The curation and consumption of these visuals injected a lightness into social justice struggles, offering Selena putatively apolitical moments amid explicitly political work: a second to breathe when taking a selfie, an excuse to ignore her quiet doubts, an easy task without legislative complexity, a reason to smile. EQUAL's selfie culture made the political bearable. It complemented the quiet indifference that Selena and her colleagues wore beneath their protest chants and public testimony, habituating them to a mode of activism that at times lacked fervor and fastidiousness.

The Brownsville Instagram post, then, offered a familiar apolitical lens for apprehending the microgrid, inviting a feeling of levity that stood side by side with Selena's more substantive curiosity regarding the microgrid's origins. I can't speculate as to whether the post altered her orientation to the microgrid in any way, but it certainly provoked in her the sort of light reaction characteristic of mindless scrolling, rendering Brownsville as something more banal than a beacon of possibility for life beyond racial capitalism: a "cool" version of normalcy. *Subjects of the Sun* interrogates this juxtaposition of normalcy and possibility—mundane visuals and transformative visions—exploring how everyday images, graphics, and digital platforms animate an energy transition that feels as habituated as any facet of our screen-based lives even as it gestures to a world beyond anything we're accustomed to. As Selena closed her browser and announced her intention to learn more, I hoped that both of these affects—normalcy *and* possibility—would coexist in EQUAL's forthcoming efforts to mimic Brownsville's success. Selena wouldn't just view MGA's residents as a visual icon of insipid inclusivity . . . right?

THE NEXT day, Selena announced her intention to schedule a tour for EQUAL to see the microgrid in-person. In the moment, this made sense to me. How better to learn about an unfamiliar space than visiting it and forming ties with the people who steward it? But, in retrospect, this sensible idea strikes me as slightly off. For in-person solidarity is not interchangeable with a sightseeing tour. If Selena had proposed to meet MGA residents to better understand how on earth they developed a microgrid, this would be aligned with the people-oriented spirit of grassroots activism that EQUAL championed. But, like Mark, she didn't mention the people. She instead focused our attention on touring the microgrid. And while the sightseeing tour is a common approach to learning about unfamiliar things in the contemporary world, it is also productive of the colonial forms of difference on which racial capitalism feasts.

Indeed, the sightseeing tour, whether pursued for imperial, empiricist, or leisurely reasons, emerged as a Euro-American colonial practice for viewing and observing the Other—usually at a safe distance from the exotic object of inquiry—positioning one's sight as a primary medium for learning about difference.[38] This ocularcentric approach to apprehension generated a one-sided form of encounter, rendering cultural difference in terms of powerful viewers and objects to be viewed. In this vein, the sightseeing tour was an integral tool in the mastery of nature, enabling colonial powers to visualize and thereby control the manifold forms of life deemed less than human, including nonwhite peoples.[39] As such, the sightseeing tour was central to the global mapping of differential value, creating the raced extractive zones constitutive of racial capitalism. While today's sightseeing tour has seemingly more benign objectives than its imperialist forebearer, its colonial history cannot be extricated from current practice, as our Brownsville tour would soon demonstrate.

At EQUAL, these colonial dynamics first reared their head when Selena suggested that the real estate management company who owns MGA could serve as our tour guide. This seemed reasonable, as said company managed the several dozen residential buildings where the microgrid was installed, and they therefore were the only party that could give us in-person access to it. But Selena was essentially proposing that a for-profit corporation run by white men would tour us around a poor Black community. None of us gave any thought as to what it would mean for us to view this community through a corporation's gaze. In Selena's proposal, then, this community would be an object of inquiry to observe at a distance from the perspective of corporate ownership, not a shared space in the struggle for environmental justice. Perhaps this tradeoff would be justifiable if seeing the microgrid would reveal something that could help EQUAL replicate MGA's success. But at no point did we consider how a sightseeing tour could aid EQUAL's efforts. We took for granted that the structure of knowledge acquisition at the core of the colonial paradigm of sightseeing was essential to gaining the information we sought. This sort of reasonable hubris is a central theme of *Subjects of the Sun*, as I explore the ways in which energy professionals employ a "view from nowhere" in destitute spaces that are imagined as a ground zero of the just transition.[40]

While the company agreed to host us for a tour, scheduling it proved difficult for logistical reasons. Furthermore, none of EQUAL's activists knew the technical ins and outs of installing electrical infrastructure, so there was, in theory, no need to physically go to Brownsville to witness the microgrid with their own eyes. But the activists *did* know a tremendous amount about solar financing programs—an important body of knowledge, as we will see—so a quick phone call with the company to discuss these programs should have been sufficient. *Seeing* the microgrid would not enable us to secure the necessary financing and technical expertise to develop another microgrid in EQUAL's community, nor would it quench our curiosity about *how* the microgrid came to be.

Yet when we discussed the possibility of simply scheduling a call with the company, we quickly dismissed this reasonable idea for reasons we never articulated. We instead affirmed that we were "excited" to *see* MGA. Perhaps we were less interested in learning how the improbable microgrid came to be and more intrigued by a banal image of it: the novel normalcy aestheticized by Instagram—this remarkable infrastructure integrated into the unremarkable texture of everyday life. Or perhaps, instead, we were intrigued by a *vision* of it: a present-day future in which the microgrid generated not only electricity but also political sovereignty and a sense of resiliency, uplifting MGA's residents out of the ravages of racial capitalism. Was our planned tour simply a way of witnessing a shallow image of "cool" power, to quote Selena, or, instead, did it emerge from a more substantive imaginary of collective power? This book troubles the line here, suggesting that solar is a particularly affective technology because it blurs distinctions that separate late liberal optics from the lens of grassroots visionaries—the status quo from the spectacular—giving form to a muddled political ideology that I will introduce shortly. I only had an inchoate sense of this muddledness at the time, unable to fully grasp our intentions as we finally locked down a date with the company and hopped on the subway to Brownsville a few weeks later, entranced by the spectacle of solar on a sunny spring day.

A SPRAWLING campus of low-income homes named for the Black revolutionary Marcus Garvey, MGA is occupied almost entirely by low-income Black people. These homes are now powered in part by solar panels, a giant lithium solar battery, and a cutting-edge fuel cell, connected underground by a complex network of circuitry that could keep the community running on solar electricity if the grid malfunctioned. While the firm that owns and manages the homes is a for-profit corporation run by white men, it has developed a reputation for doing the godly work of building quality affordable housing in a gentrifying megacity. Furthermore, Don, one of the company's white male senior property managers, went out of his way to schedule a tour with us, a group of EJ activists—a thoughtful gesture that I naively thought confirmed the company's Good Samaritan reputation. *Subjects of the Sun* explores many selfless corporate professionals like Don whose good intentions reproduce differential value vis-à-vis solar due largely to the technology's enmeshment in their screenwork, as I will show.

We sat with Don in the MGA community center, and he explained that the microgrid was primarily developed to address the city's broader infrastructural problems. The city's residential utility company, Con Edison, feared that the community's electricity demand could overwhelm the old electricity substation that powers Brownsville and spur a blackout. For this reason, they launched a "demand management" initiative that financed the microgrid. Thus, if Brownsville had had a decent substation, the utility would never have paid for the microgrid. Ever since my conversation with Mark, I had hoped that the microgrid was evidence of a marginalized community rising up from the periphery and beating the odds to secure their livelihood through innovative infrastructure, but, to my disappointment, it was just a cost-efficient, top-down operation instigated by a notorious investor-owned utility to safeguard their own lousy machinery.

At that point, there was, on the surface, no need for us to climb to the rooftops and look at the solar panels. The information we had come looking for—how to replicate this sustainable behemoth—would not be forthcoming no matter how long we stared at those panels or the lithium battery. Yet I certainly still wanted to see the internationally known solar microgrid with my own two eyes. I refused to dismiss the possibility that the community looked to it as a source of self-sufficiency, economic autonomy, and communal pride regardless of Con Ed's role in developing it—and I hoped we could sense this affective power if we inspected it in its local context. Even if MGA's residents had not fought for a microgrid, I still imagined an Afro-futurist juxtaposition between high-tech modernity and the wretched of the earth—the contiguity in the Instagram post. Regardless of whether my comrades also harbored this vision or not, we all were intent on seeing the microgrid even after we had acquired the information we had long sought.

Don proceeded to lead us toward the campus's electrical room, exposing us to local residents going about their day-to-day lives: congregating on their homes' stoops, entering their apartments with groceries, checking the mail, talking on the phone on outdoor benches. He seemed unnervingly at ease moving through these people's intimate spaces. With an entitled gaze evocative of a safari, he positioned the microgrid as charismatic megafauna—as though it overshadowed the dark-skinned natives whose homelands we traversed—bringing the colonial dynamics of our sightseeing tour into focus.[41] It was then that I asked the two local men about the microgrid, only to be met with bewilderment. As the tour progressed, I continued to peel off from the crowd intermittently to ask residents if they knew about the solar microgrid. The answer was unanimous: *No.* Our tour, then, enacted a limited form of witnessing, a "view from nowhere" putatively detached from our positionality, as sightseeing—a vestige of colonial power—denies natives the lens through which it surveils their own spaces while positioning outsiders like us as unmarked.[42]

When we viewed the microgrid's subterranean electrical equipment and rooftop solar panels, the technology just sat there, making no sound, emitting no smell, revealing nothing we didn't already know. This only confirmed my sense that there was no need to *see* the microgrid. All the while, Don talked to us about wattage, grid-interconnection rates, and countless other numerical data—information that he and his colleagues withheld from the MGA residents, reinforcing the invisibility of their solar infrastructure. As we asked questions and took notes, he responded to us like we were worthy interlocutors, but the same couldn't be said of the people living in the buildings he managed.

Unlike the solar panels and the electrical equipment, the battery caught my comrades' attention—this high-tech homage to futurity on an impoverished Brooklyn backstreet—and they started taking selfies with it for EQUAL's social media. But they weren't the only ones intrigued. The battery shed's gate had been opened for our tour, so two local residents who were walking by the battery got their first unwitting glimpse of it, poking their heads in before scooting past the point where the locked gate usually was. The two residents, Roxanne and Felicia, appeared transfixed by this giant futuristic machine sitting in their backyard. "What the hell is that?" Roxanne exclaimed.

I retreated from the shed to talk to them. "You've never seen this before?"

"Nah, that gate's always locked," said Felicia. "I never even thought about what's behind it."

"Oh, it's really cool, it's a giant battery that stores the energy from the solar panels on your roofs." Par for the course, they, too, had never heard of MGA's solar panels. Instead of outright indulging their curiosity, they slowly crept closer to the shed with an uncertainty that suggested that they felt out of place. Within seconds their caution was validated.

"There's nothing to see here, you two," said the office manager. "This is a private tour for our guests, and believe me, you don't care about this, okay?" Roxanne and Felicia got one final glance of the mysterious battery and then obligingly walked back to the sidewalk. The office manager promptly closed the gate. "If you don't set boundaries, they'll do anything," she explained to me, expecting sympathy as she concealed the battery from the residents. But I was aghast. The battery's intriguing sight could have illuminated the power of the women's place of residence, empowering them, in turn, to participate in governing the electricity that they depended on every day. By prohibiting them from even seeing the battery, the office manager effectively stymied the microgrid's transformative political potential, fomenting differential value in the process. The people who have historically been unseen in the spaces where electricity is controlled—Black people, women, the poor—remained invisible, as a private corporation run by white men determined who saw what in the governance of electrons that remained largely imperceptible to the eyes.

OUR SIGHTSEEING tour presented a dispiriting contrast to the Instagram post, offering an up-close view of how poor women of color like Roxanne and Felicia are marginalized in spaces that late liberal screenwork brands as inclusive. This marginalization demonstrates how differential value is centered on not just systems of capital but also who and what can be seen. While the Instagram post visualized solar infrastructure in ways that projected inclusivity, our tour showed how said infrastructure remains unseen to those whom landlords deem invisible, positioning their residencies as not communal spaces but instead a source of value vis-à-vis renewable energy.

The tour, then, offered a lens into the ways in which invisibility links structures of intersectional power with infrastructures of electrical power—a central theme of this book. Women, people of color, and poor folks—people like Roxanne and Felicia—have historically seldom seen and rarely been visible in the shrouded halls of power where electricity is controlled: the boardrooms and executive offices of the institutions that generate, transmit, distribute, and regulate the invisible streams of charged electrons that make modern existence possible, from federal energy agencies to rural cooperative utilities to private energy companies.[43] In this way, the structural invisibility of marginalized people coproduces the infrastructural invisibility of modern electricity. However, these invisibilities are not one and the same. The erasure of Black women like Roxanne and Felicia is of a different order than the relative obscurity of electrical infrastructure, to say nothing of the imperceptibility of electrical currents. What ties these invisibilities together, though, is the aforementioned form of visual apprehension at the heart of our sightseeing tour: the view from nowhere. This "view" refers to the removed perspective of experts whose distance from the spaces and people they observe allows them to occupy an unmarked position of power—an inscrutability in contradistinction to the discernible identities of those entities that sit beneath their gaze. Modern energy infrastructure is fundamental to this classed differentiation, as it produces and distributes the commodities that power racial capitalism—electricity, heat, and fuels—in spaces largely cordoned off from everyday life, demanding a technocratic class to manage those infrastructures *at a distance*.[44] Such distance, in turn, affords this technocratic class a panoptic view of society, a totalizing gaze of the humans who depend on those infrastructures to survive, converting laypeople into ratepayers that elites can view through demand management maps, energy consumption data, control rooms, and administrative nodes.[45] Our tour of MGA's microgrid suggested that this privileged perspective is not reserved for the dirty, industrial-scale infrastructure that has historically powered racial capitalism—that solar can similarly render everyday people as passive energy consumers prohibited from even seeing the spaces of power they depend on every day.

But while the MGA residents had been shut out from these spaces of power, there were in fact people with positionalities far closer to Roxanne than to Don who were afforded a view from nowhere: EQUAL's staff and me. Don's white-led management company went out of their way to accommodate us—a group of activists, many of whom were women of color, representing a low-income Black and Latino community—suggesting that the view from nowhere in the solar energy industry is not solely reserved for white elites. In *Subjects of the Sun*, I contend that this sort of inclusivity points to the ways in which late liberal screenwork is transforming energy governance with the rise of solar. Specifically, in an information age in which energy governance largely operates through LCD screens, visual data are creating new opportunities for people from marginalized backgrounds with cloud-based proficiencies—such as Selena—to govern an electricity production apparatus that has traditionally been controlled by technocrats at a distance. But, as I show, these opportunities are not available to similarly positioned people who lack those proficiencies, such as MGA residents. Along these lines, this book argues that screenwork is integral to the management of intersectional difference in the United States' emergent clean energy economy, as smart boards, PowerPoints, and digital media related to solar electricity blur the ideological lines that have traditionally separated the predominantly white technocratic world of energy production from the predominantly nonwhite activist world of EJ. To illustrate this core problematic, I now return to EQUAL's work in the days after our tour, considering in greater depth why EQUAL's staff and I were given the privilege of viewing MGA's solar microgrid while Roxanne and Felicia were not.

AFFECTIVE POWER

When we returned to EQUAL's office the day after our tour, EQUAL's staff convened their weekly meeting to discuss their solar work. I assumed the meeting would allow us to strategize about how we might organize on the ground for a local microgrid in light of the dispiriting things we had learned on the tour. This seemed like a well-founded assumption, as community organizing was the foundation of EQUAL's work. The meeting room we convened in corroborated this sense, replete with the trappings of an archetypically grassroots space: a slightly scratched conference table that looked like it had been repurposed from an elementary school, mismatched chairs, a tiny patch of mold on the ceiling, windows that hadn't been cleaned in years. Posters chronicling EQUAL's past EJ campaigns adorned the walls, highlighting the community's fights against poor air quality, asthma, inequitable waste disposal, and environmental racism. As nonprofit professionals, EQUAL's staff were dressed in clothes that matched the room's somewhat unprofessional appearance: a rumpled button-down, basic jeans, their organization's bright T-shirt.

But if I were a little more attentive to the room's *feel*—its affective texture—perhaps I wouldn't have assumed that our discussion would focus on community organizing. For the collective attention of everyone in the room centered on a giant smart board at the front of it—a glowing LCD surface that dwarfed everything else in that space. A dashboard beamed from the screen, projecting metrics on the solar campaign's progress: the number of participating buildings, the number of kilowatts of solar installed, projected reductions in carbon emissions. Crucially, the dashboard stylized these metrics such that they didn't appear as dull digits. They instead took the form of colorful graphics and icons that, when combined with the screen's glow, caught the gaze of anyone who walked into the room—even those who didn't seem particularly vested in the campaign. Selena presented these updated metrics on the screen every week, grounding our solar discussions in both the quantitative data of EQUAL's campaign and the fluorescent surface that projected these data. So I shouldn't have been surprised when she opened our reflection on Brownsville by sharing the technical information we had gleaned from the tour, mentioning the number of solar kilowatts that the microgrid generated and Con Edison's incentive program, first and foremost. There was no mention of Roxanne and Felicia, no discussion of MGA's residents' involvement (or lack thereof) in matters related to the microgrid, no reflection on the untapped political potential of such innovative infrastructure. After sharing the data, Selena concluded briefly that MGA did not really offer a clear model for EQUAL to replicate, and she quickly moved to the next agenda item: the numbers glowing from the smart board.

The data on-screen instantly took center stage. The staff ruminated on these numbers, contrasting their progress to date with the enumerated goals they had set at the campaign's outset, lauding themselves for the close proximity of these two sets of metrics. They spoke of large local rooftops with the potential to generate the number of solar kilowatts necessary to achieve their overarching campaign goal, focusing on quantitative output. Much like our brief recap of our tour, there was no talk of promoting justice, countering corporate power, or dismantling the status quo. While an equity agenda initially inspired this campaign, it had devolved into a crusade for optimizing metrics, animated by the dashboard's glow, creating a spectacle of technical rigor. As their eyes flitted back and forth between one another's gazes and the screen before them, the dashboard felt like our collective ground, anchoring our focus.

In this way, the room was eerily evocative of the predominantly white clean energy and property management corporations that I was also observing in the city's whiter, more affluent areas—corporations whose meetings were centered on dashboards with graphics on maximizing returns through solar. Many of those firms' employees embraced an ideology of growth and free markets that my EQUAL comrades renounced, but this ideological dissonance did not temper the similarity between the unspoken sensibilities that permeated both groups' spaces: the obsession with data, a performance of rigor, and a fixation on screenwork. Even as Selena and her colleagues mentioned "energy poverty" and "energy democracy," their screenwork endowed EQUAL's meeting with a technical sensibility unconcerned with social equity—this feeling of administrative proficiency and professional competency that is incongruent with grassroots organizing. In other words, the *affect* of the room, the smart board's fluorescent feel, emanated technocratic managerialism, not insurgent campaigning.

Affect is notoriously difficult to define. Comparable to but different from *emotions, affect* refers to the unspoken "intensities" that circulate between different forms of life—the inchoate sensibilities that permeate our relations with other bodies, spaces, things, and phenomena.[46] By *inchoate,* I do not mean "not fully formed," for affects can be thick and dense. Instead, I'm pointing to sensations that have not coalesced into a clear-cut form that we can effortlessly qualify in conventional language. As Lauren Berlant explains, affect "registers the conditions of life that move across persons and worlds, play out in lived time, and energize attachments," "saturat[ing] the corporeal, intimate, and political performances of adjustment that make a shared atmosphere something palpable."[47] As it points to intensities untethered to a particular entity or person—irreducible to a material form—*affect* is less a definitive descriptor and more an approximation of that which it signifies. Sianne Ngai aptly theorizes this indeterminacy: "Affects are less formed and structured than emotions, but not lacking form or structure altogether; less 'sociolinguistically fixed,' but by no means code-free or meaningless; less 'organized in response to our interpretations of situations,' but by no means entirely devoid of organization or diagnostic powers."[48]

To say, then, that EQUAL's meeting had a *technocratic affect* is to attempt to capture something elusive—to begin to qualify the room's feel as the activists engaged with metrics, performed technical competence, and reveled in the smart board's glow. In *Subjects of the Sun,* I argue that the affective resonances between a grassroots campaign for renewable energy in a low-income community of color and the predominantly white, cleantech corporations that I simultaneously explored point to the power of datafied renderings of solar to not simply affect how social spaces *feel* but, more significantly, shape the intersectional politics of energy transitions. Specifically, I contend that the technocratic affect of EQUAL's smart board is symptomatic of a broader political project to mitigate climate change vis-à-vis the so-called free market—*a project that, paradoxically, many of its purveyors consciously oppose.* As I will show, the affective power of dashboards, PowerPoints, and cloud-based platforms inspires people who reject this project's neoliberal ideology to enthusiastically enact it in the name of environmental protection, infusing free market dogma into our everyday spaces, professional lives, and social movements. Consequently, anticapitalist women of color at an EJ organization, for instance, can unwittingly employ the market logics of efficiency and growth when they're fighting for local solar. Along these lines, screenwork affectively animates imaginaries of a late liberal world in which metrics help overcome structural inequalities, as visualizations of demographic data often occlude the complexity of racial difference. As I will show, an affect of technical rigor emanates from EQUAL's screenwork, steering my comrades away from their community organizing work.

Yet it would miss the mark to suggest that their digital dashboard deluded them into abandoning their radical politics—that it necessarily signifies compromised values. Instead, this book argues that the materiality of energy infrastructure *requires* regimes of quantification such that even the most progressive, democratic effort to solarize communities necessitates a rigorous engagement with datafied screenwork. This quantification imperative is directly related to energy's in/visibility. Specifically, kilowatt-hours (kWh), renewable energy tax credits, energy prices, and carbon metrics are the primary medium through which electricity generation becomes discernible to most people who don't directly work with the power sector's physical systems. When most of us *consciously* engage with our electricity generation infrastructure—when that infrastructure moves to the forefront of our considerations—we do not kinesthetically interact with the wires, cables, electrical rooms, and power plants that enable energy to effortlessly flow throughout the Global North. When most laypersons flip a light switch or charge their phones, they do not consider the violent sound of refining fossil fuels, the sprawling spatiality of transmission lines, the mechanical convulsions of petroleum extraction, or the corrosive plumes of smoke puffing out of their homes. Instead, when most of us consciously engage with electricity generation, it is through our electricity bills, reports on utility rate hikes, proposals to fight climate change with renewable energy, air-quality data, and activism grounded in such information. And in all of these instances, we primarily apprehend electricity through the medium of numbers. As such, most of us do not touch, smell, hear, sense, or taste the vast material substrate of our electrified existence. While we similarly do not directly *see* this substrate, metrics enable us to indirectly register it in a visual form that digital platforms aestheticize. This visual imperative is even more acute with solar electricity, for the process of generating energy from the sun does not entail the forms of extraction and combustion that most other electricity generation requires—intensely physical practices that one can touch, smell, taste, and disrupt with one's body—underscoring the centrality of metrics in engaging with this process. Because solar electricity is not merely quantifiable like the human body or any other datafied object but also, more specifically, apprehended *primarily* through data, any activism to address it must engage closely with screenwork—a point this book expands on extensively.

As such, just-transition activism employs the lenses, media, and affects of the corporate energy sector it opposes, generating a shared field of action that muddies, but never eradicates, the intersectional divisions that uphold racial capitalism. Specifically, this book contends that sustainable energy screenwork affectively cross-pollinates radical climate justice politics with technocratic ideology, diluting the normative political poles that separate the white-collar expert from the anticapitalist activist. Out of these blurred boundaries emerges what I conceptualize as the *equicrat*, short for the *equity-minded technocrat*: a liberal subject who reflexively mobilizes their technocratic prowess in service of social equity. The equicrat heuristic calls attention to the counterintuitive compatibility of grassroots governance and rule by experts under late liberalism. In theorizing this compatibility, the equicrat illuminates the performative terrain through which subjects at once reject and reproduce differential value. The NYC-based EJ organizations and sustainable energy corporations that I discuss in *Subjects of the Sun* collectively comprise an ideal site for exploring this equicratic reconfiguration since they both offer insights into how solar works with late liberal screenwork in ways that destabilize the normative distinctions between nonprofit activism and for-profit commerce.

Our Brownsville tour underscored the significance of this equicratic reconfiguration. For the MGA management company offered my women-of-color colleagues access to the microgrid while denying this access to their women-of-color tenants due to a late liberal elitism in which data proficiency operates as a common ground across raced, classed, and gendered lines. Put simply, Selena could connect with the company because they all trafficked in the same energy management screenwork. But when the company viewed the microgrid through a dashboard lens, focusing on metrics to reduce their energy costs, the microgrid did not appear as a force for an *equitable* clean energy future or a means of empowering the poor people of color who occupy their properties.

So if you are one of the residents who are theoretically benefiting from this electrical upgrade, your feelings toward the microgrid are irrelevant to the company. Layperson sentiments are impertinent to renewable energy when electricity is just a set of numbers for a managerial class to optimize on a smart board. Yet if you approach the company and demonstrate some basic proficiency with their dashboard's data points, if you take on a technocratic tone and inquire how they secured the investment to bring a microgrid to the 'hood, if, in sum, you cultivate the *affect* of an energy professional, they will give you the view of their beloved infrastructure that they deny to Roxanne and Felicia—even if you're an outsider like my EQUAL comrades and I. This equicratic convergence, then, points to the affective power of the solar dashboard: its capacity to transform the perspectives and performances of people who have traditionally been excluded from energy governance, empowering them in the process. But the marginalization of Roxanne and Felicia suggests that this affective power is a double-edged sword; while it generates new late liberal solidarities between activists of color and white corporations, it also keeps intact the structures of racial capitalism that render poor Black women invisible.

Thus, the Brownsville case study begins to demonstrate the ways in which visual renderings of solar infrastructure (whether on Instagram or a smart board) work in tandem with solar's material properties (its diffuse spatiality and datafied power) to evoke a late liberal order in which professional work can seemingly overcome the differential value that it is in fact embedded in. These dynamics evince solar's *affective power*: its capacity to shape our political aspirations, everyday comportments, and unexamined predilections in ways we lose sight of when we view infrastructure only through an Instagram post, dashboard, or sightseeing tour. *Subjects of the Sun* aims to illuminate this affective power, uncovering how solar inspires late liberal subjects in ways that at once upend and uphold the relations of racial capitalism.

Throughout this book, I take seriously the aforementioned insight from API's propaganda: "There's energy everywhere. But sometimes it can be hard to see." For in a society that privileges sight and is powered by an electrical force undetectable to the human eye, transforming our energy system requires us to see not that invisible phenomenon but instead its physical conduits all around us: the machinery, landscapes, and bodies that generate the streams of charged electrons that otherwise elude our vision. I therefore ground my analysis in the five aforementioned material properties of solar infrastructure: sunshine, a decentralized terrain of residential rooftops in dense urban landscapes, the modularity of solar panels, the quantifiability of electrical currents, and humans' physical labor in the sustainable energy industry. In attending to these material properties, I aim to ethnographically uncover how electricity infrastructure produces not only physical power but also political possibilities, exposing how our politics are configured by the very technologies that our politics are supposed to control.[49]

Chapter 1, "Shine," focuses on shiny images of pastoral solar farms in a place that couldn't be further from pastoral: NYC. These images depict anthropogenic technologies as a natural outgrowth of the nonhuman world, situating solar infrastructure in an imaginary of pure nature. As such, they suggest that solar collapses one of the foundational divisions of racial capitalism: the nature/society divide that renders the nonhuman world as extractable terrain. I show how this collapse informs late liberal visions of high-tech commodities that can paradoxically "return" alienated communities of color to a premodern state of purity, obfuscating solar's extractive supply chains and racialized production processes. However, I attribute this obfuscatory naturalization not only to the fetishistic workings of capital but also to the affective power of the sun. I suggest that the sun's material properties—its shine, glimmer, and interactions with the city's built environment—affectively animate apprehensions of solar infrastructure as a transcendent natural force, giving form to pastoralized solar images. Chapter 1 thus challenges Marxian orthodoxy on alienation and commodity fetishism under racial capitalism, foregrounding the affective power of a biospheric phenomenon in an analysis of technological transformation.

Chapter 2, "Space," introduces the corporate energy equicrat by exploring how NYC-based sustainable energy technocrats use cloud-based platforms to affectively connect with marginalized spaces. It opens with a discussion of Google Earth simulations that visualize solar panels on rooftops in poor communities of color. I argue that these visual interfaces and other similar platforms work in tandem with the decentralized spatiality of sustainable energy technologies to enable (predominantly) white energy experts to work with poor communities of color, demanding that they engage closely with everyday people in intimate, personal spaces. This, in turn, transforms the political contours of energy expertise, fostering an ideologically muddied form of technocratic care that conjoins free market practice with socialist and leftist principles. As such, the cloud-based culture of energy experts taps into a late liberal imaginary that conceptualizes solar energy and energy efficiency commodities as tools for ameliorating the structural inequalities of racial capitalism.

Chapter 3, "Modules and Metrics," theorizes how the modularity of solar panels and the quantifiability of electrical currents affectively incubate the grassroots equicrat. I ground this discussion in an ethnographic analysis of two of EQUAL's graphics: an online flyer for a grassroots community solar campaign that misleadingly aestheticizes racial diversity, and a spreadsheet comparing solar installation contractors that focuses narrowly on market values. These graphics point to the ways in which justice-oriented work often operates through late liberal representations of inclusivity and idealizations of market-based equality, revealing the equicratic character of EJ activism. I suggest that solar's modularity and quantifiability accentuate this equicratic character by centering economies of scale in a mode of activism that has traditionally rejected the economizing imperatives of racial capitalism. Specifically, I contend that the power to connect and disconnect solar from the centralized grid, and the physical imperative to quantify solar's electrical inputs into the grid, enable EJ activists to participate in electricity production. This, in turn, shifts EJ activism away from the immediate bodily concerns of low-income communities of color and toward market-based energy governance, as visions for energy democracy often focus more on the equicratic desire to produce more solar in such communities than on the stated needs of the people whom solar is supposed to serve. This has the effect of deepening EJ activism's dependence on what is often called *philanthrocapitalism* and orienting EJ activism toward neoliberal mandates for cost efficiency.

Chapter 4, "Bodies," focuses on the corporeality of solar work, offering a blueprint for practically transforming the relations that constitute energy infrastructure and thereby moving beyond the paradigm of differential value. I ground my discussion in social media posts that aestheticize solar installation labor as a force that can counter racial capitalism. I argue that these aestheticizations actually reveal the political constraints of late liberal imagery. As a corrective, I call for an ecosocialism that centers multisensorial relations with solar, focusing less on screenwork than the book's other three chapters. I argue that the development of solar infrastructure necessitates we see, touch, sense, consider, and work on our surroundings with a certain care that is absent in many forms of industrial-scale, fossil fuel energy production. This care is not simply an ideological concern for "the environment" but also a *corporeal feel* for the spaces we inhabit. As such, I suggest that the work of transforming the landscape vis-à-vis solar can cultivate a bodily attunement to the built environment. Theorizing the political power of this corporeality, I contend that the everyday labor of energy transitions can better align care for ecosystems with an intersectional pro-worker politics.

I attempt to uncover this potentiality by exploring the corporeal experiences of both blue-collar and white-collar workers in the sustainable energy industry. A labor politics that foregrounds corporeal relations with the environment can shift the work of energy transitions from monetary *returns* toward what is often understood as ecological *return*: an ecocentric ethos absent in both corporate environmentalism and EJ. Yet an overemphasis on *return* can neglect the fact that many marginalized workers have no place to return to—that dispossession is the precondition of Black and brown labor. As such, a pro-worker approach to energy transitions must be attuned to the complex ways in which marginalized workers navigate the space of dispossession through their physical labor, leaving room for the contradictions of environmental conservation in a megacity powered by racial capitalism. I therefore call for a just transition that brings together an intersectional labor politics and an ethos of ecological return through a closing discussion of two Black and brown worker-owned, solar-powered businesses, exploring how they balance the ecological and economic dimensions of energy transition.

1

I wasn't surprised when I saw Danica in her "No One Is Illegal" T-shirt standing next to a giant screen projecting a PowerPoint cover slide with the image shown in figure 1.1. An icon of an inclusive clean energy economy, this oft-repurposed photo suggests that solar can empower those most marginalized under racial capitalism. While it features job trainees in Richmond, California—*not* NYC—NYC-based activists like Danica often use it because it simulates a late liberal ideal that they hope to manifest locally: happy workers of color driving the emergent solar industry!

This late liberal ideal inspired the event that Danica had prepared the PowerPoint for: a technical tutorial on the nuts and bolts of solar. The tutorial was geared to her fellow environmental justice (EJ) activists who wanted to better understand the specifics of the "clean energy technology" they had long championed. They hoped that this knowledge would strengthen their efforts to promote solar as a means of equitable economic empowerment. Her organization, a Brooklyn-based affordable housing nonprofit, had convened this event, which she opened with an impassioned pitch linking solar to broader social justice struggles, gesturing to the image while sharing her vision for an "inclusive clean energy future." But the image, like her pitch, presented a vision that was far shallower than the political commentary she usually shared with me over after-work drinks. A tenant organizer with an urban planning degree, Danica often praised radical policy interventions like prison abolition, espoused sophisticated understandings of racial capitalism, and critiqued liberal representational politics. Yet she consistently returned to the image in figure 1.1 and ones like it—images that eschew structural

FIGURE 1.1. A frequently repurposed photo of Black solar installers—an icon of the just transition. CleanEnergyAuthority.com, "Mosaic Uses Crowdfunding to Finance Solar Projects," December 20, 2012, https://www.cleanenergyauthority .com/solar-energy-news/mosiac-uses-crowdfunding-to-finance-solar-projects -122412.

critique to promote inclusive labor markets—seeking to distill her desire for climate action and antiracism into a single accessible visual.

This distillation imperative is baked into PowerPoint—the primary medium through which she shared information with her organization's constituents. While she created presentations for campaigns that she characterized as "leftist, not liberal" (e.g., a campaign for sex worker housing), PowerPoints traffic in a late liberal aesthetic that can detract from such leftist organizing. With an air of authority that social media lacks, an effective PowerPoint presentation makes complex ideas palatable through appealing visuals, graphics, and simplified text.[1] As such, it can distill structural injustices into clear-cut data points (a graph with the percentage of minorities living by toxic facilities, for instance) rendering these injustices manageable—a matter of technocratic policy—and thus not cause for agitation.[2] Unconcerned with intractability or subtlety, it often propagates inoffensive, unambiguous visuals—a photo of smiling people of color signaling inclusivity, for instance—as "politics becomes entangled with the affordances of a software package."[3] PowerPoint is therefore amenable to "diverse" representations that visualize difference as a straightforward spectacle, offering little space for the contentious discourse that powers social movements.

But despite Danica's "diverse" cover slide, the rest of her PowerPoint had nothing to do with social justice; it simply shared information on the technical ins and outs of solar. The image in figure 1.1 was only intended to underscore the political stakes of her otherwise didactic presentation. Yet when this presentation shifted to technical matters, the cover slide's late liberal aesthetic did not subside. For her subsequent slides presented digital renderings of solar farms with subtle racial undertones, depicting renewable energy not in an urban environment (a notable oddity given that we were in NYC) but, instead, in an idyllic pastoral setting—"nature" at its most aestheticized. Solar panels appeared on grass beneath a blue sky and a bright sun without any trace of urban life or human beings (see figures 1.2 and 1.3). If late liberalism refers to the management of racial difference vis-à-vis private markets, then the term does not, at first glance, seem applicable to these images. But in idealizing a pristine nature that is, paradoxically, bereft of humans *and* partly constituted by anthropogenic commodities (i.e., solar panels), these images do in fact grapple with difference, naturalizing the core ontological divisions of racial capitalism. That is, these images offer a twenty-first-century twist on a trope that has long informed differential value: the allegedly pure outdoors—a space mythologized in ways that erase its Indigenous inhabitants and thus catalyze colonial extraction.[4] Yet, as we will see, these

FIGURE 1.2. A reproduction of my interlocutors' PowerPoint visuals. Graphic by Cynthia Zhang.

FIGURE 1.3. An internet image that aestheticizes the pastoralized solar panel and its affective shine through hyperreal digital graphics. Limitless Energy, accessed November 25, 2022, https://www.limitlessenergyllc.com.

images also tap into desires to "return" to a space unmarred by extractive racial violence vis-à-vis technology that is produced *through* extractive racial violence—a late liberal fantasy in which markets solve the discord they sowed.

Danica, a self-described anticapitalist, had no interest in promoting such fantasies; she was simply trying to populate her PowerPoint with palatable visuals. Yet I want to suggest that her images begin to shed light on how EJ activists and other just-transition advocates often unwittingly reproduce—through their late liberal renderings of solar infrastructure—the very extractive, exploitative logics that they reject.

In this chapter I show how the aesthetic project of figures 1.2 and 1.3— its natural affect—gives form to the racial justice sentiments of figure 1.1 by focusing specifically on a central visual element of the former: sunshine. As I show, the images in figures 1.2 and 1.3 illustrate how the sun naturalizes solar energy in ways that suggest it can overcome the differential value, extractivism, and alienation of racial capitalism. Put simply, the sun obfuscates solar's dependence on racialized exploitation, mineral extraction, and even fossil fuels. In this way, sunshine affectively animates aestheticizations of solar as a "clean" energy. I suggest that these aestheticizations are symptomatic of solar advocates' alienation from solar's exploitative supply chains, maintaining differential value by illusorily simulating a space unmarked by differential value.

But the sun's affective power cannot be reduced to late liberal screen-work. Across historical contexts, the sun has animated numerous social, political, and cosmological imaginaries that celebrate its awe-inspiring nature—what English speakers often characterize as *the sublime*.[5] Drawing from William Cronon, I argue that racial capitalism vis-à-vis settler colonialism "domesticated" the sublime, transforming sunny landscapes into a spectacle to gaze at from a distance in the American settler colony.[6] I locate this domesticated sublime in New Yorkers' affective encounters with the sun in urban spaces as they gawk in awe at the shiny veneer of the city's sun-drenched skyline. Through this discussion I suggest that the sun is a spectacle in a megacity often imagined as the antithesis of "nature" in ways that the sun is not in spaces where racial capitalism doesn't scrape the sky. I contend that this sunny spectacle coheres with late liberal screenwork to naturalize solar infrastructure—aestheticizing it through the imagistic registers of figures 1.2 and 1.3. Thus, the aesthetic naturalization of solar infrastructure is inextricable from a settler colonial orientation to the more-than-human world that, as I show, foments differential value even in spaces committed to solar-powered EJ.

In considering how sunshine links long-standing settler colonial optics with late liberal visualizations of an "equitable, clean energy future," this chapter offers a nontraditional framework for understanding the entanglements of renewable energy and racial capitalism. While many have called attention to the ways in which renewable energy can reproduce the extractive order it is supposed to supplant, I more specifically zero in on the affective processes through which solar is problematically imagined as the natural antithesis of differential value.[7] This naturalization of solar matters because it shapes on-the-ground climate justice activism, hiding from view the exploited workers and ecosystems that solar depends on—obfuscating the contradictions of the just-transition framework.

Yet in calling attention to this chimeric affect, I challenge Marxian orthodoxy on the delusory effects of commodified technologies. Karl Marx argued that everyday people fetishize commodities—see them as possessing a life of their own, a character independent of human production—when capitalism alienates us from the natural world that we depend on to survive, obscuring the exploitative social relations and material processes that transform nature into modern goods and services.[8] Therefore, contemporary students of Marx might suggest that images that render manufactured solar commodities as natural are symptomatic of *commodity fetishism* and, by extension, of their curators' *alien-*

ation from nature. In other words, to depict factory-made technologies in a way that suggests they are almost the natural outgrowth of the biosphere (as figure 1.2 does) is to ostensibly fetishize such technologies, overlooking the unjust extractive conditions under which they were produced. Yet I challenge this prominent understanding of commodity fetishism by illuminating how the *sun*—not simply capital—endows solar commodities with the appearance of having a life of their own. This paradoxically positions a commodity as a medium for attenuating the alienation from nature often attributed to commodities under racial capitalism. Put differently, sunshine bestows a technology that is produced through an extractive political economy with the affective power to redress the existential unease caused by said extractive political economy. This suggests that the sun can shape our imaginations in ways comparable to capital precisely because it is affectively *unlike* capital.

While the entirety of this book engages closely with visual forms, this chapter is focused more on images than other chapters in order to illuminate the affective texture of environmental work in a city that is often imagined as the antithesis of "the environment." How does urban sustainability rely on an imaginary of the nonurban? Why would cityfolk render solar in terms of the pastoral? And how does the pastoral affectively animate a program for a more equitable city vis-à-vis renewable energy? The imagery discussed in this chapter is a helpful starting point for grappling with these questions, introducing the paradoxical structure of my interlocutors' projects, careers, and aspirations, which I take up in greater ethnographic detail later in the book.

DISSIMULATING SOLAR EXTRACTIVISM

The smiling faces and triumphant postures in figure 1.1 suggest that solar is not enmeshed in the structures of labor exploitation that have traditionally befallen Black and brown workers under racial capitalism. If nothing else, Danica and others often presented this image in Power-Points to illustrate the idea that solar can provide the edifying employment usually denied to poor people of color. In this imaginary, solar has the power to create pathways out of poverty, giving marginalized people greater autonomy over their livelihoods in an industry that is saving the planet.

The image, though, focuses on workers who *install* solar infrastructure —not workers who are involved in the *production* of solar infrastructure.

And while solar installers are certainly exploited (as I discuss in chapter 4), when we look beneath the shiny veneer of solar panels and consider the conditions under which this infrastructure is produced, we can detect not merely the violent labor practices and environmental desecration endemic to racial capitalism but also, more specifically, the most brutal instantiation of differential value: a racialized, genocidal system of enslavement. Put differently, as figure 1.1 aestheticizes solar in a way that suggests it can redress the economic vestiges of the transatlantic slave trade, solar is nonetheless produced through a contemporary iteration of racialized slavery.

I'm referring to solar manufacturing in the colonized lands of northwestern China. Almost half of the world's production of polysilicon—the central raw material in most solar panels—occurs in China's Xinjiang region.[9] Xinjiang is the indigenous homeland of the Uighur people—a predominantly Muslim ethnic minority in China—who have endured decades of racist exploitation and genocidal governance (such as forced sterilization) under the Chinese Communist Party.[10] Through coordinated acts of racial violence, the Party transformed Xinjiang into a robust site of industrial-scale mining and a global manufacturing hub powered by coal, imprisoned labor, and state-sponsored corporations—all of which are central to solar supply chains under racial capitalism.[11] Specifically, renewable energy markets demand *cost-effective* alternatives to fossil fuels. (*To save the planet, we must bring down the costs of solar!* or so say many renewable energy advocates.)[12] Xinjiang has several attributes that make solar production there attractive from this economistic perspective, including a criminalized Indigenous population whose high rates of unjust incarceration enable low-cost forced labor programs to operate across the region's factories; racialized economic discrimination that forces Uyghurs and other precarious minorities to seek employment at the lowest rungs of unregulated industries; a lack of political and economic sovereignty among Indigenous peoples, which prevents them from contesting extractive projects on their homeland; and coal-fired power plants unbeholden to environmental and public health regulations.[13] Under these conditions, state-sponsored corporations have pioneered an extractive apparatus that feeds quartz and other minerals into solar supply chains; developed a racialized base of precarious workers to mine and process these raw materials, often under dangerous conditions; built a prolific site of polysilicon production through a racialized penal system that amounts to modern-day slavery; and established cost-efficient factories that are paradoxically producing the world's green infrastructure with coal-fired electricity.[14] This horrific operation works directly through differential value. Han Chinese are in positions of power at every node of the system, as colonial, nationalist, racist, and Islamophobic discourses legitimize the violent persecution of Uyghurs and other Xinjiang-based Indigenous people in the space of solar production. Put differently, the extraction of surplus value vis-à-vis solar occurs through calcified racial differences between Han and Indigenous peoples—differences built into the legal, social, and political structure of an internal settler colony. Racial capitalism, then, is at the core of solar manufacturing, embedding "sustainable" technologies in a system of fossil-fueled slavery.

The resonances with the transatlantic slave trade are perhaps even more acute in the mining of cobalt, a central component of solar battery storage technologies. Workers in the Democratic Republic of the Congo produce over half of the world's cobalt under unregulated, hazardous, and often deadly conditions—work that is largely powered by fossil fuels.[15] Congolese miners risk their lives to extract a mineral whose demand has grown precipitously with the rise of renewable energy, working without basic protections at the bottom rung of a poisonous, fossil fuel–intensive multinational industry intent on reducing the costs of a green energy transition.[16] Many of these miners are children, whose small hands are valued due to the manual precision required to extract cobalt.[17] And all of these miners are rural Black Africans who lack any meaningful control over their labor in one of the poorest countries in the world, essentially forced into precarious work by a global system of differential value established through anti-Black colonization and the transatlantic slave trade.

It's no mystery why images of solar don't depict racialized de facto slavery. No industry is going to promote their product by highlighting their supply chain's human rights abuses and environmental degradation—especially an industry with a reputation centered on sustainability. And consumers of *any* commodity often know little about who made said commodity and how it was made, relating to it primarily in terms of its utility and social capital. Indeed, modern humans are alienated from the systems of production we depend on, and images of commodities are integral to this alienation, giving everyday goods the appearance of having a life independent of the workers and resources through which they came to be.[18]

Yet this alienation is particularly problematic in the case of solar due to its political profile. Solar is often imagined as the antithesis of extractivism, allegedly offering a true alternative to the environmental destruction, geopolitical violence, and corporate hegemony of a fossil-fueled world.[19] To suggest otherwise is to not simply contest its reputation but to more specifically depress hopes for a better tomorrow, creating an existential conundrum that is not on par with any discord that might emerge from an exposé of a clothing brand's sweatshop, for instance. Solar arguably depends on a public reputation that is more at odds with its complex material realities than that of most other commodities.

Yet the popular understanding of solar as an unproblematic climate solution is as symptomatic of alienation as any commercial narrative that elides the labor and environmental conditions of commodity production. While solar certainly has substantial positive impacts, this does not override the alienating implications of visuals that obfuscate solar's entanglements in Xinjiang factories, Congolese mines, and fossil fuels—images like those of figures 1.1, 1.2, and 1.3, which aestheticize solar as both a social and environmental good divorced from the modern-day slavery and extractive economies that it relies on.[20]

This aestheticization is especially illusory in the context of Danica's activism. While she presented the images in figures 1.1 and 1.2 to educate her comrades about the technical specifics of solar, this educational effort was part of her broader work to mobilize an activist base to incubate *solar markets* in low-income communities of color. Although she and her comrades were intensely critical of racial capitalism, they knew they needed to work with private contractors and private investment if they wanted to transition their communities to solar, for reasons I explore in chapters 2 and 3. So she never championed solar to make a profit for herself, but she *did* support local projects that enabled solar companies to profit off her community. Thus, her PowerPoint's aestheticization of solar aimed to affectively animate activism centered on growing a commodity market, demonstrating how late liberal visuals can indirectly buttress corporations even as they are curated to support worker rights (figure 1.1) and evoke natural purity (figures 1.2 and 1.3).

I want to suggest that these images distance their object from racial capitalism in ways that other promotional images don't—a contrast that points to the unique power of solar to dissimulate differential value. This might sound like a specious claim because these images do not deviate from countless others that misleadingly aestheticize commodities in terms of their alleged contributions to society. Numerous brands adopt, for instance, greenwashing aesthetics to suggest that their goods help the planet.[21] I conceptualize greenwashing as an aesthetic that discursively severs commodities from the material and social worlds that produce them by paradoxically simulating an *overcoming* of such alienation. Specifically, to greenwash is to posit corporate products as *un*alienated from the natural world—as a part of nature or as a tool for reconnecting us to nature—by calling attention to their putative environmental benefits. At first glance, the images in figures 1.1–1.3 appear exemplary of greenwashing in that they desituate solar technology from its corrosive means of production by simulating the antithesis of such extractivism—a fantasy that solar both redresses the inequality of racial capitalism (figure 1.1) and emerges from a pristine pasture unmarred by the extractivism that produces it (figures 1.2 and 1.3). Put differently, *the images present a commodity in a way that suggests it can ameliorate its structural conditions of possibility (i.e., environmental degradation and differential value).* This is particularly notable in figures 1.2 and 1.3, which situate solar in an idyllic landscape bereft of humans as if it sprouted organically from the grass, rooting renewable energy in nature's soil, not racialized toil.

But greenwashing also, by definition, aims to protect corporate profits—which couldn't be further from Danica's intention in presenting her PowerPoint. While her presentation was, as stated, part of a broader effort to incubate solar markets in low-income communities, she was not sneakily misbranding products to pad her bottom line or promoting commodities to a consumer base. If not for the purpose of accumulating surplus value or targeting potential consumers, why did she curate images that suggest that solar is not simply a tool for addressing climate change or saving money but, more significantly, a countervailing force to the exploitation, extractivism, and alienation that it in fact relies on?

In response, one might hypothesize that Danica didn't really care about these images. Perhaps she simply sought random avatars for the topic at hand and presented whatever images Google turned up. But, to the contrary, I want to suggest that her curation choices bespeak EJ activists' utopic understanding of solar. For as I show, the absence of NYC's landscape in her presentation is symptomatic of a naturalist imaginary in which solar commodities can transgress the industrial order that produces these commodities. Specifically, I argue that the sun—a central aesthetic component of figures 1.2 and 1.3—affectively transforms solar from a commodity produced under racial capitalism into an antidote to racial capitalism, tapping into a visual orientation to nature in a late liberal urban context. To unpack this claim, I now explore those images with greater depth, illuminating the sun's affective power before returning to my interlocutors in NYC.

NATURALIZING SOLAR

Perhaps the oddest thing about Danica's PowerPoint was the preponderance of images lacking in urbanity. Why would an urban activist give a community-focused presentation that situates solar not in a city but in a pasture? A close look at these images reveals meaningful clues.

In figure 1.2 the solar panels appear embedded in a grassy landscape, as though they emerged from the field, not a factory. These images, then, *naturalize* solar, positing commodities produced through racialized supply chains as the organic offspring of the nonhuman world. This suggests that the image positions the pastoral setting as not merely the scenic backdrop for solar but also as its ecosystemic container. Put differently, the image doesn't just place solar in a rural space; it aestheticizes a *sensibility* that solar is derivative of that space. Over the course of my field-work, I came to understand that this sensibility shaped my comrades' decision to visually situate solar in rural settings, simulating a world in which solar can return us to a nature that predates the ravages of racial capitalism.

I first detected this sensibility when I was canvassing for a community solar project led by EQUAL (Environmental Quality for All) in the organization's predominantly low-income Black and Latino neighborhood. This canvassing work was part of my broader efforts to support EQUAL's just-transition initiative as an engaged ethnographer and volunteer, building on my years of partnering with the organization as a sustainable energy policy advocate earlier in my career. On my first day of canvassing, I was hanging a flyer in the lobby of a low-income co-op building when Renée, a middle-aged Black woman, approached me suspiciously, displeased to see a solicitous stranger. But before I could explain what I was doing, she saw "SOLAR" in big letters on my flyer, and her comportment immediately changed. "Oh, you're doing solar energy? Oh, that's real good, we need that." After I told her about the community solar campaign and invited her to join, she nodded affirmatively and explained:

> I support what y'all are doing in the community because I support the environment. We need the environment, we don't have the

environment here, and it's killing us. People here don't realize how important the trees is and how important the air is. Many people don't go outside, we need to get outside, and go to the mountains to find ourselves. . . . That's why I support . . . what y'all are doing. Solar energy, it's good for the earth, it's good for the planet. We need to be taking care of the earth, get outside, go to the mountains. We need to get back to the land. . . . That's why I support your solar energy campaign.

Reneé speaks to a sense of alienation from nature that is shaped by her experience in the city—a space that she posits as the antithesis of nature. She suggests that a return "back to the land"—to the mountains and the outdoors—can ameliorate this alienation. Solar is a medium for such existential realignment, a conduit to the nonhuman world, and thus a part of "the environment" that her community lacks. The "back" in *back to the land* is key in her assessment, for it suggests that she and her community had once been a part of nature, such that solar is a means of returning to a purer past. In this way, she renders an energy commodity not simply as a gateway to a mythic nonhuman environment but also as a salve for the alienation of Black urbanity. Solar here is therefore affectively positioned in opposition to the necropolitics of environmental injustice, counterposing the purity of the land with the poverty of the city.

I had many interactions with local residents when canvassing for solar that resembled my conversation with Reneé. Tellingly, none of the residents I spoke to—almost all of whom were low-income, over the age of forty, and Black or Latino—raised the matter of carbon emissions or fossil fuels in lauding solar, and while they often alluded to "the planet," they did not explicitly cite climate change as an impetus for action.[22] Instead of rendering solar as a climate solution, they affiliated it with an idealized environmental purity counterposed to the toxicity of their own spaces—a naturalist imaginary that predates contemporary carbon discourses.

For instance, Tanya, a Black woman in her mid-forties, spoke at length about her affinity for solar in the city by evoking her purer past life in the country. Specifically, she ruminated on how you can't see the stars in the city at night because there's so much pollution. She grew up in rural North Carolina watching the stars, and she wants her daughter to do the same thing in the city. Solar, in her view, can enable this by "purifying" the air, returning her to her beloved stargazing—bringing the purity of the backcountry to NYC. At work in her assessment is the sense not simply that solar will replace dirty energy but also that it can recuperate a relationship with nature. This recuperation is not centered on solar's carbon emissions profile, as Tanya instead imagines solar as a force to transform her relationship with the cosmos. While solar's measurable impact on pollutants is key here, it's no surprise that Tanya focuses her assessment of solar not on a generic discussion of air quality but rather on her intimate physical experiences with the nonhuman world. In linking solar and bucolic spaces, these testimonies speak to a naturalistic sensibility among environmentally minded urban subjects—people with a reflexive conception of their alienation from the so-called planet—that is aestheticized in the PowerPoint's images, as I will show.

The nature/city dichotomy at the heart of Reneé's and Tanya's testimonies distinguishes their affiliation of solar with a pristine nature from a more general characterization of solar as environmental. For it is no mystery why one might affiliate solar with a generic green. As an infrastructure for reducing humans' emissions, solar technology is an integral part of a contemporary naturalist ethos that views the environment as something that must be saved from fossil fuels. Solar has long been conceptualized as a means of minimizing human impact on the nonhuman world, cohering with a dualist metaphysics that conceptualizes nature as distinct from human society. Thus, Reneé's and Tanya's bucolic rendering of solar is significant because it affiliates the panels not with a *broad* conception of nature (e.g., the climate system) but rather with a *particular* pastoral vision of a less alienated past. The PowerPoint illustrates this particular vision, depicting solar panels not as something that mitigates humanity's impacts on an already-degraded environment but instead as a part of a pristine environment that's independent of humans' destructive whims.

This signifying chain becomes all the more notable when we consider *nonsolar* strategies to minimize fossil fuel consumption. While both energy efficiency and solar reduce buildings' carbon emissions, I never encountered ruminations on energy efficiency that resembled Tanya's solar narrative or visuals that pastoralized energy efficiency in the ways that Danica's PowerPoint pastoralized solar during my eight years advocating for energy efficiency policy in NYC. Energy efficiency was never rendered as a means of returning us to nature or affiliated with bucolic spaces. Of course, the images in figures 1.2 and 1.3 resemble solar farms (which are clusters of solar arrays installed on grassy fields), so they do have some real-world correlate. Even so, these images barely approximate the density and configuration of solar farms, and this disjuncture, coupled with their digital graphics, suggests that the images were intended to be more like a naturalistic fantasy than a depiction of real-world infrastructure. Furthermore, if Danica wanted to share a photo of a solar farm—instead of this digital rendering of nature—she could easily have done so.

Why, then, do visuals like those of the PowerPoint and colloquial discourse like that of my interlocutors depict solar not only as the natural cure to extractive structures of differential value but also as an organic offspring of the nonhuman world? As I now show, the answer to this question lies in the sun, which sheds light on why NYC activists curate images that render solar as the antithesis of the extractivism that it in fact relies on.

SETTLER NATURE AND SOLAR ROMANTICISM

I met Randall, a young housing activist, when I volunteered for his organization's campaign to bring solar to one of the city's poorest, Blackest communities—a place *with no significant green space*. As we were canvassing his community, singing the praises of the residential solar market, he explained to me his affinity for solar: "*Solar panels are a part of nature because they're using the sun*. We need things that are natural. We can just leave them [i.e., natural things] alone, and they can sustain themselves. A tree sustains itself; it takes the sunlight, and it takes the soil from the ground and the rain, and it sustains itself. When you leave nature alone, it sustains itself. *Solar panels are the same way; if you leave it alone, it works from nature*" (emphasis added). Here Randall (unintentionally) exposes the affective linkages between sunshine and the pastoralization of solar. In his analysis, the sun endows solar with an air of nonhuman nature, positioning an industrialized commodity as a biospheric phenomenon that can operate without human ingenuity. In rendering solar as an organic life-form by virtue of its interactions with the sun, his words suggest that the sun can draw attention away from the human relations of racial capitalism that produce solar. Put differently, the sun is metonymic of nature and can thus naturalize that which it powers, obfuscating the exploitation and extraction that allow us to tap this power.

Perhaps this is why solar commodities are so often pastoralized in images like those in figure 1.2. For both Randall's conception of a self-sustaining nonhuman ecology *and* pastoral solar landscapes that similarly appear self-sustaining suggest that in the modern environmentalist imaginary, the sun can imbue anthropogenic technology with an affect of nonhuman nature. Consider the aestheticization of sunshine in figure 1.2. Here the solar panels glisten from the sun's rays, representing the power of the sun being converted into electricity. In other words, energy production is visualized through a hypershininess in the atmospheric space between the force of nature in the sky and the force of power in the grass. This shininess naturalizes such energy production by suggesting that the panels tap into the sun without the assistance of engineers, transnational supply chains, permits, installers, contractors, plastics, chemicals, fossil fuels, enslaved workers, or anything else involved in the production of a solar farm. Thus, the sun's aestheticized affect—its shine—disentangles solar from the racialized means of production. This shine, then, is the centerpiece of a visual fetishization of an industrial technology that transforms said technology into an offshoot of the natural world. As such, it gives an imagistic form to Randall's observation that "solar panels are a part of nature because they're using the sun," obfuscating racial capitalist production. By extension, the shininess simulates a "postindustrial nature," accentuating a pristine nonhuman landscape that is nonetheless marked by humans' modern innovations.[23] While the sun is evocative of nonhuman power, the shine that the sun generates gestures to this human/nonhuman hybridity; across visual genres, shine is "a dynamic medium through which the organic and the inorganic fuse," in Anne Cheng's terms—"an active mode of relationality."[24] We can detect this fusion, this relationality, in the imagined point where sunshine meets the panels—where timeless nature and modern energy are one and the same.

Notably, this paradoxically timeless-yet-modern landscape is reminiscent of a romantic genre of settler landscape imagery made possible by Indigenous genocide. In other words, we can situate the shiny solar image in a lineage of nature visuals that pleasantly aestheticize environmental shifts that were in fact caused by settler colonial violence. Specifically, the dispossession of Indigenous people's ancestral homelands and the subsequent "conservation" of nature in these spaces gave rise to an aestheticization of pristine landscapes—images that upheld an imaginary of an ideal environment to be preserved at all costs (see figure 1.4).[25] In this way, romantic visuals of American settler landscapes at once erased Indigeneity and constructed a normative baseline of environmental qual-

FIGURE 1.4. A nineteenth-century aestheticization of a "pristine" settler colonial landscape. Asher Durand, *Pastoral Landscape*, 1861. From Kantrowitz, "American Beauty and Bounty."

ity. In both its dearth of humans and pastoral serenity, the shiny solar image evokes these romantic visuals, promoting an ecocentric ideal that is insidiously rooted in differential value. While the digitally produced landscape of figure 1.2 sharply contrasts with the nineteenth-century pastoral aesthetic of figure 1.4 in many respects, this aesthetic is still part of a hegemonic visual genre in the American colonial imaginary that depicts a settler nature that is devoid of humans and thus coterminous with genocide—a genre that the image in figure 1.2 adapts for the twenty-first century.[26] As such, the solar images erase differential value doubly: first, in the naturalization of solar panels, an act that conceals racialized supply chains; and, second, in the reproduction of the settler trope of pristine landscapes that hides the brutal otherization it depends on. In other words, the naturalization of solar vis-à-vis the aestheticized sun/shine at once hides the violence of racial capitalism and perpetuates the visual registers of settler colonialism.

This settler naturalization would perhaps not be so significant if it was limited to a PowerPoint's shticky images. But I want to suggest that Danica's aestheticization of nature is instead evocative of the broader affective texture of solar energy transitions among many modern

FIGURE 1.5. For the spectators who take part in "Manhattanhenge," the sun congeals with the built environment to evoke a force beyond the human that is nonetheless embedded in an anthropogenic landscape. Photo by Michael Appleton, New York City Mayoral Photography Office, posted on Instagram, July 11, 2016, 8:50 p.m., https://x.com/NYCMayorsOffice/status/752666332924153856 /photo/1.

urbanites—the feeling that solar can return us to a pure nature that is, in actuality, imbricated in differential value. Randall unwittingly spoke to this insidious feeling in his rumination on solar's naturalness, which I now quote at length to illuminate its contiguity with settler ideology:

> *Solar panels are a part of nature because they're using the sun.* . . . We need things that are natural. . . . When you leave nature alone, it sustains itself. *Solar panels are the same way; if you leave it alone, it works from nature.* We need nature; if the planet goes, so do we—we have to think about the next generation. In . . . [rich white] communities they think about the next generation, and we need to think like that in our LMI [low- and moderate-income] communities. I don't want to hear these lies that we can't do this or that. . . . We need to change the narrative about police officers being violent, we need to bring back a vocation into the school so that people have a

job and have something to do when they graduate, we have to educate about ownership, about not throwing bottles on the floor and getting glass all over the street. We have to bring back the family structure into our community. The number-one thing is to change the narrative. . . . I want my community to be an epicenter of solar, homeownership, vocational training in schools. . . . I don't want to hear that we can't do whatever we set our sights on.

In Randall's commentary, (1) solar technology is metonymic of a self-sufficient nature that is necessary for long-term survival; (2) privileged communities are committed to sustainably regenerating like this self-sufficient nature; and (3) poor communities of color don't yet regenerate sustainably because they eschew self-sufficiency, blaming their problems on things like police brutality. In other words, the purity of nature is counterposed to the impurity of poor communities of color, as Randall amplifies conservative critiques of the Black "family structure" and racist "broken windows" theory (i.e., the supposed sociological problem of glass all over the street) in characterizing the alleged failure of those communities to parallel nature's organic reproduction.

In the process, he locates differential value at the intersection of two axes of difference: (1) nonhuman nature and human society and (2) a white culture of preservation and a Black culture of depravity. He theorizes the first axis of difference when he posits nature as a passive domain of life on which the future of humanity depends. In this analysis, the human is contingent on but still discrete from nature—an ontological separation that bears the imprint of the settler pastoral imaginary aestheticized in figure 1.4. When this nature takes the form of solar panels, it can generate human wealth as part of a broader project for upward mobility (inclusive of homeownership initiatives and vocational training). Nature vis-à-vis solar is thus an extractable terrain that can improve human society. At the same time, Randall theorizes a second axis of difference in valorizing a white culture of preservation in which privileged people recognize humans' dependence on nature in contrast to Black LMI communities like his own. (While he doesn't say the words *white* and *Black*, these categories are contextually implicit in the euphemistic binary of "LMI communities" and "other communities" that lack the former's pathologies.) At the intersection of these axes, Randall endorses a local environmentalism that doesn't simply acknowledge the importance of nature but more specifically extracts value from it, building prosperous futures by tapping the sun's natural power. In naturalizing solar, then, he dually reproduces differential value, separating both an extractable terrain of nonhuman life from the humans that extract it, and the beneficiaries of this humanistic extractivism from underprivileged people of color.

Randall doesn't describe the sort of shiny settler landscape in Danica's PowerPoint, but in linking a self-sufficient purity with the conservationist ethos of privileged communities, he reproduces the colonial environmentalism that the PowerPoint aestheticizes. In other words, both Randall and the PowerPoint draw from colonial imaginaries of conservation in naturalizing solar, at once celebrating a mode of environmental preservation affiliated with white settlers and upholding a romantic vision of an untouched nature enabled by Indigenous genocide. As such, the PowerPoint visualizes the loose concept of naturalized solar that Randall attests to.

I want to therefore suggest that the sun's affective power—which Randall unwittingly highlights—partly inspired Danica to aestheticize solar as a conduit to a pure nature outside of racial capitalism. For if the sun enables Randall to see solar panels as an organic form, then perhaps it played a similarly affective role in the PowerPoint's pastoral aesthetic. To explore this hypothesis, I now consider the sun's interactions with the PowerPoint's viewers and intended audience in a megacity that lacks any resemblance to settler ideals of purity. How does their orientation to sunshine evoke a pure nature predicated on differential value in a multicultural metropolis viewed as the anthesis of natural? In considering this question, I hope to show how human relations with the sun enable New Yorkers to naturalize solar in the struggle for a just transition.

THE DOMESTICATED SUBLIME

A few weeks after Danica's tutorial, I spoke with Sean, the vice president of engineering at a large solar and clean energy technology corporation in the Financial District, who, to my surprise, had attended the tutorial. As I discuss in chapters 2 and 3, sustainable energy muddies the normative political divisions separating corporations and anticapitalist activism, such that it wasn't unusual that a white corporate vice president would attend a meeting geared toward community activists who disdain racial capitalism. But when I spoke with Sean, I was still trying to wrap my head around this peculiar corporate/grassroots alliance, attempting to understand not his thoughts on the PowerPoint's bucolic images but the reasons he was even there viewing the PowerPoint to begin with. Like everyone else in attendance, Sean hadn't batted an eye at the sight of a hyperreal solar farm detached from the spaces he sought to solarize; he explained that he had attended the tutorial not to view a slideshow but instead to cultivate political solidarity with other solar advocates who, unlike him, represented the city's low-income communities of color.

Yet the pastoralized solar panels came to mind when Sean shared his out-of-place politics with me as we chatted in his company's sterile employee lounge on the thirty-first floor of a skyscraper. In spite of this corporate sterility, I was enthralled by the expansive shiny sight before me: the Downtown Brooklyn skyline, the Lower Manhattan skyline, the river separating these skylines, and, out into the distance, the Statue of Liberty—these monumental icons shimmering in the late afternoon light as the sun prepared for its descent. Here, in front of my eyes, was the shine that the PowerPoint *attempted* to visually capture—sunlight on a glassy infrastructural surface—but the buildings' near-twilight shimmer was arresting in ways that pastoral solar simulations could never be. Specifically, the shimmer's movement and variegated luminosity gave it a liveliness and depth that can't be translated to a shiny two-dimensional still, as shimmers produce a "surge of affect," "transform[ing]" the visual field into a permeable field of flux and sensuous connection," "a field of attraction and joy," as Jennifer Deger puts it.[27] The sight before me, then, was precisely the sort of lively shine that the PowerPoint image attempted to simulate: the luminous effect when sunlight congeals with an industrial glass facade (most solar panels resemble skyscrapers in that both are encased in glass)—a sight that is ubiquitous in developed urban spaces in spite of the image's pastoral setting. So, for all its bucolic conventions, the image reflected a modern, often *urban* affective experience: the feeling of a nonhuman life force in an anthropogenic infrastructure—a presence that exceeds the built environment it nests in. Perhaps, then, the pastoralized solar panel emerges from an unease with the alienating excesses of the modern, evocative of a "naturalist machine."[28]

But I registered no such unease when Sean, whose back was to the skyline's shimmering glass, stopped himself midsentence as the sun began its descent and asked, "Do you mind if I sit next to you so I can watch the sunset?" After repositioning himself accordingly, he took in the landscape for a beat and said, "Isn't this an amazing place on earth? It just makes me feel alive, you know? If [my company] didn't have this office,

I probably would have left New York. . . . Here it's so open; you get the sun everywhere . . . this is a place where you can look up and know that life exists. . . . I don't know what's so enlivening about it"—he shook his head, at a loss for words—"it just literally is enlivening." In Sean's indescribable sense of life pulsing through skyscrapers at sunset, I detected not merely the awestruck affect affiliated with encounters in nature—*the sublime*—but, more specifically, the indescribability of a force of nonhuman life in a landscape marked by modern man's mastery of nature[29]—a metropolis infused with the more-than-human. This inspiring encounter with a space that is as larger-than-life as it is engineered is reminiscent of what William Cronon calls *the domesticated sublime*.[30] In contrast to the "bewilderment" and "loneliness" of early Euro-American settlers traversing the perilous vastness of uncolonized land, the domesticated sublime points to the experience of encountering the pleasing vastness of "tamed" forests excised of their Indigenous stewards. The sublimity of this colonial nature is immanent in its expansiveness: dimensions that evoke a godly force in spite of their anthropogenic origins. Along these lines, the domesticated sublime is experienced in spaces curated *by humans* to admire from a distance, a managed site whose awesome scale is pleasant to the eye.[31] Although Sean gazed at a built environment that does not pretend to be "nature," as the sun's rays danced on its vast horizon, shimmering within an expansive array of glass and brick, this luminescent landscape epitomized shine as Cheng conceptualizes it—the organic/inorganic fusion "through which the visual spills into the sensorial," that is, an affect that breaks down the boundary between the natural and man-made, engulfing its viewer in the process.[32] This shine, then, exemplifies the domesticated sublime: an anthropogenic environment that, in its visually alluring expansiveness, gestures to a force beyond the human, inviting the sort of awestruck reaction that Sean enacts. In other words, the sun endows the built environment with this indescribable sense that it is not merely a captivating feat of human ingenuity—a *technological sublime*, as David Nye conceptualizes it—but also, to quote Sean, "enlivening" in ways that engineering alone couldn't achieve.[33] Specifically, the skyline's shimmer is the point where a mastery over nature in the form of skyscrapers meets an omnipotent force of nature in the form of the sun, animating experiences with shine that are particular to the modern urban dweller.

Consider Manhattanhenge: an annual event in which New Yorkers form crowds to gawk at the sunset on the one day of the year when it's perfectly aligned with Manhattan's street grid. This shiny alignment evinces the power of the sun to infuse a sense of life beyond Anthropos in the man-made skyline. Specifically, as the image in figure 1.5 suggests, the shine's organic/inorganic fusion attunes New Yorkers to their built environment with a ritualistic sense of excitement that stands in contrast to the more passive spectatorship of the Empire State Building's light show or the city's Fourth of July fireworks.

Yet Manhattanhenge is but a more extreme version of the daily shine-gazing ritual that animates life around the vast glassy surface of buildings scraping the sky. Consider the visual apprehension of the sun on the Brooklyn Bridge and in a waterfront alcove in the park beneath the bridge—spaces with unobstructed views of the Manhattan skyline and the Statue of Liberty. Locals and tourists alike take photos of and selfies with the larger-than-life landscape from these spaces at all times of the year, but on any remotely clear day at sunset, the size of the crowds *at least* triples from earlier in the day (by my estimation) as throngs of people show up primarily to marvel at the mass of buildings across the river. They angle themselves to take in not just the sight of the sun's descent but also the sight of the towers and buildings glistening in the sun, an effect magnified by the descent, even in the dead of winter. This is to say that people are more transfixed by the monumental landscape when it is infused with the sun's piercing rays just prior to twilight than at any other point in the day—their spectatorship a testament to the indescribable force that Sean spoke to. Furthermore, smartphones have transformed this shiny encounter into an intensely imagistic experience, visualizing sublime shimmers in ways that resemble the PowerPoint's image. Consider figure 1.6, one of countless photos that my father regularly takes of the NYC skyline at sunset, capturing its shininess. While sunset gazing is not at all specific to NYC, it takes on a unique form in the dense glassy skyline, producing a sheen that is not necessarily any more or less affective that that of a shiny mountain, for instance, but that is particular in its ability to infuse the sublime into structures that usually instead testify to man's dominance over nature.

As a *domesticated* affect, this alluring elusiveness is thus commonplace: a familiar feeling in manufactured spaces designed with a considerable degree of predictability, such that the presence of something seemingly in excess of the human is simultaneously unthreatening and a sight to

FIGURE 1.6. One of my dad's many sunset photos of the glistening NYC skyline, 2019. Courtesy of Llewellyn Lennon.

behold. Indeed, Sean regularly pauses his workday to gaze at the skyline at sunset; its awesomeness does not preclude its everydayness. And this everydayness, the fact that the shiny domesticated sublime is just a normal part of NYC life, begins to shed light on the curation of pastoralized solar images in a space that couldn't be less pastoral. For the sunny sheen of a factory-made surface is a uniquely modern phenomenon, fusing the organic and inorganic, signaling the long-standing tendency of contemporary urbanites to encounter the sublime in industrial infrastructures (in contrast to rural folk and to workers familiar with those infrastructures' ins and outs).[34] The shiny conjoining of nature and machine in Danica's images illustrates this phenomenon, naturalizing an affective encounter that is in fact historically specific.

Of course, history offers numerous examples of human monuments and built environments that sunshine "enlivened"—their glossy surfaces enchanting their human spectators while gesturing to life beyond the human.[35] But encountering more-than-human liveliness in skylines that were built to solidify modern man's mastery over nature is still a particular experience. Modern infrastructure has traditionally been understood as the antithesis of the natural world, so when the sun infuses modern infrastructure with the natural world's affective affordances, it disrupts a dominant conception of the built environment.[36] And this particular semiotic linkage between nonhuman nature and human infrastructure vis-à-vis shine—the sun's capacity to highlight an anthropogenic technology's more-than-human liveliness—informs the PowerPoint's naturalization of solar. Put differently, the domesticated sublime of modern urban spaces—that which Sean and I experienced while admiring the skyline's enlivening glare—can give affective shape to the sort of luminescent naturalist machines shown in figures 1.2 and 1.3. Indeed, as David Nye argues in reference to early industrial infrastructures in the United States, urban as opposed to rural people were unique in their capacity to perceive nature in modern technologies.[37] By extension, I want to suggest that the PowerPoint's pastoral setting is ironically conditioned by the city's sensibilities. While I would imagine that Danica did not choose images that consciously reminded her of watching NYC sunsets, my point is that for those of us who live in modern spaces that were built in contradistinction to nature, sunshine has naturalizing affordances that her images attempt to aestheticize. This aestheticization, then, naturalizes solar—visually situating it in the domesticated sublime.

This suggests that the sun partly enables urbanites to *fetishize* solar commodities—to see solar as possessing a life of its own—obfuscating the racial capitalist relations that produce all modern energy infrastructure. Thus, the sun can play a similar ontological role as that which a student of Marx might assign to capital, mystifying modern humans such that we can see an alien vitality in mass-produced technologies manufactured in dirty, brutal conditions like those of Xinjiang. Of course, capital *also* alienates technology consumers from the exploitation and extraction through which their technologies are produced. Racial capitalism partly functions by keeping the atrocities against the Uighurs and the Congolese out of consumers' sight and mind, promising cheap goods whose utility and price veil the human labor and environmental degradation that make those goods possible. What I'm contending, then, is that the sun affectively accentuates this consumer estrangement in the case of solar, enabling us to see solar as not merely a life force independent of its production but also a sublime presence that ameliorates the structural conditions of its production—as though it can return us to the purity preceding industrial modernity. I now expand on this thesis, exploring how the sun's sublimity works hand in hand with racial capitalism beyond the PowerPoint images.

THE DESERT IMAGINARY

Amita, a solar energy engineer at Sean's company, was inspecting the steel frame of what would become a giant 150-kilowatt photovoltaic system on the top of a new building at a loud construction site in one of the city's populous working-class communities of color, in the absence of anything resembling wilderness. Unprompted, she turned to me and said, "Whenever I see solar panels, it just makes me happy; I don't know why." She grinned as she said this, hinting—in her tone and facial expression—at an indescribable joy reminiscent of the sublime, the sort of happy incredulity that Sean displayed in the face of the shimmering skyline. When I asked her why solar panels made her happy, she didn't say a word about her urban rooftop solar work; she instead shared her dreams for a future world nourished by massive *solar farms in deserts*—thousands of contiguous solar panels in the sunniest places in the world *uninhabited by humans*, a point she emphasized—enabling us to, in her words, "live in harmony" with a "natural resource." In attempting to describe the ineffable goodness of a technology she knew intimately, a technology she worked with every day, her thoughts went not to her lived experience with said technology in the city but instead to a fantasy of a harmonious nonhuman nature in the desert—a space saturated with sunlight.

The sun is obviously central to this imaginary, as she naturalizes a human-made technology by situating it in a sunny environment bereft of humans. Amita's imagined desert, then, evinces the sun's alienating affordances: its capacity to affectively disentangle solar from the commodity form and human labor. The sunny fetish here operates through a domesticated sublime: an awe in the presence of human-made life that settlers like Amita affiliate with a pristine world beyond the human.[38] While there could be other signifying chains that led Amita to affiliate an urban commodity she knows well with nonurban spaces, my point is that this imagined terrain is defined by its sunshine and operates as an ecological trope, enabling her to attest to the sublime presence of technologies that are, in fact, made by enslaved Indigenous people in fossil-fueled factories.

The sun's power to inspire fetishistic imaginaries is not particular to spaces in the grip of racial capitalism like NYC; sunshine endowed anthropogenic infrastructures with an alien vitality in past social contexts that were not beholden to the commodity form. I want to suggest that theorizing this *transhistorical* affective power can actually elucidate how the sun, in the *present moment*, obfuscates solar's enmeshment in racial capitalism through a domesticated sublime.

Historically speaking, the sun's affective interactions with infrastructure were perhaps most palpable in ancient Giza. Much like NYC's glistening skyline in the twenty-first century, Giza's massive sun-kissed pyramids captivated their human progenitors' sight, animating the ontological perceptions of awestruck onlookers. While these pyramids are paragons of human innovation, Lynn Meskell contends that they actually evince "one of the earliest and most concrete attempts to overcome the limitations of the human condition."[39] Operating as a tomb "for the great living god, pharaoh," the pyramid was "beyond the scope of mere mortals," as its colossal dimensions were intended to invest a transcendent power beyond the human in its supreme occupant.[40] Yet the pyramid's power lay not only in its dimensions but also in its proximity to the sun. As George Bataille explains, "In the eyes of the Egyptians, the pyramid was an image of *solar radiation*," so "in the person of the dead king, death was changed into a *radiance*, changed into an indefinite being" (emphasis added).[41] Indeed, the pyramids' glimmering limestone would reflect light, which is speculated to have been visible from space. In this way, the anthropogenic pyramid simulated the nonhuman power of the sun, "embod[ying] . . . light and shadow."[42] Through such radiance, "the pyramid entraps and engages its observers," highlighting their ephemerality and minuteness—the existential chasm between a more-than-human transcendence and its mortal onlooker.[43] As such, the pyramid's entanglement with the sun conditioned a visual apprehension of the built environment that shaped a broader conception of nonhuman life. Designed with careful attention to the sun's orbit, the pyramids were an anthropogenic infrastructure that appeared to be touched with something beyond the human.[44]

The sun, then, can endow the products of human ingenuity with an affect that suggests they have their own lively presence, inviting a mythical relationship with the built environment—a fetish, of sorts—that is not reducible to racial capitalism's ontological perversions. For the Giza pyramids and the NYC skyline similarly interact with sunshine in ways that demand quiescent observation, inspiring humility in the face of monumental matter that dwarfs mortal human existence. Thus, amid anthropogenic landscapes, the sun can induce a visual apprehension of a world beyond the human across historical contexts. Of course, modern New Yorkers and ancient Egyptians relate(d) to their built environments in vastly different ways. Sean, for instance, does not believe that the skyscrapers before him are the manifestation of a particular God. Furthermore, as stated, a shiny modern skyline is affective in ways that similarly shiny spaces in other times were not due to the omnipotent status of the human in the context of urban modernity. Even so, we can locate, across contexts, a common visual experience with the sun as it inflects anthropogenic landscapes at a distance—an infrastructural liveliness not particular to the commodity fetishism of racial capitalism.

In this liveliness, the sun conceals as much as it reveals, centering our attention on a surface that shimmers as though it has a life of its own—divorced from the exploitative relations that made both the pyramids and the NYC skyline. There is, of course, a degree of concealment in any finished product, but the dimensions and shine of these infrastructures inspire a gaze that is particularly transfixed by the surface—indifferent to that which is beneath. In other words, the sun's infrastructural shimmer is mystifying regardless of whether racial capitalism or a prehistoric kingdom made that shimmer possible.

But this mystifying force is remarkably compatible with the particular fetishistic impulses of racial capitalism, as it can endow *commodities* with the sense they are something else. Put differently, the sun—across time and space—anchors mythmaking with an affective power that racial capitalism latches on to vis-à-vis urban modernity. This affixation is nowhere more apparent than in Danica's PowerPoint, which simulates the sun's affective power in promoting her energy commodity of choice. The PowerPoint, then, reveals a form of commodity fetishism that resonates with a time *before* the commodity form—this cross-cultural timelessness suggesting something, dare I say, *natural*.

On a gray winter day, I was walking through the dirty remnants of the previous week's snow on Seventh Avenue in Manhattan when my gaze met the domesticated sublime aestheticized in an advertisement plastered on a commercial storefront (figure 1.7). The ad's cloudless sky, grassy field, and carefree blond boy presented a stark contrast to the gray horizon, slushy urban streets, and diverse New Yorkers that the ad hovered over. It was as if this pastoral image was trying to remind us of everything NYC is not. In reality, it was a corporate social responsibility ad for Amalgamated, a NYC-based bank whose clientele consists primarily of local union workers. New Yorkers regularly encounter ads featuring lush places that couldn't be further from their hometown; subways and buses are covered with images of sunny tourist destinations, urging us to find some outdoorsy respite from the city. The Amalgamated ad, then, was notably discordant with the space it hovered over not because of its outdoor aesthetics but rather because of its central object: a solar farm. Unlike Bermuda's beaches, Appalachian hiking trails, and every other touristy nature that New Yorkers gawk at during their subway commutes, the solar farm is not something that most New Yorkers could ever directly experience in real life; while they can subscribe to a community solar project that pays for a solar farm's electricity production, the farm itself is not something that most city residents will ever encounter in the way the blond boy is.

Furthermore, in contrast to countless companies whose ads sell nature to New Yorkers, Amalgamated is not promoting their outdoor scene as a service or good that their clientele can purchase from them. Instead, I want to suggest that Amalgamated displayed this image to choreograph an affect: a *feeling* of clean energy solutions that takes form in the dissimilarity between the pastoral scene and the megacity that their clientele live in. For while Amalgamated invests in renewable energy, these investments target *rooftop* solar—not solar *farms*. Unsurprisingly, though, Amalgamated visualizes these investments through a pastoral scene, tapping into the naturalist imaginary that Amita, Randall, and my other interlocutors invoked when characterizing their affinity for solar. The image, then, is not simply a picture of a solar farm; it is a hyperreal take on a pastoral genre, resonant with a rural imaginary that oddly inflects my interlocutors' work for a just transition *in the city*. Foregrounding a boy whose positionality and playfulness suggest something pure,

FIGURE 1.7. Amalgamated Bank's pastoralized solar panels, which faced New York City streets. Amalgamated Bank, "Go paperless or spend >$10 on ur Donate theChange debit card & we give to @100isNow to support #renewableenergy," Twitter (now X), May 24, 2017, 12:44 p.m., https://x.com/AmalgamatedBank /status/867421177236643840.

the image situates solar in a landscape that pops with floral color, aestheticizing it not as an anthropogenic technology but rather as bucolic life. The solar panels' surface shimmers in the sun, simulating a liveliness on par with the photoshopped dandelions and white skin, evoking a nature that is so untouched it exceeds reality. This hyperreal nature doesn't simply position a financial institution as a good environmental steward. More insidiously, it choreographs the antithesis of the capitalized relations that enable the solar panels, obfuscating differential value through the master signifier of whiteness in the form of a blond boy. In the process, solar is rendered as a utopic "solution"; sun, skin, shine, and grass conspire to simulate a future that resembles a mythic past preceding racial capitalism.

I initially smirked at this corporate aestheticization of the domesticated sublime broadcast on a slushy city street; the contrast between the image and my location felt absurd. But after a beat, this amusement morphed into a mild annoyance when I considered the very reasonable desires of my friends, family, and comrades for greener pastures and sunny skies among our concrete environment's corrosive hustle and climatic precarity. It felt as though the bank was trying to profit off the naturalistic imaginary of many cityfolk, cynically evacuating the substance of this imaginary by reducing it to a banal purity detached from differential value.

While this might sound like an overblown reaction to a photoshopped solar farm, I believe my critique is warranted in light of New Yorkers' lived experiences with environmental injustices epitomized by NYC's unjust terrain of sun exposure. As Tanya's testimony demonstrates, many marginalized New Yorkers naturalize solar as a response to their sense of environmental alienation, rendering the energy transition as a medium for connecting with the forms of life that the city's corporate elites commodify and sell to the highest bidder. In addition to green space, clean air, fresh food, and biodiversity, the sun is one such form of life. For NYC's ever-expanding skyline and housing affordability crisis have made *sunlight*, of all things, a precious commodity denied to the most marginalized, underscoring the need for some semblance of nature in gentrifying communities of color with a dearth of open space. Specifically, "luxury" residential high-rises are rapidly being built in those communities, casting literal shadows on people who can't afford to live on the upper floors and penthouses of tacky glass towers. At the same time, real estate companies aestheticize and sell the sun to the city's burgeoning bourgeoise, as the ads in figures 1.8–1.10 suggest. The sky in these ads appears light and empty, a space inhabited only by sunlight, upscale clientele, and skyscrapers glowing on the horizon. Sunlight is a commodified amenity here, as the bourgeoisie are promised special access to the sun's liveliness, warmth, and daily descents—but only if they're willing to pay the right price. These new skyscrapers cast shadows that threaten proposed rooftop solar projects on adjacent buildings, while droves of poor undocumented immigrants live in illegally converted basements without even a slice of sun exposure in different communities. Indeed, many of the city's most precarious workers—many of whom install solar panels for a living—dwell in spaces entirely deprived of natural light.[45]

Yet New Yorkers are organizing on the ground to address these sun-exposure injustices, evoking imaginaries of nature that offer important context for deciphering the affective texture of Amalgamated's solar ad. Consider Save Our Sunlight, a multiyear (and ultimately unsuccessful) campaign to stop the development of a seventy-three-story behemoth that could cast debilitating shadows over a cherished community garden. At the intersection of two high-traffic streets, the garden is a source of well-being in a neighborhood marked by commercial development, vehicular pollution, and modern alienation (the last of which is

Sun-Drenched Modern
Living Spaces

Wrapped in a radiant glass façade and reinforced
with a double-height Indiana limestone base, BKLYN
GOLD was built from the ground up to surpass any
expectations of what rental living should be.

FIGURES 1.8–1.10. Luxury housing ads promise the bourgeoisie privileged
access to the sun. Figure 1.8 from Realtor.com, listing for 555 Tenth Ave.
Unit 20G, Manhattan, NY 10018, accessed December 2, 2022, https://www
.realtor.com/realestateandhomes-detail/555-10th-Ave-Apt-20G_New-York
_NY_10018_M49015-13429. Figure 1.9 from BKLYN Gold, accessed November 19, 2024, https://www.bklyngold.com. Figure 1.10 from the Douglas
Elliman website, accessed December 2, 2022, https://www.elliman.com
/newyorkcity/fra/associer/527-a-538-lgg/lisa-gustin.

an approximate truncation of my interlocutors' lengthy ruminations on the existential unease of life amid concrete, cars, and skyscrapers). The campaign, which included direct action and citizen lobbying, framed the imperative to "save our sunlight" as a matter of "nature," casting the sun that beamed on their green enclave as an organic salve to alienated urbanity.[46] For the "nature" they wished to preserve was not a pure non-human presence but instead immanent in their human practices. Their summer solstice celebration exemplifies this immanence. There they wear Venetian sun masks and collectively marvel at the protracted sunlight, enacting gratitude to the supreme giver of life hovering in the sky amid the soil, trees, vegetable beds, and flowers they tend to. This sunny nature is a partner in celebration inscribed on one's physiognomy—a divine collaborator in the maintenance of life. Their naturalistic discourses, then, pointed not to a facile romanticization of the outdoors but rather to a timeless, cross-cultural practice of celebrating the sun.

Indeed, we could situate the gardeners' sunny nature in a multimillennial history of sun worship, deification, and mythology that goes beyond solstice celebrations. When we consider, for instance, the righteous Mesopotamian god of the sun, Utu, working toward justice on earth from the heavens, or the Etruscans celebrating their solar deity for the gift of fertility, or Vishvakarma, "the divine sculptor of the Hindus, [who] chiseled the human form . . . out of the orb of the sun," or the pharaonic manifestations of the sun god Ra in ancient Egypt, we see that sunshine has consistently inspired imaginaries that don't begin to conform to a human/nature split.[47] Humans have historically viewed the natural source of vitality in the sky as a catalyst of creativity, a paragon of morality, and a partner in life making.[48] But in the context of a contentious fight over a seventy-three-story tower, the community's evocations of nature are notable because they situate the sun in broader struggles against racial capitalism. In this schema, nature-qua-sunshine represents the final frontier in an ongoing battle for space in a dense megacity plagued by gentrification and inequality.

It is in this context, over treeless city streets, that the Amalgamated ad dangled an idealized nature centered on the sun's bounty. So the ad felt like a seductive tease; it minimized the potent imaginary of a sunny otherwise that the gardeners enacted with their Venetian masks, reducing our understandable desires for something beyond the city to a mythic solar farm that New Yorkers couldn't actually access. While I can only speculate why Amalgamated chose this image, I sensed that they were trying to tap into urbanites' craving for nature by aestheticizing something that most of us will never have—as if our desire for a sunny pasture is just fodder for corporate gain and not rooted in the real deprivation of sunlight that marks the gentrifying tumult of a constantly expanding skyline.

Yet my fellow pedestrians who passed the image on the street only seemed indifferent to it. Their nonreactions bespoke not only the habituated presence of advertisements in urban spaces but also the hackneyed nature of depictions of environmentalist ideals that are, at their core, rather profound. For Amalgamated's ad presents a shticky take on a holistic futurity that is not, in and of itself, insipid. From the forward-looking practice of wishfully blowing on a dandelion to the child's playful presence—a metonym of a brighter tomorrow—the image's objects position solar as a means of a lively future in which we cultivate loving relations with the nonhuman world. Standing in the urban cold, I wondered what it would feel like if this spirit was not photoshopped by a tentacle of racial capitalism (i.e., a bank) but instead enacted by Tanya, Reneé, Amita, and other urbanites enthralled with a solarized domesticated sublime. What if solar's presence could attune us to the life concealed by the city skyline, helping us practice gratitude for sunlight and soil while repairing the multispecies relations that racial capitalism has ravaged? Perhaps a fetishized solar farm could paradoxically tend to our alienation from nature in the city. Or perhaps I had lost my mind with these nonsensical musings on the cold NYC street while gaping at the window of a bank overrun with retouched dandelions and solar.

THE AFFECTIVE POWER OF THE SOLAR FARM

Both the Amalgamated ad and Danica's images collapse the differences between putatively pure nature and modern society without challenging the structures through which racial capitalism extracts value from these differences. But I have also suggested that in the context of solar transitions, the sun similarly troubles the nature/society distinction without challenging differential value. So what is the culprit? Reductive images in a "society of the spectacle" or the closest star to the earth?[49] I want to suggest that the distinction here between natural sunlight and human imagery is negligible with regard to solar, as the sun invites a fetishistic gaze that resembles the alienating lens through which racial capitalism visualizes commodities. I encountered this fetishistic gaze not only as a spectator of the Amalgamated ad and Danica's PowerPoint but also in the real-life presence of that which those visuals aestheticize: the solar farm. Indeed, when witnessing a solar farm in-person—not a photoshopped simulacrum—I began to better grasp the entanglements of imagery and sunlight as they relate to solar, underscoring the intractability of the commodity fetish in otherwise laudable efforts for a just transition.

My solar farm encounter was certainly not planned. There are no solar farms in NYC—the densest metropolis in the United States. But I unexpectedly encountered one when I traveled from NYC to Buffalo for a statewide just-transition strategy retreat with eight other renewable energy activists. Johanna, an effervescent community organizer, drove us in a rented van on what at first glance appeared to be an austere interstate highway. Johanna, though, had a very different impression of our setting. "Oh, I love this, *look*, it's so beautiful!" she gushed as the highway ran adjacent to a colorful mountain range. "Beautiful . . . you just have to ignore all traces of humanity." By *humanity* she meant the fossil fuel infrastructure all around us. In addition to the highway, the mountain range loomed over a strip mall with multiple gas stations and ran parallel to electric transmission lines and a commercial railroad track. Furthermore, at that point on the highway, we were surrounded by four gas delivery trucks—hulking machines that hummed as if they aimed to hammer home the point that the breathtaking landscape before us would offer no respite from fossil fuels. Johanna, though, adhered to her own directive to "ignore all traces of humanity" as she chirpily ruminated on the flora that blossomed at the mountain's base, displaying an encyclopedic knowledge of the local ecosystem bordering the otherwise monotonous highway. She then commented on the clouds' contours, the residual smells of an early morning rain, and the liveliness of nearby cornfields, with profound appreciation. Amid an anthropogenic landscape, she sensed a nonhuman nature, accessing life in excess of fossil fuel infrastructure.

When we weren't driving by alluring fields, lakes, or trees, her enchanted monologues on natural beauty would morph into horror stories of fracked gas, as she'd recount in harrowing detail how new planned pipelines threatened the biodiversity along the highway. She railed against investor-owned utilities and pipeline-praising plutocrats, assailing a for-profit energy system for wreaking havoc on the beauty before us while casually lambasting "capitalism" more broadly. For Johanna, then, free market fossil fuels were the antithesis of the natural world, but their presence on the road (in the form of pavement, vehicular traffic, fuel delivery trucks, and transmission lines) didn't prevent her from dwelling in this natural world. As she broadened her attention beyond the "traces of humanity" in her immediate line of vision, the feel of the air on her skin, the scents of the space we traversed, and the sense of the mountains' vitality gave affective form to a nature that was, in her estimation, both distinct from the anthropogenic environment *and* an intimate part of it.

Johanna's nature-philia was contagious. The other activists in the van (including I) marveled at the beauty her monologues illuminated, as her various edicts to note the nature immanent in the endless stretch of concrete lanes, metal signs, fast-food rest stops, and other vehicles moved us to look with reverence at the highway's nonhuman backdrop—that which often gets taken for granted. For it's easy to grow complacent with the gray blur that saturates your sight in the enclosed space of commodified transport—to overlook the nonhuman world around you. While this world appears as the other to our anthropogenic landscape, Johanna's orientation to our surroundings underscored the fragility of such a distinction.

This distinction practically dissipated as we drove through rural landscapes and stumbled on a technology we often simulated on screens in our urban environs: a massive solar farm. "Oh my god, look, wow, oh my god, this is everything!" exclaimed Johanna while the rest of us instinctively rolled down our windows and cocked our heads outside, not merely to take in the expansive sight of the farm, but also to position ourselves closer to it—to *feel* it. Whatever awe the mountain inspired multiplied considerably amid these technologies, as my comrades photographed the industrial-scale infrastructures like ecotourists capturing the wonders of the natural world. "The whole thing is such a beauty," Johanna added, taking in the built landscape with the soft smile and humble gratitude that she usually reserved for vegetation and wooded trails. "OMG, look at those sunflowers!" she bellowed, noting the flora at the panels' base.

Her constant directive to *look* encapsulated the sensorial texture of this domesticated sublime. For our experience with the farm was first and foremost visual; we were less like the blond boy playing in the grass in the Amalgamated ad and more like the ad's spectators: urbanites observing the pastoral. The PowerPoint visuals had seemingly jumped off the screen and come to life on a stage, as we commented in awe on encountering in-person that which we had only dreamed about: the cipher of sustainable futurity we had seen in Danica's presentation—a reminder that visual apprehension co-constitutes broader social imaginaries.[50] At the same time, when the farm's panels glistened in the sun, the bright blue sky framed the farm not just in terms of futurity but also as a sight to behold irrespective of its political significance. Resembling the sunny configuration in figure 1.2, this glisten worked in tandem with the infrastructure's embeddedness in an expansive, lush landscape lacking humans to suggest an ecosystemic autonomy evocative of Randall's analysis: "When you leave nature alone, it sustains itself. *Solar panels are the same way; if you leave it alone, it works from nature.*" The sight before me affectively instantiated this claim, as the solar farm felt like the sovereign nature that circulates in ads throughout NYC: images of colonized green spaces that position the spectator outside of this enclosed purity, paradoxically "enabl[ing] new forms of experience and engagement with natural environments" that you can't actually touch.[51] As a productive landscape that requires not human stewardship but plentiful sunshine, the farm didn't invite our haptic care. It was, for us, a spectacle to view at a distance—not a space of embodied leisure or labor. I want to suggest, then, that this space's anthropogenic solar infrastructure counterintuitively endowed the landscape with a nonhuman presence because it positioned us humans as spectators from afar—not as an intimate part of the ecosystem—directing our visual attention to a grassy field that we could engage with kinesthetically *if said infrastructure was not there.*[52] Put simply, solar transformed a space that one can steward into a space in which one can spectate. If solar electricity production were a more hands-on process, then perhaps our spectatorship would feel different— perhaps the sight of this farm would inspire us to learn new land stewardship skills. But the sun and farm interact to produce energy in a way that requires no human intervention, generating a breathtaking sight and nothing more.

This domesticated sublime was conditioned not only by the farm's solar dynamics but also by our drive's social dynamics: the attunement to space that Johanna had cultivated in the van. Through her meditation on the highway's overlooked ecosystem, we had conjured a collective effervescence: a shared ecocentrism that was both a visceral response to the life outside the van *and* a communal response to the life within it. The cool breeze, crisp dew, sunshine, and one another's presence coalesced with our political commitments to form a self-conscious relationship with the landscape that inflected our sight of the farm. Put differently, a site of electricity generation in which the sun—not the human—is the star of the show worked closely with Johanna's ecocentric affects and our anti-petro politics to domesticate the sublime.

Crucially, this affective entanglement did not activate a coherent environmentalism. When a disdain for fossil capital, the imagined futurities of a just transition, the domesticated sublime of a sunny landscape, and the naturalist resonances of pastoral images congeal in a rented van, the resultant experience does not translate well into a political agenda. For in the course of our drive, we troubled the boundary between nature and society while simultaneously reifying that very boundary through our gaze at a distance—and this boundary, in turn, allowed us to simultaneously critique capitalism *and* celebrate (naturalized) high-tech commodities, training our outrage on the exploitation of the fossil fuel economy, while overlooking that of its clean energy counterpart. This simultaneity, then, bespeaks a lack of ideological clarity—a muddied political project—not a thoughtless commodity fetishism or an obliviousness to racial capitalism. I want to therefore suggest that the domesticated sublime of sunny landscapes works closely with naturalist images/imaginaries, sustainability politics, and an alienating urbanity to inspire fetishistic impulses among people with a critical consciousness of capital. At the affective intersections of image and sunlight, energy justice activism is blinded by the solar spectacle—a dearth of lucidity accompanying enlightened intentions.

This ideological muddiness came to the fore for me when we arrived at the strategy retreat at a Buffalo-based college. Which is to not to say that the retreat was without a clear stated purpose. To the contrary, the retreat organizers—EJ activists from all over the state—explicitly stated that their goal was to map out a political strategy for transitioning to 100 percent renewables statewide, dismantling investor-owned utilities, and "democratizing" energy so marginalized communities can control the electricity they depend on. By rejecting investor-owned utilities and calling for everyday people to own the means of energy production, this platform offered an unequivocal rebuke of racial capitalism, setting the tone for the retreat. Policy oriented and city based, our gathering felt nothing like our sublime drive: no pastoral scene, no solar farm, no socionature on the highway periphery. Instead of ruminating on beauty and life, we talked about efforts to lobby state legislators and the amount of solar power we could generate in different parts of the state. Infrastructure thus took shape not as a physical life force as it did during our drive but, instead, in terms of gigawatts—abstractions we could plot on a screen.

Yet despite this technocratic orientation, the sun's affective power could be felt at the retreat, muddying its ideological coherence. Specifically, the sun emerged when the retreat organizers presented a just-transition graphic that contrasted what it called "the extractive economy" with "the regenerative economy" (figure 1.11). This image proposes an ideological schematic for the just transition, centering equity, well-being, and ecological regeneration in visualizing a clean energy future—linking energy to life beyond metrics and policies. When the retreat organizers showed this image, the room affirmatively lit up, as the graphic presented a clear vision for change that made their politics legible. Crucially, the "resources" part of the schematic visualizes energy with two icons: a power plant, which symbolizes fossil fuels, and a sun, which symbolizes solar. As an icon, the sun here obviously lacks the rich affective texture of a NYC sunset or the shininess of a sun-kissed solar farm; it is simply iconic. Yet this signifying move reduces solar to the sun, resembling the obfuscatory shine in Danica's PowerPoint by rendering solar as a biospheric force and not an anthropogenic infrastructure rooted in racial capitalism. Furthermore, by characterizing the sun as a resource, the graphic traffics in racial capitalism's extractive grammar. Along these lines, the differential value at the heart of the solar supply chain undercuts the regenerative/extractive dichotomy that the graphic employs, as racial capitalism roots a so-called regenerative resource like solar in the fossil-fueled exploitation of the so-called extractive economy. The sun icon, then, presents solar as a natural force untethered to the differential value that links extractive industry with infrastructural work for a just transition.

The experience of viewing a tiny signifier of the sun on a graphic at a retreat is obviously different than experiencing the domesticated sublime at a solar farm. But in both instances, solar is visually apprehended as a part of nature vis-à-vis the sun, and this apprehension can attenuate one's consciousness of solar as a commodity produced under exploitative conditions. For when we encountered solar alongside the road *and* in the retreat's conceptual schematic, we saw it metonymically, regarding it with little consideration for the corporeal practices that brought it into being: the haptic extraction of minerals, the physical strain of fabrication, laborious transnational transport. Indeed, solar is relatively unique in that it is a mass-produced consumer good that most consumers never touch, hold, smell, taste, or hear, and it is thus particularly amenable to the

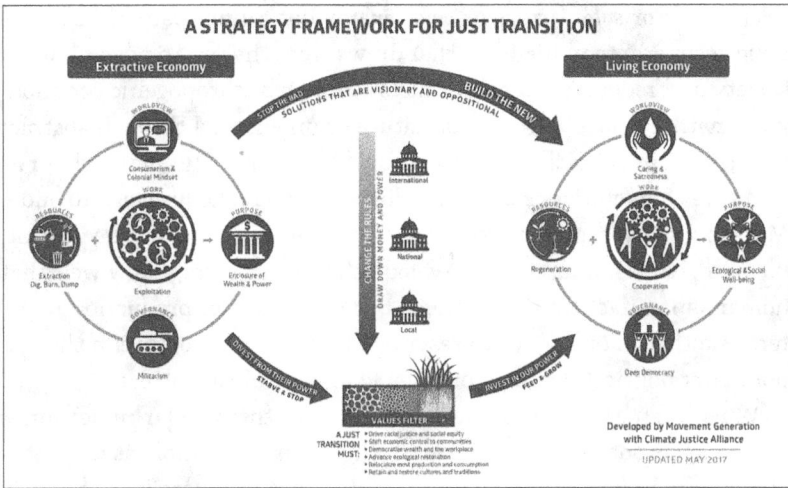

FIGURE 1.11. The extractive/regenerative economy dichotomy—a popular heuristic for the just transition. Movement Generation, *From Banks and Tanks to Cooperation and Caring*, 16–17.

commodity fetish. While there are numerous commodities that consumers have no proximate tactile relationship with (e.g., insurance plans), solar is a *material* consumer good—not a service—that its consumers primarily relate to at a distance: a physical form that is largely intangible for those of us who don't work with solar directly. The naturalization of solar is therefore not informed by a physical proximity that is characteristic of human relations with many other material goods. That said, I cannot speculate as to whether my interlocutors would naturalize solar if they touched instead of simply gazed at it. But I am suggesting that this absence of tactility creates the affective context in which the sun—a biophysical phenomenon that all humans have a sensuous relationship with—can coalesce with the naturalist ethos of someone like Johanna or the political imperatives of our energy democracy retreat to endow solar with a life of its own. Whether this endowment involves the sunny sheen of a real-life solar farm or a two-dimensional icon of the sun, urban environmentalists *see* a natural force of life in a technological commodity across contexts, attributing meanings to solar that an enslaved Uyghur in a solar factory likely wouldn't.

Along these lines, in both the technical renderings at the retreat and the lively impressions on the road, solar appeared as just that: *solar*, not *solar panels* or *solar arrays* or *solar energy* but a univocal yet nebulous *thing*, a superhuman life force that draws from the sun to power human society and reduce human emissions—not an anthropogenic commodity. Whether lauded as a part of nature or understood through abstract metrics, solar "reflect[ed] the social characteristics of [human] labour as objective characteristics of the products of labour themselves," to quote Marx on the commodity form.[53] Of course, everyone at the retreat, including those of us who marveled at the solar farm, knew well that humans manufacture solar panels. But the means of production registered neither in our affective reactions to the infrastructure in the van nor in our political strategy for a just transition at the retreat.

When seen as a sun-infused, superhuman thing, solar undermines efforts to uproot racial capitalism through a just transition, as it appears as the antithesis of the differential value that produces it—clean and organic, not derivative of racialized extractivism. The idea that solar is the inverse of racial capitalism informed the political strategy that emerged from the retreat, which included incredibly pragmatic and principled plans for ensuring communities of color could access solar without relying extensively on corporations. As such, this strategy positioned solar as the centerpiece of efforts to redress racial capitalism by endorsing the mass deployment of commodities produced under racial capitalism.

In this way, solar occupies an ideologically murky space even when it is paired with a value-driven framework and a specific political strategy for a just transition; its natural affect and quantifiable benefits situate it as an ethical antidote to the structures of extraction and exploitation that bring it into being.[54] As I have shown in chapter 1, this contradiction is not just attributable to the mystifications of capital but also to the sun's affective power, which both naturalizes solar and enables us to see it as a thing independent of its relations of production. While we could attribute this to greenwashing, what I am instead suggesting is that solar's relations with the sun in the context of racial capitalism obfuscate differential value in ways that enable people who actively oppose racial capitalism to valorize a product made through racial capitalism. Thus, solar's misleadingly green character is not simply the product of conscious dishonesty or greed; it is also attributable to the sun-powered technology itself. For the sun has long been a potent source of myth—a progenitor of affective power that inflects our social imaginaries—shaping how we apprehend the material forms it interacts with, including the solar panels it powers.

2

I peered at the screen of Celeste, a white solar professional, to view the Google Earth rendering of roofs in a working-class Black community that she was planning a solar installation for (figure 2.1). Lacking in vegetation and permeable space, the scene in this view illustrates how concrete and other man-made surfaces dominate marginalized urban areas—a metonym of environmental injustice. Indeed, poor Black and brown communities are home to a greater share of impermeable, treeless spaces, which, in turn, gives rise to inequities related to air quality, flooding, urban heat island, and food access—the ecological imprint of racial capitalism.[1]

Yet my thoughts went to the concrete not when I first viewed Celeste's screen but rather when she said that very word, *concrete*. The concrete she spoke of, though, wasn't the paved streets that her dual-monitor display projected. Instead, she was talking about the paradoxically *tangible* qualities of her *virtual* screen-based labor: the ways in which her work at a desk felt more real, more impactful, more *concrete* than her erstwhile policy advocacy work. "I chose solar because . . . it feels like a more concrete positive thing that you can achieve . . . for the [low-income] communities [of color] I used to organize," she said, contrasting her present technocratic work coordinating the early stages of solar installations with her previous, more idealistic career leading campaigns for affordable health care. Yet this "concrete" work occurred not through a medium she could touch but instead via Google Earth—a *cloud-based* platform that limits one's apprehension of the built environment to a pixilated aerial view. Beaming from screens in an open-layout office, Celeste's Google Earth portal was nowhere near the truly concrete spaces she aimed to touch.

FIGURE 2.1. A Google Earth simulation of rooftops used to scope a forthcoming solar installation. Photo of the simulation by Myles Lennon, December 2017.

Her employer—a solar technical assistance provider—works with installation contractors and community groups to do the upfront technical and administrative grunt work for solar installations, focusing primarily on marginalized communities. As Celeste and I were talking, she was conducting a preliminary site analysis on Google Earth—an assessment of rooftops to (1) identify if they have pipes, vents, bulkheads, or other so-called obstructions that could impede an installation; (2) rule out any rooftops that are near shadow-inducing buildings; and (3) map out the placement and dimensions of solar arrays for unobstructed, sun-exposed roofs. Crucially, if Celeste were assessing any old building, I'm not sure she would call her Google Earth work concrete. But she was specifically simulating solar arrays for a giant affordable housing cooperative in a community of color—solar arrays that, she explained, could yield *immediate* savings for people burdened by high energy bills, protecting a demographic that is vulnerable to the city's affordable housing crisis. These savings—their unambiguous impact on poor people's lives—rendered her screenwork *concrete* in her analysis. *Concrete*, then, stood in contrast to the often impractical, more idealistic world of social justice policy; instead of righteous sloganeering and lofty visioning, solar could address people's precarious realities in the here and now in clear-cut ways.

Even so, I found it interesting that she called virtual screenwork *concrete*, removed as it was from the actual people and spaces that she sought to impact. Without knowing the communities she was indirectly engaging with—without understanding their needs or experiences in the buildings simulated on her screen—how could she be so sure that the two-dimensional solar arrays that she affixed to virtual rooftops with her mouse would truly catalyze concrete change? Given both the preponderance of actual concrete in the community she viewed from a distance and the environmental injustices that these impermeable surfaces manifest, was her screenwork really the best way to address the community's "concrete" infrastructural problems? How does virtual solar transform our relations with space when abstract energy bills feel more concrete than on-the-ground organizing in those communities that concrete often symbolizes?[2]

This chapter grapples with these questions by exploring the digital platforms of equity-minded technocrats like Celeste and the ways this screenwork interacts with the marginalized "concrete" spaces that they remake virtually. I contend that solar's decentralized topography connects this screenwork with those spaces in a way that situates white-color professionals in low-income communities of color that they otherwise have no relationship with. This, in turn, transforms the ideological landscape of energy management, as Celeste's "concrete" calculus illustrates. For Celeste is a self-identified Marxist, ecofeminist, and intersectional activist committed to "structural transformative change," but she is not so naive to think that her solar screenwork is the transformative change she desires. Instead, the technical assistance she provides to building owners and tenants who are considering going solar is intended to fill alleged "gaps in the market" (according to her boss), as it entails doing the preliminary analytical work that solar companies don't want to do before signing a customer contract. In this way, her company enables energy transitions centered on consumer markets as opposed to public ownership, upholding a specific formation of racial capitalism—what is often called *neoliberalism* (which I discuss later in the chapter)—that she consciously opposes. "I don't necessarily see revolution to be part of our mission," she said, pointing at the word Google splayed on her simulation. Concrete screenwork, then, is a sort of compromise: a pragmatic middle ground between her political ideals and racial capitalism's intransigent structures.

Specifically, her anticapitalist ethos does not, in her mind, preclude the possibility that capitalized solar can challenge a system of differential value in which marginalized communities spend a disproportionate share of their income on energy while shouldering a disproportionate share of its environmental burdens.[3] By enabling her to work "in" such communities, the screen is the primary medium through which Celeste addresses this energy inequality. Her eyes glued to Google Earth, she identifies rooftops in low-income communities of color with potential for solar, efficiently navigating the city's skyline without leaving her desk. This screenwork, then, allows her to work with "the market"—that is, private solar contractors and renewable energy investment instruments—to intentionally situate solar in spaces that have traditionally been encumbered by energy generation. Put differently, Google Earth gives her the practical ability to instantaneously visit the city's rooftops, redressing the differential value of urban space by steering the market toward the most marginalized.

This chapter elucidates the late liberal affordances of software platforms that compress space in ways that enable equity-minded technocrats like Celeste to rationally manage racial disparities vis-à-vis the market. The "concreteness" of this virtual work is rooted in a particularly agential orientation to space that these platforms cultivate: a sense that one can navigate and transform the environment through a technocratic gaze.[4] As Celeste shows, Google Earth empowers its viewers to feel that they can remake space from their bird's-eye perspective, rendering one's environment ripe for industrialization by virtue of its "machinic" view.[5]

But Celeste contends that Google Earth's agential lens can only redress differential value when it specifically interacts with what I conceptualize as the *diffuse spatiality* of solar infrastructure—the focal point of this chapter. Whereas fossil fuel energy is generated in power plants and then distributed to buildings through a centralized network of substations, solar electricity generation is far more diffuse, occurring on the rooftops of people's everyday spaces: their homes, community centers, schools, and places of worship. Furthermore, this diffuse spatiality isn't particular to solar but instead characteristic of a broader ecosystem of sustainable energy technologies designed for domestic spaces, as I discuss shortly. This diffuse spatiality *decentralizes* energy production, enabling individual homeowners, residential cooperatives, commercial property owners, and community organizations to govern the electrical power that they previously could only purchase. But this more "democratic" form of energy governance is just as reliant on technocratic corporations as centralized energy generation, requiring advanced design, contracting, permitting, financing, and grid-interconnection services, for instance.[6]

As such, this chapter argues that sustainable energy's diffuse spatiality inspires technocrats to form intimate relationships with everyday people and their personal spaces, transforming the social landscape and political contours of energy expertise. I suggest that this dynamic at times motivates people who consciously oppose racial capitalism (like Celeste) to pursue market-oriented careers, while motivating more traditional energy experts to embrace a leftist politics. As a result, the diffuse spatiality of sustainable energy fosters an ideologically muddled approach to technocratic care that conjoins free market practice with socialist and antiracist principles. As I will show, solar and other sustainable energy infrastructures have affectively cross-pollinated the radical politics of the climate justice movement with the technoscientific ideology of "cleantech" corporations, diluting the normative ideological poles that traditionally divide these spaces, as the technocrat and the activist, the neoliberal and the anticapitalist, coalesce through the screenwork of infrastructural transformation. Celeste exemplifies this boundary-blurring subjectivity, which I conceptualize as *the equicrat*, short for *the equity-minded technocrat*: a liberal subject who reflexively mobilizes their technocratic training toward social equity. The equicrat heuristic calls attention to the counterintuitive compatibility of equitable, grassroots governance and rule by (corporate) experts under late liberalism, illuminating the performative terrain through which subjects simultaneously reject and reproduce the differential value of racial capitalism.

This chapter focuses on the corporate equicrat: the white-collar energy expert like Celeste who embraces a social justice ethos at odds with their corporate habitus; chapter 3 focuses on the grassroots equicrat: the community-based activist who adopts the technocratic values and practices of the electricity generation sector—ideologies at odds with their social justice habitus. These chapters show how the *decentralization* of energy counterintuitively powers what we can think of as the *centralization* of ideology: an erosion of the phantasmatic lines separating conflicting political visions (e.g., free markets versus ecosocialism).[7] More specifically, solar's diffuse spatiality broadens the range of subjects who can participate in electricity governance, which, in turn, dilutes the normative ideological poles that traditionally divide these subjects.

As Celeste demonstrates, this equicratic shift largely takes shape through digital platforms. For these platforms' "machinic panopticism" enables their viewers to transform myriad spaces from afar, enacting the decentralized governance that diffuse energy generation demands.[8] This alignment of virtual and concrete spaces animates Celeste's equicratic efforts to solarize communities that fossil capital traditionally neglects; it is precisely the potential to equitably decentralize energy governance through screenwork that draws Celeste to an office job she otherwise would be skeptical of. This suggests that Google Earth and similar virtual interfaces represent a break in the ways that the technocratic gaze operates as a mode of governance. In contrast to the modernist cartography and surveillance that comprised the *centralized* power of twentieth-century statecraft, Celeste and other equicrats remotely survey space in support of a *decentralized* regime of power that ostensibly redresses differential value, reappropriating the technocratic gaze toward late liberal ends.[9] This, again, evinces the compatibility between, on the one hand, sustainable energy's diffuse spatiality and, on the other, virtual interfaces that allow their viewers to transform space via screens.

While this screenwork taps into a late liberal imaginary that views sustainable energy markets as an antidote to the inequalities of racial capitalism, this chapter shows that it also obfuscates the differential value embedded in energy generation systems. Specifically, this screenwork alienates energy equicrats from the communities of color they wish to save, choreographing "concrete" change that is divorced from labor dynamics on the ground while disingenuously simulating racial progress with regard to energy.

Before discussing this insidious screenwork, this chapter explores how sustainable energy's diffuse spatiality affectively incubates the corporate energy equicrat. In the process, I theorize a mode of technocratic practice that simultaneously sustains a long-standing regime of equity-driven expertise and disrupts the corporate world of energy experts. I then return my analysis to screenwork, showing how virtual interfaces constitute energy equicrats' spaces to demonstrate the symbiotic relationship between diffuse energy generation and screen-based management. I argue that this symbiotic relationship positions the work of energy equicrats as late liberal governance, deceptively branding neoliberal practice as social justice—effectively undercutting aspirations for a just transition unbeholden to racial capitalism.

THE ELUSIVE MARKET FOR DECENTRALIZED ENERGY

In 2016 I joined SunLight—a national energy efficiency and renewable energy (EERE) corporation that specializes in solar—as a pro bono anthropologist tasked with helping the then predominantly white male company diversify and foster inclusivity. SunLight had built a successful brand by conducting specialized energy assessments of large buildings but was rapidly expanding their portfolio of solar installation projects in predominantly low- and moderate-income (LMI) communities of color. John, the company's CEO, took great pride in making solar accessible, and while he rejected dogmatic free market ideology, he regarded SunLight's LMI projects as evidence that under the right circumstances the market can be a force for social good. SunLight was one of several solar companies expanding into Black and brown spaces that are not usually regarded as paragons of sustainability. These companies' principals often cited their LMI work as evidence that markets can decarbonize the world while prioritizing the most marginalized. That they grew their business by solarizing affordable housing was, in their mind, proof that capitalism can be kind.

This pro-capitalist imaginary hinged on solar's diffuse spatiality. As an infrastructure that can generate electricity at the same site it's consumed—in homes, schools, houses of worship, and communal spaces—solar can transform any functional roof with sun exposure (including those in poor communities) into a site of energy production. This diffuse spatiality brought SunLight's energy experts into communities they had never before had a relationship with. The company's engineering team had once primarily worked on reducing commercial properties' energy consumption, but the emergence of NYC's solar market created demand for their expertise in sunny spaces in the city's outer boroughs—away from Manhattan's shadow-inducing skyscrapers. As such, the solar market created an opportunity for these experts to directly support those

NYC residents who are disproportionately vulnerable to high energy bills and climate change. This would suggest that John's assessment had merit; the market was in fact transforming electricity generation in those communities most in need of affordable, sustainable energy. In this schema, solar's diffuse spatiality corroborates a moral case for market-based energy governance.

But Martha, a counterintuitively *anti*corporate SunLight employee, saw the situation differently. She viewed solar's diffuse spatiality as the basis of a just transition that counteracts racial capitalism. By enabling everyday people to generate electricity in their homes, a just transition, as Martha and others imagined it, can distribute power to the people most burdened by extractive industries, catalyzing energy worker cooperatives that reject profits and prioritize community well-being.[10]

Thus, solar's decentralized terrain is at the heart of both anticorporate imaginaries of community power *and* pro-corporate narratives of market power. This would suggest that John and Martha read their own competing agendas into solar's diffuse spatiality—that this terrain did not shape their ideological persuasions.

Indeed, a more holistic view of the solar market suggests that this diffuse spatiality validates neither platitudes about kind capitalism nor leftist imaginaries of ecosocialist economies. For starters, the corporate deployment of solar in NYC's LMI communities is the product not of a "free" sphere of private commercial exchange (as some "kind capitalists" contend) but instead of *government* programs that expand renewable energy access. In the United States, the public sector incubates renewable energy markets through policy that enables consumers to, for instance, contract with a private company to install solar panels on their private homes. Through renewable energy tax credits, incentives, and promotional resources, government aims to move the market, bringing down costs, investing in new businesses, and creating a competitive environment in which consumers and businesses can collaboratively ramp up renewable energy production. While market focused, this approach is not dogmatically "free market": it provides *considerable* subsidies to LMI residents so they can purchase sustainable energy upgrades they couldn't otherwise afford. Thus, this market-based approach to energy transitions includes socialist dimensions with equity-oriented aims, troubling any claim that a private market is helping the poor.[11] As such, just-transition campaigns in NYC are partly focused on expanding this equity-oriented marketplace, enabling poor people of color to access for-profit EERE corporations' services even as these campaigns are led by activists who strongly oppose racial capitalism (as I discuss in chapter 3). These EERE corporations were happy to participate in this highly subsidized LMI marketplace, as it expanded their customer base while building their brand as community-responsive. But, again, this approach to decentralizing electricity generation is rooted in a redistributional public sector—not a private market.

At the same time, the diffuse spatiality of solar in NYC is *necessarily* enmeshed in an even more diffuse network of private industries, problematizing Martha's aspirations to disentangle renewable energy from racial capitalism. Specifically, solar infrastructure consists of numerous materials produced by transnational extractive industries, including plastics, silicon, and lithium.[12] This makes it nearly impossible to procure solar that has not been made in some way, shape, or form by extractive multinational corporations. Solar installation cooperatives in NYC, for instance, deploy solar panels that not only were made by corporations but, more crucially, include materials that have *never* been incorporated into goods without capital and private industry. The transnational spatiality of solar commodities demonstrates that a just transition via solar will always depend in some way on private corporations.

Yet solar's decentralized terrain—its deployment in marginalized communities—is marshaled as evidence that solar can disentangle electricity generation from racial capitalism, or, alternatively, that it bears the imprint of a kinder capitalism, competing conclusions that similarly overlook the broader political economy of renewable energy production. That John and Martha—people of different ideological persuasions—draw competing conclusions from solar's diffuse spatiality suggests that this spatiality is not in fact the primary progenitor of their respective renewable energy visions.

But as I got to know more EERE professionals, I found that solar's distributed imprint did in fact affectively animate ideological sentiments—yet not anything as clear as a pro- or anticorporate politics. At SunLight, in particular, I encountered a range of ideological orientations to the market, the diversity of which troubled the anticapitalist conception of *the* corporation—a mythically monolithic entity focused myopically on shareholder value.[13] There was nothing approaching a consensus at Sun-Light that capitalism is "good," as many employees, like Martha, offered intersectional critiques of private industry. These critiques pointed to the energy sector's equicratic shift—an ideological transformation that,

I contend, is tethered to the spatialized decentralization of electricity generation. Along these lines, SunLight's investment in diversity—the ostensible reason I had been granted access to the company—was not just motivated by the late liberal politics of corporate diversity, equity, and inclusion (DEI) but also by EERE's diffuse spatiality, as I now show.

CORPORATE INTERSECTIONALITY

As part of my initial assessment of SunLight's diversity challenges, I worked with the company's head of human resources (HR) to organize a series of open-ended conversations with employees at every level of the company. In our very first conversation—with Sean, a millennial white male vice president (VP)—I encountered equicratic ideas that were at odds with John's ideology of kinder capitalism. "I bet if Adam weren't a white man, he would've lost his job," proclaimed Sean, startling me. While Sean's statement would suggest otherwise, Adam, the privileged white man he mentioned, was his employee, and I didn't expect him to offer such a blunt assessment of SunLight's raced and gendered proclivities. "People like Adam literally have the privilege, because of how he looks, to choose office work or physical labor even though he's really not the best fit for the office stuff." Adam had worked for several years doing blue-collar solar work for a small solar company before making the jump to a white-collar solar engineering position at SunLight. Sean, though, was suggesting that a candidate for Adam's position with similar work experience but without his race, class, and gender privilege would not have been hired and that if they somehow were hired, they would have lost their job once their lack of office skills surfaced. This intersectional analysis grew even more surprising when Sean implicated himself, contending that although he doesn't believe that he discriminates, he obviously isn't the best person to assess his own implicit biases and his culpability in a company culture of hiring and promoting educated white men like him and Adam.

I was more than satisfied that Sean—a white man in a senior position—had shared such a candid assessment of the company, and I would have been happy to leave the conversation there. But Sean didn't simply want to discuss intersectional inequality at SunLight; to my surprise, he also wanted to problematize the company's white-collar character more broadly. "It's unfortunate that we value office work and give people opportunities to grow and to actually earn a living wage in the office when we barely pay our [solar installation] crews enough to get by, but in fact the really important work that engages your imagination—the work that we should be valuing—is happening outside the office in the field." ("The field" entailed the rooftops and electrical rooms where the company's solar installation crews worked—the blue-collar carve-out of an otherwise technocratic corporation.) Here Sean revealed SunLight's classed identity crisis. On the one hand, SunLight prided itself on its elite credentials; at the time, *two* employees had doctorates in astrophysics, and their website branded them as "energy experts." On the other hand, the company had seen remarkable growth in their lower-paying solar installation work. As a result, John had recently decided to dramatically expand SunLight's roster of blue-collar solar installers despite the company's technocratic brand. Yet, as Sean suggested, this shift had not changed how SunLight valued labor; solar installation revenues did not translate into rising wages for solar installers.

For Sean, this income inequality was especially troubling because it was linked to the broader existential matter of the meaning and value of work:

> Unfortunately, in our society we don't value people who do work outside. As a society, we value office work where people sit behind screens more than we value hands-on physical work. We forget that working with your hands allows you to think critically about how systems operate and allows for you to form a critical understanding of systems that you can't get by just sitting behind a computer all day. Sitting behind a monitor . . . is bad for our minds and our bodies, and we need to learn to do work that actively engages our bodies and to value labor that actually makes the world work. . . . Office work just kills the mind and soul. . . . This is what happens when we, as a society, value elite status over real-world issues. It's terrible how we don't value people who actually make things work, who actually know how things work, the people we're most dependent on to function.

This analysis surprised me more than Sean's critique of white male privilege. After all, Sean was a VP at a national technology firm headquartered in NYC's Financial District—a company whose employees spend their days at screens, crunching numbers, running reports, inputting data into cloud-based platforms, talking on Slack, eating lunch with their eyes glued to their phones, unable to pause for a moment with sales targets to meet; why was he so cavalierly sharing his distaste for corporate values and his affection for corporeal labor? He offered other similarly confounding commentary on white-collar ideology, expressing skepticism toward SunLight's colorful graphs that suggested that better profits could improve the company's response to climate change. In this regard, he rejected John's belief that the corporation could be a force for good, wedding this commentary with a critique of the tech sector's differential value (epitomized by SunLight's hiring and promotional practices, and the broader privileging of white-collar over blue-collar labor). In other words, Sean offered a holistic takedown of what he called "corporate society," problematizing the very structure of twenty-first-century racial capitalism.

It's not uncommon to find disaffected white-collar workers in the tech sector, but I had trouble understanding how someone who vocalized unapologetic critiques of the existential order of contemporary corporations could climb the ranks at a technocratic corporation.[14] Unlike many disillusioned professionals, Sean did not strike me as someone who had sold his soul to corporate America. In his stinging commentary, I detected values that he didn't seem willing to compromise, even if his position would suggest otherwise—views that were more particular and less hackneyed than the social justice platitudes of the "'woke' workplace."[15] To understand Sean's seemingly out-of-place orientation to racial capitalism, I would need to look not simply at his place of work but also at the spaces his work touched—the decentralized terrain of sustainable energy upgrades.

THE HVAC VP

One evening after work, Sean and I went out for a drink, and halfway through my beer, I felt an itch to ask him something that had been troubling me since our "diversity" discussion. How could he square his critique of the elevated status of corporate screenwork with his critiques of SunLight's egregious white male privilege, given that many of the Black, brown, and women workers he sympathized with actively *prefer* the comforts of screenwork over the exhausting physical realities of blue-collar jobs? Indeed, countless marginalized people look to computing, data, and office work as a way of transgressing the structural oppression of a labor market that unevenly relegates them to the realm of somatic work.

When I raised this, Sean's response began to shed light on why someone with his *particular* positionality would unambiguously espouse an intersectional, blue-collar politics in his workplace. While he respectfully affirmed marginalized workers' aspirations for upward mobility and criticized his own intersectional privilege, he explained that for many years he had himself done hands-on blue-collar work in residential buildings. But he acknowledged that even when he was a low-waged HVAC technician on small-scale energy efficiency projects, he still had more privileges than his Black and brown coworkers—people of color without college degrees who did the same labor he did and had the same skills he had. Unlike them, Sean was given the opportunity to make the jump to office-based work and to eventually land a VP position—a move that he directly attributed to his raced, classed, and gendered privilege. In contrast to many other white male executives, Sean had a professional pedigree that enabled him to intimately understand the ways in which hands-on work is debased and people with his intersectional positionality often rise above that work.

At the same time, there seemed to be another element of his argument unrelated to intersectional positionality. He was concerned not only with *whose* work is undervalued but also with the *types* of work that are undervalued. He explained:

It's extremely important for the spirit to actually interact with people and to connect with people. The people that I get to work with is why I do this work . . . [but I got closer with people] when I was actually doing things with my hands, in buildings. It's very rewarding to fix things for people with your hands, and then they aren't freezing in the winter. . . . And they would look at me and say, 'You fixed my heat, I haven't had heat in thirty years. My kids grew up without heat.' . . . And I had this real drive in the depths of winter to go somewhere in the middle of the night to help people because this is a skill I have, and that's what I used to do. . . . It was very rewarding and also incredibly abusive. People would say things to me that you shouldn't say to anybody, much less the guy who comes to fix the heat, but even that was really special because to be out there and see what kind of conditions make people this abusive—I don't think people naturally wanna be that abusive. . . . I don't have those same interactions with people now that I'm a VP. . . . It feels better to do something *with* people . . . than trying to do something in a more fractured way, which is unfortunately how we usually do things [in the office].

Here Sean unpacks his orientation to SunLight's workplace by analyzing the affective interrelations of his erstwhile haptic labor and his moral/political values. In the process, he offers a dialectics of infrastructure and ideology, identifying three aspects of working with HVAC systems that, he contends, shaped the ways he values labor: (1) visiting people's personal spaces, (2) helping people access energy, and (3) bonding with people in the flesh. In Sean's analysis, when you're physically present in a person's home, when you get to know them face-to-face, and when you improve their bodily conditions, you cultivate a holistic work ethic that connects the "spirit[ual]" and material dimensions of living and livelihood. In this way, the corporeality of residential HVAC labor contrasts sharply with the "fractured" world of office work, entailing certain proximities—a physical closeness to laypeople and their homes—that partly enabled Sean to sympathize with the less fortunate, understand his actions' real-world ramifications, and value the various forms of connectivity that constitute life. Or so he claims. I am not trying to adjudicate the alleged causal relationship between his labor and his interpersonal values. Instead, I suggest that when he enunciates this relationship, he also articulates an ideology of work in which hands-on skills facilitate social relationality and tap into "the spirit."

At the core of this ideology is not only haptic labor but also the infrastructure that such labor acts on. In contrast to centralized electricity generation infrastructure, residential HVAC systems are predominantly located in people's places of residence and, as such, are predominantly controlled by people who have a personal relationship with those places—not by technocrats who have never been to those places. As a result, such infrastructure requires labor often rooted in certain interpersonal connections that are missing from labor focused on transmission lines, power stations, and electricity usage data. In Sean's ideology of work, such interpersonal connections endowed his labor with an in-person intimacy that is necessarily absent from labor focused on, for instance, an industrial-scale electricity storage apparatus, as the latter does not entail communing with people in their homes. Put differently, work on energy systems in people's homes is, by its very nature, affective in ways that work on the centralized and nonresidential nodes of industrial-scale energy is not, engendering a domestic solidarity—a "special" type of communion.

Sean suggested that this solidarity informed his intersectional assessment of SunLight's problems. Specifically, he attributed his white-collar colleagues' inability to act sympathetically across lines of difference to a "fractured" mode of screenwork where people focus more on their phones than connecting with one another in the flesh, in contrast to his hands-on work "in the field."[16] In this way, Sean positioned haptic labor as the grounds on which he cultivated an intersectional consciousness of white-collar workplace relations—an awareness he further spoke to in the following disclosure:

> We've got a big group of people here who . . . started [at the company] around the same time and became very close and . . . they're all white, college-educated males, and all of them can make a *Family Guy* reference, and they all laugh, and they have a certain comfort with each other, and cliques form, and all of this is shaped by their gender and race. There's just a certain white male culture here that's not, like, overtly exclusionary, people are really nice, but I don't think they, or most of us, really get how—it's hard to put into words. Like, a lot of people here are always focused on heavy drinking in their free time, and that translates into the culture here, but . . . many of our new women employees don't drink for religious and health reasons so we need to curtail that . . . 'cause when you think of [SunLight], you think of a bunch of white dudes drinking beer. . . . And I'm not saying that alcohol [has] a racial or gender group, but there's a [white] bro-y aspect to things . . . that I find disconcerting. . . . We will always hire the best person for the job, but all sorts of biases go into that equation. Do we support diversity? Yes. But . . . we usually already know who we want to hire when we have a budget line—people in our networks—and it's a lot harder to hire racially diverse if most of your networks are white.

This rumination on the insidiousness of racism and misogyny in his office bespeaks not only an intersectional ethos but also an attunement to the particularities of workplace relations.

Sean proceeded to attribute this attunement to the consciousness he cultivated when he performed hands-on work in people's homes—to "my experiences with hurt people up-close." The haptic labor of residential energy infrastructure, then, shapes the ways that Sean positions himself intersectionally, informing a performance of enlightened white male subjectivity at odds with that of elite professionals in the ethnographic literature on white-collar energy industries. Looking broadly at this literature, we can locate a trope of the energy professional—a composite figure that is, to be sure, a simplification but that nonetheless helps to highlight the ideological discontinuities between energy professionals affiliated with decentralized energy production (e.g., Sean) and those affiliated with centralized energy production. For in numerous ethnographic works on energy, experts are quite often white men predisposed to a narrow scientism, white-collar elitism, growth-oriented fundamentalism, free market deregulation, overt misogyny, racial paternalism, a hostility to diversity, and/or an alienation from the everyday world that their technologies, research, commodities, policies, and markets intimately touch. Countless lab-based research scientists, economists, consultants, bureaucrats, policymakers, development practitioners, and corporate executives in energy industries consistently demonstrate these problematic traits.[17] Significantly, such energy experts perform their expertise in isolated offices, labs, and conferences—elite, inaccessible spaces. As such, their parochialism is considerably shaped by the *relational spatiality* of their labor: their physical remoteness from the energy infrastructures they control and manage, their disconnection from the laypeople who rely on those infrastructures, and their distance from places burdened by those infrastructures' environmental impacts. In Sean's self-assessment, the relational spatiality of his labor similarly shapes his political orientations, but he rejects the parochialism of traditional energy experts because he previously performed haptic labor in people's homes, forming intimate, cross-class relations that elude the elite. While this explanation obviously simplifies the ideological workings of subjectivity—Sean's former occupation is not the singular cause of his deviation from the energy professional trope—it does suggest that infrastructural spatiality shapes the imaginaries through which technocrats position themselves as such, as the anthropology of infrastructure has consistently shown.[18]

More specifically, the differences between the low-income Black and brown spaces that Sean previously worked in and the more sophisticated energy infrastructure that SunLight initially worked on (before shifting to solar) allow him to account for his ideological deviation from the "bros" he derides. *Bro*, in this context, refers to one of Sean's affable, white male junior colleagues who wear backward hats with their button-downs at their desks, who work beside cartoon memes hanging over their computers, who keep goofy toy weapons by their keyboards, and who congregate in the office on Wednesdays after normal work hours to drink from a keg and talk about sustainable energy innovations, a ritual paid for by the company. The moniker *bro*, then, refers to a youthful masculinity distinguished by its "chill"—part and parcel of a young, straight, white habitus at many tech startups that conceals its constitutive privilege by foregrounding a certain affability derivative of the "default" raced and gendered slot that its purveyors occupy.[19] While corporate bro-ness is not particular to energy firms, this affect is unmistakably shaped by energy infrastructure at SunLight. Specifically, the young men whom Sean characterizes as bros perform this particular masculinity through their confident displays of proficiency in online energy management platforms; their infrastructural jargon, which they exchange with a restrained bravado during their weekly tribal keg ritual; their gritty forays out of the office to work in scorching boiler rooms across the city (as part of SunLight's energy efficiency upgrades); and the solar and HVAC puns they affix to their company T-shirts. These practices demonstrate a gendered, jovial affection for infrastructure—a lighthearted confidence that bros can change the world through EERE—enacting the tech-startup culture that Sean bemoans. While Sean knows he's not immune to SunLight's bro-ness, the spatiality of his previous labor is, again, his point of differentiation, as he highlights his experience working on *low-income residential* energy systems in contrast to his bro-y colleagues' experiences with sophisticated large-scale commercial properties. This spatiality is a point of pride that, in his telling, shapes a particular political consciousness that he brings to the upper ranks of the white-collar corporation—intersectional and anticorporate sensibilities that are uncommon in technocratic spaces.

But this political consciousness wouldn't mean much if Sean just went about his day without actually addressing the tensions between his principles and his work. After all, there are plenty of white-collar workers with critiques of their employers.[20] That he has misgivings about his office's intersectional character, labor protocols, and screenwork is not, in and of itself, so notable. But I want to suggest that the diffuse spatiality of EERE affectively animated a workplace where Sean could counterintuitively act on his critiques of the corporation by occupying a privileged perch within the corporation. As I will show, this diffuse spatiality paradoxically transforms Sean's anticorporate sentiments into corporate assets. Put differently, EERE's decentralized terrain gives affective form to a certain discontent with racial capitalism that is actually quite compatible with SunLight's market-based work—a strange ideological configuration that I now explore.

THE DECENTRALIZED TERRAIN OF EQUICRATIC NEOLIBERALISM

Sean wasn't exactly circumspect about his politics, no matter how out of place they might have seemed in a corporate office. While he was always polite and mild-mannered, he didn't hold back from sharing his views with other members of SunLight's leadership: his sense that the company should stop focusing on growth and instead focus on quality, his conviction that the company overvalued office work at the expense of blue-collar work, his belief that true value comes from doing things with your hands and that corporations overvalue screenwork, his critiques of the company's tendency to hire mediocre white men. Why would a white-collar energy technology and data firm give a VP position to a person who's intensely critical of not only racism, sexism, and elitism in the corporate workplace but also capitalist values more generally? A person who claims he'd rather be working in people's homes with his hands than staring at a screen in an office?

Sean hinted at a partial answer to this question when he called attention to his own raced and gendered privilege. The white male nepotism of the corporate world—and at tech startups, in particular—enables many people with his intersectional profile to secure powerful white-collar positions irrespective of their *particular* personality traits.[21] We can locate this nepotism in Sean's close relationship with SunLight's founder and CEO, who is also a straight white man and who hired Sean and other straight white men whom he knew personally when the company was still a small startup many years earlier. In some regards, then, it is not at all a given that Sean's anticorporate values could hinder his corporate success.

But white male nepotism isn't the only reason a technocratic, white-collar firm would, contrary to common sense, empower someone who disdains white-collar technocracy. The energy systems that SunLight specializes in were also key. Specifically, over the past few decades, "sustainable" electricity and heating infrastructures transformed the ideological landscape of energy expertise, generating an equicratic politics that gave affective form to Sean's position at SunLight. In particular, the *diffuse spatiality* of residential electrical and HVAC systems and of rooftop solar projects began to alter the political climate of NYC's energy ecosystem in the mid-2000s, injecting anticorporate and antiracist sensibilities into the technocratic realm of energy governance, as I discuss shortly. Put differently, new efficient electricity, heating, and solar technologies positioned *residential and community spaces* at the center of energy policymaking, infusing the energy profession with those spaces' left-leaning politics. This, in turn, created a corporate environment amenable to Sean's anticorporate perspectives. Specifically, the diffuse spatiality of EERE technologies aligned with similarly diffuse forms of governance by the market in ways that paradoxically amplified activism *against* "governance by the market."[22] This form of governance is frequently known as *neoliberalism*—a word that often imprecisely connotes a crude economism that is at odds with Sean's leftist views.[23] Yet neoliberalism is actually quite compatible with anticorporate just-transition activism, as twenty-first-century racial capitalism and its critics oddly align in the decentralized space of EERE.

Unpacking this point first requires clarity as to what *neoliberalism*—a notoriously capacious term—actually is.[24] Across a range of scholarship, *neoliberalism* often refers to a particular formation of racial capitalism in which "market terms, . . . market metrics, and . . . market techniques and practices" proliferate beyond commercial activity to "everyday life," transforming how contemporary humans understand themselves and govern one another.[25] Put differently, neoliberalism generally describes the application of economization and spheres of commercial exchange to things that were once understood to be outside the locus of the market: life essentials, liberal entitlements, human affects, group identities, and/or public sector services (including environmental conservation and electricity generation). In this schema, for-profit corporations can apply commercial imperatives or economizing rationales to the provisioning of water, the securing of happiness, or the empowerment of minoritized subjects, for instance.[26] At the same time, one can engage in neoliberal governance without necessarily working toward the goal of extracting surplus value; environmentalists can, for instance, eschew a profit motive while promoting climate justice on the grounds that it can yield a good "bang for the buck."[27] This proliferation of economization supplants and undermines the modes of liberal governance that took form and gained prominence throughout the world for the better part of the twentieth century. Put differently, the enhanced role of private firms in the provisioning of social services and nonstate actors' efforts to secure and optimize well-being through, for instance, utilitarian metrics of value, have challenged and attenuated the model of the strong central state—a liberal vision of comprehensive, centralized governance. This does *not* mean that "the state" has withered away or even shrunk in the shadows of the market but rather that "the economization of life" undermines a philosophy and approach to governance focused on the ideal of a robust public sector.[28] Thus, whether *neoliberalism* is used to theorize governance by the World Bank or governance by Twitter, it refers to both an expansion of economization and forms of decentralized or more diffuse governance that are at odds with the twentieth-century ideal of the Keynesian and "high modern" state.[29] In other words, no matter how neoliberalism is construed in the wide range of literature, this literature generally supports the claim that political, economic, and cultural shifts over the past fifty years expanded the scope of commercial forms of valuation— whether this entails the commoditization of social goods or the subsumption of everyday life under economizing techniques—which has, in turn, tasked more nonstate actors with the management of life vis-à-vis commodities and commercial exchange.[30] Neoliberalism, then, represents a more diffuse form of racial capitalism—a racial capitalism that extends not just its commodifying imperatives and extractive practices but, more crucially, its utilitarian ontology from the marketplace to domestic and communal spaces or to any other traditionally noncommercial sphere of life, such as the environment. At times it even rebrands everyday people as the marketplace itself (as I discuss in greater depth below—see figure 2.5). By subsuming life under commercial logics, neoliberalism is a radically decentralized project. Even as it clearly has (raced, classed, and gendered) winners and losers (like any other form of racial capitalism), it obfuscates this differential value by creating an ontological order in which *anyone* can ostensibly improve themselves through economization techniques—in which investments in solar can improve poor people's carbon footprint and local environment—economizing projects that, on the surface, are not centered on surplus value.

This decentralized program of economization maps onto the decentralized program of EERE, as the environmentalist imperative to expand electricity generation from centralized power plants to people's personal spaces is congruent with the neoliberal imperative to expand the locus of economization from designated marketplaces to people's personal spaces. This congruence creates the ideological conditions for Sean to act on his political views through a technocratic position seemingly at odds with those views. Specifically, I argue that public policy transformed this spatial congruence into a political formation that catalyzed the emergence of an equicratic ideology in which racial capitalism and anticapitalist sentiments aligned. In the next section, I connect the dots between neoliberal policies, sustainable energy infrastructure, and anticapitalist activism to elucidate the energy equicrat's emergence over the past three decades before returning to Sean's particular politics at SunLight.

DECENTRALIZING ENERGY GOVERNANCE

In the 1990s and early 2000s, private industry and governments improved and expanded the production of a range of energy efficiency technologies (including spray foam insulation, LED lighting, blower doors, and highly efficient home appliances and boilers) and rooftop solar technologies, creating the material conditions for *consumer-centered* energy governance. Specifically, these technologies were primarily designed and marketed to reduce electricity and heat consumption in private homes and buildings. As such, these technologies set the stage for energy policy focused not only on large-scale nodes of energy infrastructure, such as power plants, transmission lines, and pipelines, but also on small, consumer nodes of energy infrastructure, such as boiler rooms, building envelopes, lighting fixtures, and rooftops. These technologies, then, transformed the spatiality of energy governance, magnifying the capacity of *individual consumers* to manage energy infrastructure in their homes and community spaces. In this way, the spatiality of energy efficiency and renewable energy technologies in the 1990s and 2000s was conducive to neoliberal energy policy: legislation that decentralized energy governance by incentivizing laypeople to manage electricity and heat through a consumer market that was largely developed to mitigate climate change.[31] Before the emergence of such technologies (in addition to new grid-management infrastructures), laypeople were not explicitly enlisted into local, state, and federal energy governance. In sum, EERE technological innovations created the material conditions for neoliberal reforms that positioned individual consumers as active agents in energy governance vis-à-vis the market.

The first set of such reforms aimed to transform electrical, heating, and building infrastructures by incentivizing consumers to acquire and install the aforementioned technologies (including solar) through tax credits, rebates, feed-in tariffs, and other financial mechanisms.[32] By creating an EERE consumer market, these neoliberal policies enabled subjects who had never before played a formal role in energy governance to do so through economization, often in the comfort of their own home. Instead of mandating building upgrades or changing building codes as part of comprehensive, top-down efforts to reduce carbon emissions, these state and federal policies focused on *individual choice*, as they created programs predicated on the principle of market competition and the conception of a *homo economicus* homeowner who carefully navigates the marketplace to acquire EERE goods and services particular to their individual building's needs. The significance of this approach to governance is worth underscoring; for the first time, individual laypeople were effectively recruited (or, in market terms, *incentivized*) by government and industry to serve as market-based agents in the management of energy—a neoliberal move predicated not merely on its proponents' philosophy of governance but also on the diffuse spatiality of new EERE technologies. Thus, this decentralized terrain of sustainable energy aligned with the decentralized terrain of neoliberal markets to generate new subjects of energy governance.

As such, this process of subject formation relied heavily on the *affectivity* of EERE technologies, that is, their capacity to make individuals *feel* like "empower[ed] energy consumers" in the decentralized space of their homes.[33] By lowering energy bills, reducing fossil fuel consumption, and mitigating heat leakage in the country's sprawling, privately owned housing stock, these technologies gave many homeowners and tenants the sense that they could save money, save the planet, and enhance their physical well-being by participating in residential EERE markets (as I discuss further in chapter 3).[34] Advertising, graphics, and sales pitches also highlighted these technologies' material impacts on private housing, rendering nonindustry laypersons as empowered energy consumers capable of improving their home economics, bodily comfort, and positive contributions to the world through EERE (see figure 2.2). In this way, decentralized energy technologies worked hand in hand with corporate promotions and neoliberal policies to shape a new conception of the advanced, individuated energy subject optimizing their bodily well-being and sense of domestic control through the market—a feeling that *anyone* could govern their own energy infrastructure and improve their lives in the process.

By interpellating consumers as agents of energy governance, neoliberal EERE policies capitalized on the diffuse spatiality of sustainable energy technologies in a way that rapidly expanded the industry of small EERE companies focused on residential buildings. In other words, policy-

FIGURE 2.2. Ads like this not only market solar commodities but also recruit laypeople to participate in the governance of energy—part of a broader neoliberal shift to decentralized energy management. Positive Energy Solar, "Go Solar and Save the Planet."

makers' neoliberal focus on commercial instruments to address *individual consumers'* homes (as opposed to building codes or other policy instruments to address an *aggregated* building stock) catalyzed the formation of startups that vied for a piece of the new consumer pie—a *fragmented* market-based response to society's broad energy challenges, as opposed to a top-down governmental response. This fragmented response, in turn, transformed the social landscape of energy expertise, as the emphasis on consumer economization created new technocratic spaces unmoored to the hierarchical, centralized politics that are commonplace at electric and gas utilities, public utility commissions, extractive industries, and other traditional energy institutions—a move that had implications for Sean's unlikely positioning at SunLight, as I discuss in the next section. Again, the decentralized spatiality of EERE technologies was at the root of this neoliberal shift to consumer economization.

Crucially, several political developments in the mid-2000s forever transformed energy governance, accelerating this neoliberal shift by counterintuitively linking market-based policies with the leftist politics that Sean espoused. These political developments include the second US-Iraq War, which amplified calls for "energy independence" and "no blood for oil"; the reinvigoration of public concern regarding climate change, shaped by, among other things, the release of Al Gore's film, *An Inconvenient Truth*, and high-profile "natural disasters," notably Hurricane Katrina; a rapid increase in the cost of gasoline, which magnified energy independence discourses; and a global economic recession that catalyzed large-scale government investments in EERE.[35] It is in this political context that leftist and progressive activists who had traditionally opposed neoliberal reforms began to embrace neoliberal EERE policies. These activists—who were affiliated with labor unions, environmental justice (EJ) organizations, liberal think tanks, and Democratic legislators—regarded EERE as the basis of a *realistic* policy platform for addressing the United States' energy, environmental, and economic problems amid a conservative political climate that was hostile to the "big government" straw man pejoratively affiliated with such activists. Specifically, this left-leaning coalition put together reports, campaigns, books, websites, videos, and conferences that rendered EERE policies as *politically pragmatic, commonsensical* mechanisms for mitigating dependence on "foreign oil," stopping climate change, reducing everyday Americans' energy costs, and creating jobs—*lots* of jobs—especially in postindustrial, low-income, and working-class communities pummeled by the recession and corporate deregulation.[36] Instead of calling for aggressive building codes that mandated EERE or for EERE installations fully subsidized by the government, many of these activists proposed to expand or create tax credits, financial incentives/rebates, public-private loan funds, and other government investments in EERE markets—a capitulation to the political currents of the moment. These market-oriented proposals included equity provisions, notably EERE workforce development services, marketing, and small-business investments targeting low-income communities of color—measures that aimed to grow EERE markets in places burdened by environmental injustice.[37]

In the process, many of these activists unwittingly championed a new form of neoliberal governmentality that directly leveraged the diffuse spatiality of EERE technologies in support of social equity. For it was the *decentralized* deployment of these technologies—the fact that they could be installed in people's homes—that aligned EERE with the rhetoric of *community empowerment*, as these activists envisioned society's most marginalized people leading a movement to stop climate change and grow inclusive economies by reducing their energy consumption through cost-efficient upgrades of their everyday spaces.[38] This equitable ethos, then, endorsed a form of governance that was neoliberal in the sense that it tasked laypeople to address climate change through personal economization, positioning consumer EERE markets as a means of social change.

Thus, the neoliberal political terrain of the 2000s fused the hopes and aspirations of the then-marginalized left wing of American politics with market-based policy tools, capitalizing on the spatial congruencies between decentralized EERE technologies and decentralized community-driven activism. Put differently, neoliberal tax credits, rebates, and consumer marketing connected the politics of the left with "sustainable" energy infrastructure precisely because these policy tools, politics, and infrastructure centered on *decentralization*—a diffuse terrain of power.

In calling attention to the neoliberal attributes of EERE activism, I do not mean to downplay its progressive impacts. The slate of policy proposals focused on fostering equitable EERE markets over the past two decades infused questions of economic and racial justice in the traditionally technocratic realm of energy governance, shifting the political discourses of electricity, heat, and fuels. Centering equity, EJ activists and allied leftists advanced specific legislative provisions and unapologetically antiracist discourse in forums traditionally dominated by predominantly white credentialed experts. But in the process, these activists stimulated an equicratic shift, as market-based legislation focused on LMI communities of color had the effect of creating jobs for white-collar professionals who had not traditionally worked in those communities. Thus, the decentralized spatiality of EERE infrastructure complemented both the decentralized governance of neoliberalism and the decentralized organizing of EJ politics to animate an *equity-minded* EERE marketplace that, in turn, created a new institutional niche for the energy equicrat. Let's return, then, to one such equicrat, Sean, whose worldview emerges at the ideological intersection of EERE markets, leftist activism, and decentralized energy infrastructure.

SPATIALIZING THE EQUICRATIC TURN

The juxtaposition of leftist and neoliberal politics in NYC's EERE markets created the institutional conditions under which Sean could act on his anticapitalist, intersectional sensibilities as a corporate VP. For SunLight emerged out of this contradictory juxtaposition, growing to become one of the city's largest EERE companies while embracing a politics that is oddly compatible with Sean's views.

Specifically, SunLight participated in the "equitable" neoliberal programs that financed EERE in poor communities of color, providing expensive energy expertise to some of NYC's poorest residents under the guise of the market. SunLight's leadership and many of their employees took pride in this work, characterizing their solar and energy efficiency projects as components of broader struggles to combat NYC's housing affordability challenges and climate injustices. In this way, they cultivated corporate discourses and a company identity that celebrated social justice, championing an ethos that aligns well with the spirit—if not always the specific substance—of Sean's critiques of racial capitalism and white privilege. While Sean's views were more radical than many (not all) of his colleagues', they were at home in a company that was brought to life in a political climate preoccupied with expanding EERE markets in marginalized communities. In this climate, many SunLight employees abhorred a corporate structure of power that ravages low-income communities of color (through, for instance, gentrification) while paradoxically reproducing this structure *with the goal of empowering those communities* (as I further discuss below).

This contradiction is indicative of a professional environment amenable to Sean's organizational critiques: a place where a curmudgeonly leftist (like him) isn't marginalized for their views, where said leftist can find some ethical kernel in the otherwise extractive world of the private sector. When I raised Sean's critiques with some of his colleagues—the idea, for instance, that "corporate society" undervalues blue-collar work and overvalues white-collar work—most of them didn't disagree, sharing their moral misgivings about serving corporate clients. However, when they were assigned to work not with those clients but instead on residential buildings in LMI communities, when they had to venture out of the office to upgrade those buildings' infrastructure, when they formed personal relationships with those buildings' superintendents and encountered the smiling faces of residents who had just learned about their rooftops' new solar panels, and when they received effusive feedback from building managers about their reduced energy bills, the fact that this work was done through a neoliberal incentive program didn't really matter to them.

As Sean's liberal colleagues radiated pride during these community visits, their mien bore the imprint of EJ activists who advocated for equitable EERE markets—those activists who organized the very communities that SunLight worked in vis-à-vis LMI solar subsidies. Thus, those activists' pragmatic politics had generated a political economy of climate mitigation well suited for white-collar professionals who prided themselves on caring for the most marginalized—equity-oriented technocrats, or *equicrats*, in short. I characterize them as such to highlight not only their ideological orientation to work but also the spirit of their labor: the sense of political solidarity they cultivated with residents of LMI communities; the concern, wry humor, and even outrage they brought to their craft in neglected buildings; the intentness with which they often powered through the boring administrative aspects of their job by staying focused on their work's justice-oriented goals; the pride they took in memorizing their projects' energy savings, knowing these savings would meaningfully impact people struggling with their bills—an affect of care that resembled that of the activists who had championed the market-based government programs that subsidized their work. As these programs respatialized SunLight's project portfolio, the company attracted more credentialed professionals who were critical of the free market status quo yet strove for "meaningful" careers—employment shaped by a *market-based* approach to social change—offering technocrats pragmatic opportunities to enact their social equity politics. In this space, Sean's critiques of white male privilege and white-collar elitism were resonant, despite the fact that SunLight was an elite white-collar company run by white men. For all of SunLight's progressive bona fides, their equicratic habitus emerged from a market-based approach to selling expertise to property owners on a building-by-building basis—an approach that capitalized on EERE's decentralized spatiality.

These spatialized dynamics also shaped SunLight's division of labor and therefore its classed composition, demanding a white-collar position that was counterintuitively well suited for a professional with pro-labor politics, namely, Sean. Because SunLight and its competitors install solar on a building-by-building basis (and not on a municipal scale, for instance), their solar projects' division of labor differs from that of other energy infrastructure projects. Specifically, the development of pipelines, transmission lines, power plants, and other centralized infrastructure traditionally relies on a fairly calcified division of labor among different firms that can be classified according to the normative white-collar/blue-collar distinction: on the one hand, there are white-collar engineering, architectural, design, and legal firms, and, on the other, there are blue-collar construction trade firms (e.g., electricians) and general contractors. At first glance, this division of labor appears applicable to rooftop solar projects in NYC, which also require both physical labor *and* advanced technical expertise (with regard to building codes, fire codes, light exposure, rooftop conditions, design, and financing). These diverse labor needs reflect the material properties of NYC's skyline. For NYC is a giant metropolis with a very diverse building stock—a hodgepodge of commercial skyscrapers, multifamily walk-ups, luxury towers, single-family homes—with roofs of variable size, condition, and sun exposure, governed according to a patchwork of legal regulations. As such, the city's rooftop solar projects are more complex than those of suburbs, rural areas, and smaller cities, demanding a broad set of proficiencies comparable to that of many traditional, centralized energy infrastructure projects. But unlike such projects, the installation of rooftop solar in a neoliberal market works in service of *individual property owners*, not according to a public mandate that requires large-scale service provisions. Consequently, many firms like SunLight often eschew the complex webs of contractors that accompany those projects, opting, instead, to do both white-collar *and* blue-collar work. This dynamic creates a fascinating juxtaposition of differently classed subjects that is absent in many of the firms who work on traditional energy infrastructure projects.

Specifically, in the course of my experience as an embedded researcher at SunLight, I encountered, on the one hand, the wonkish milieu characteristic of the technology and design sector (complex cloud-based data systems; nerdy engineering humor; employees with advanced degrees; NPR playing in the background; a predilection for finding solutions to complex problems; a love for producing charts, graphs, and other visualizations of data; an affinity for the Democratic Party; casual allusions to podcasts; a measured, hushed work environment; workers eating lunch by themselves at their desks; predominantly white male leadership; frequent coffee breaks) *and*, on the other hand, the less-polished, working-class habitus of blue-collar construction companies (hip-hop blasting from speakers on the jobsite; shared lunch breaks with no shortage of comradery; cavalier attitudes toward dirt, debris, rain, and residue; stained, ripped, and baggy clothing; masculinist humor; a dearth of women; more references to religion than to formal politics; piercing loud noises; physical exhaustion; a bottle of piss lying dormant in the corner of parapet walls; hourly wages and not annual salaries).

Yet as employees of the same company, the subjects who occupy these two disparate worlds worked closely together, and while their classed and raced differences often generated conflict (as I discuss in chapter 4), their relationships were habituated and familiar in ways that are lacking in comparable blue-/white-collar collaborations that connect two or more different firms. Even when those cross-class relations grew fraught, they were rooted in social bonds that you won't find among laborers and engineers at separate firms—a common ground located, for instance, in the workers' shared experiences: breaking bread, hiking, and strategizing together at SunLight's all-company meetings; making friendly chit-chat and getting to know one another when blue-collar workers would occasionally come to SunLight's main office and when white-collar workers would venture into the field; reading the company's weekly newsletter, which highlighted the skills, interests, and successes of both office workers and field workers as though there was no distinction; and working with the company's logo on their cards and T-shirts, an emblem simulating corporate unity. Furthermore, the classed distinctions between workers were often not so legible; SunLight employs many white-collar experts who spend a good portion of their time engaged in hands-on labor—in crawl spaces and in scorching boiler rooms, physically wrestling with incredibly complex HVAC systems, and on rooftops, setting the stage for solar installations.

For these reasons, Sean's labor politics were actually well aligned with his corporate position. On the one hand, his technical expertise, professional credentials, and intersectional positionality fit the mold of the paradigmatic energy expert, and, on the other, his background performing hands-on labor and his proclivity for blue-collar work kept him attuned to the needs of SunLight's solar installer crews. When blue-collar workers expressed grievances regarding occupational safety, low compensation, and SunLight's flawed systems for ordering jobsite materials, Sean strategically leveraged his position to support them, at once informally representing them in leadership meetings, lashing out at another VP in ways that led to pro-worker resource shifts, insulating the workers from cumbersome HR protocols, securing new hires to lighten their load, informally advising them about labor organizing tactics, and performing interpersonal solidarity through small acts of kindness. Self-deprecating to a fault, Sean did not see himself as a savior and remained critical of his own positionality, but he felt a moral imperative to use his relative power to make the corporation more hospitable for workers without his privileges. His sense of purpose working for a corporation that pragmatically supported poor communities of color did not inure him to SunLight's hypocrisy, as his equicratic positioning emerged not from righteousness but from a place of profound ambivalence. Sean was clearly a thorn in the side of many of the company's other leaders, a couple of whom openly clashed with him and many of whom rolled their eyes at what they perceived to be his sanctimoniousness. Nonetheless, there was a clear understanding that not too many people were qualified to do Sean's job, as his deep understanding of SunLight's technocratic systems coupled with his blue-collar credentials, intricate knowledge of NYC's residential buildings, and managerial experience made him the ideal person for bridging the blue-collar/white-collar divide. They realized, too, that his pro-worker advocacy could help keep the company's precarious blue-collar solar operations intact—a business imperative given that this work generated greater profits than their white-collar energy management services. So Sean's upper-level colleagues tolerated and even at times appreciated his pro-worker sentiments. In this way, his working-class critique of racial capitalism, when paired with his white-collar skill sets and positionality, helped him carve out a niche in the upper ranks of a technocratic corporation that relatively aligned with his political values—a sense of purpose that was nonetheless freighted with self-doubt. At the root of this apt positioning was the company's broad division of labor; their dual white-collar/blue-collar character minimized (but never eradicated) the classed hierarchies that are foundational to contracting networks of multiple firms. And this division of labor was shaped by the decentralized spatiality of EERE projects catalyzed by neoliberal policy, as the material and ideological contours of solar gave affective form to the equicratic slot that Sean occupied at SunLight. When we consider this spatiality, then, we begin to see the ways in which solar, as a physical infrastructure, can animate novel subjectivities, transforming the character of technocratic positions, including that of a white-collar VP.

While it's not exceptionally common, there are in fact white-collar VPs at fossil fuel corporations who, like Sean, have blue-collar work experiences on their résumés. Unlike Sean, though, these VPs generally lack anticapitalist sensibilities, and while it would be wrong to isolate one material cause in explaining this discrepancy, the spatiality of solar is key. Solar's impacts on the classed division of labor in the energy sector, its compatibility with a leftist politics centered on decentralization, and its interventions in traditional energy governance attest to the affective power of its decentralized landscape—its capacity to reconfigure the ideological orientation of electricity generation. This is not to suggest that all VPs in the solar industry share Sean's politics. My point, instead, is that the diffuse spatiality of solar transforms the ways that energy expertise is performed with significant implications for the politics that it's aligned with. Sean, then, is but an example and not totemic of the emergent energy equicrat—a figure I now explore in greater depth.

EQUICRATIC ACTIVISM IN THE PRIVATE SECTOR

On a sweltering summer day, a group of two dozen EJ activists and allies convened outside the governor's downstate office in Manhattan to protest a new statewide energy policy that was negatively impacting low-income communities of color. At first glance, this demonstration resembled other EJ protests, marked as it was by colorful political signs, boisterous chants, a multiracial and multigenerational group of participants, and bright T-shirts with nonprofits' logos. But if you looked closer, two things stood out.

First, unlike most energy-focused protests, the activists weren't contesting a new pipeline, inequitable power plants, or fossil fuels. Instead, they were focused on an opaque *algorithm*—a technocratic matter that usually doesn't draw the impassioned attention of grassroots activists. Paradigmatically neoliberal, the algorithm in question—known as the Value of Distributed Energy Resources (VDER)—was intended to transform how the state valued electricity so as to incentivize more nonstate actors to participate in the governance of energy infrastructure through the "distributed electricity market." This (state-facilitated) "market" compensates owners of solar arrays for injecting some of the electricity they generate (and don't need) into the electricity grid, with compensation varying based on a range of factors, such as the time the electricity was generated. This variability was intended to incentivize decentralized electricity production in ways that optimally relieved pressure on the grid and minimized expensive power generation. New York State's infamously inaccessible Department of Public Service developed VDER in conjunction with electricity utilities and solar industry insiders in forums that were largely inaccessible to the public it was supposed to serve. But when VDER created extreme price volatility, EJ activists and allies began to closely monitor—and attempted to shape—VDER's development because they knew that this volatility would stymie private investment in community solar projects that serve LMI people. The protest at the governor's office was their latest attempt to change or abolish VDER as part of their broader fight for an electricity pricing mechanism that would enable private developers to invest in LMI community solar projects.

Specifically, they envisioned a complex, recalibrated algorithm that incentivized a robust "community solar market" in places wrecked by environmental injustices. The protest, then, was an on-the-ground social justice action focused not on expanding the social safety net, regulating industry, or securing subaltern rights but instead on *stimulating investment through a technocratic tool*. It was therefore decisively equicratic, fusing the leftist principles, activist performances, and counterhegemonic affect of social justice movements with a technical mechanism for neoliberal governance. Even so, all of the participating activists explicitly disdained racial capitalism and posited their work in opposition to it. Yet, for reasons I have only begun to explore, the decentralized spatiality of solar aligned with the decentralized governance of neoliberalism to situate a capitalist algorithm in an activist world that embraced a different form of decentralized power: grassroots organizing.

This points to the second unusual aspect of the protest: in a sea of rowdy activists was Brian, a mild-mannered, white-collar white man in his late forties who clearly didn't fit the protestor mold. You could sense this difference on the surface: while his fellow protestors wore shorts and T-shirts, Brian wore khakis and a button-down shirt. But what really set him apart was his professional experience navigating renewable energy markets. As a senior sales representative for New Grid Power, a prominent solar corporation in the northeastern United States, Brian has detailed knowledge of financing, design, permitting processes, and customer recruitment for solar projects. Although he cared about environmental policy, he had not before involved himself in EJ politics, as his orientation to solar had been defined primarily by his expertise and marketing prowess. Brian, then, was not simply a little out of place; he was, in many respects, the antithesis of the people he was protesting with. While he lacked their comprehensive analyses of structural power, they knew little about the *specifics* of infrastructural power. While he thrived in the corporate world, they had long worked at nonprofits opposed to the corporate world. While he focused on cultivating commercial expertise, they focused on cultivating communal cooperation.

To be sure, there has historically been no shortage of alliances between grassroots activists and technical experts. From W. E. B. Du Bois's pioneering sociology of race at the turn of the twentieth century to radical women's health clinics in the 1970s, technoscientific experts and credentialed elites have long worked hand in hand with grassroots movements in the United States, achieving social change through bureaucratic fora and technocratic governmentality.[39] In fact, such alliances were foundational to the US EJ movement.[40] But Brian's solidarity with EJ activists was notable because it represented an ideological shift in the politics of commercial energy expertise—an emergent equicratic ethos. As discussed, such expertise has historically worked in service of extractive infrastructural projects and is often indifferent or hostile to the needs of marginalized communities. Even when energy experts have worked to close structural disparities in energy access, they have traditionally done so through top-down approaches—not through street-based demonstrations with anticapitalist organizers. While numerous other solar industry experts had serious problems with VDER—not just Brian—none of them participated in the protest. Most of them instead opted to air their grievances through state-run backroom channels in which government and industry worked together to furtively craft policy. Why, then, did Brian align himself with activists, and, more important, how does this alliance point to a broader equicratic shift?

These questions' answers lie less in the abstractions of the algorithm that Brian and the activists opposed and more in the spatiality of the infrastructure they both supported. As discussed, in contrast to centralized fossil fuel infrastructures, solar requires expertise centered on residential dwellings and community spaces that are controlled by the people who own them, operate them, and/or live there—not the utilities that control the energy those spaces traditionally relied on. As such, solar projects necessitate professionals like Brian: experts responsible for convincing everyday people to generate electricity on their roofs or to subscribe to community solar projects in their neighborhood. In other words, solar projects' spatiality animates expertise focused on empowering ratepayers in a residential energy market that previously gave them no choice but to passively adhere to the local utility. Under a traditional system of electricity generation, there are no Brians—no one who is responsible for transforming everyday folks into codevelopers of energy infrastructures. Solar's decentralized landscape, then, demands a technocratic form of the affective labor that EJ activists have long engaged in: personal appeals to people's emotions regarding their environment and well-being. Like those activists, Brian cherishes this affective labor more than any other facet of his job:

> I love sitting down at someone's kitchen table or in a co-op's community room, telling them how they can be a partner—a *leader*—in the clean energy economy. It's the most fulfilling part of my job—seeing people get excited and realizing they can save a bunch of money and save the planet too. I absolutely love it. And it's why I do this work: *to connect with people.* I could probably be making more money somewhere else. . . . But it's important for me to do something that's challenging, something that uses my skills, something that I care about where I get to connect with others.

I could detect the edifying affect of Brian's work—his sense that he was doing something important—when I accompanied him to meet LMI co-op boards in their communal spaces, basements, and apartments. His earnest voice would grow alive with urgency as he affirmed the importance of taking climate action "for our kids and our grandkids," looking his prospective customers—his "gracious hosts"—straight in the eye, raising the emotional stakes of the seemingly rational matter of diversifying one's electricity consumption before delving into the cold logistics of financing and incentives. It seemed like he thrived off the in-person intimacy of these convenings, performing a folksy charm congruent with the domestic space that served as his stage.

VDER had imperiled this person-to-person labor—and it was this labor's *affective* dimensions, *not* its monetary compensation, that drove Brian to protest with EJ activists. For Brian had recently been transferred from New Grid Power's languishing portfolio of community solar projects to its thriving commercial facility operations—a more profitable line of work. Thus, VDER didn't directly impact Brian's source of income or seriously threaten the company, giving him little economic incentive to participate in the protest. Furthermore, publicly targeting the governor with EJ activists was a risky move. The governor was notoriously vindictive, and many solar companies decided it made more sense to try to change VDER through the state's official channels than through public agitation. Brian protested not because he financially needs community solar but because his affective labor shapes his professional self—a calling defined by the interpersonal process of sitting at a kitchen table or in a community center, bringing new people into the world of distributed energy.

The spatiality of solar, then, affectively animated his common cause with activists whom he barely knew, giving his work a sense of purpose that paralleled his comrades' outrage that poor New Yorkers were being shut out from the community solar market.

While Brian was outraged too, his emotional investment in this political matter was shaped less by his long-standing ideological commitments and more by his in-person experiences in poor people's personal spaces—affective encounters that have no place in the centralized administration of fossil fuel infrastructure. Unlike Sean's pro-labor maneuverings, Brian's participation in the protest was not rooted in an intersectional critique or an interest in dismantling the systems that have empowered him in the corporate world. Instead, Brian subscribes to a white liberal politics that understands protest as a matter of free speech and democracy as opposed to a matter of Black and brown survival. Furthermore, while Brian is not a free market ideologue and does not endorse unfettered capitalism, in his mind the market has an important role to play in the functioning of modern society; he believes that corporations are essential to the provisioning of life-sustaining goods and services. Put simply, Brian lacks a critical analysis of racial capitalism.

But this lack does not make him any less of an exemplar of the ideological transformations instigated by decentralized electricity generation. For solar's diffuse spatiality—the homes and community centers where Brian worked—prompted his unlikely alliance with activists who define themselves in opposition to his corporate milieu, informing protest that blurs the normative boundaries distinguishing corporations from EJ activism—the ambitions of a movement engulfed by algorithmic improvement. These blurred boundaries, the shared space of the protest, and the affective labor of community outreach collectively comprise an equicratic shift in the space of electricity generation, evincing changes to the ideological landscape of energy expertise even as Brian lacks Sean's radical politics. Thus, the equicrat, as I'm conceptualizing it, is less an ideologically coherent figure and more a personification of these blurred boundaries—evidence of the ambivalence of a neoliberal approach to racial capitalism in which market mechanisms like pricing algorithms are imagined as an antidote to differential value. In the absence of decentralized electricity generation, it is unlikely that Brian would ever protest in solidarity with Black and brown people demanding racial, economic, and environmental justice, no matter his political commitments. In this way, the spatiality of solar affectively animates an equicratic politics that troubles the lines separating corporate power and grassroots protest—a capacious liberalism that creates a unified space for both people of color who disdain racial capitalism and white people who happily take part in the workings of the market.

I want to therefore suggest that the *decentralization* of energy counterintuitively powers a sort of *centralization* of ideology: an erosion of the phantasmatic lines separating, for instance, faith in private markets and ecosocialism. As residential dwellings in LMI communities became sites of electricity generation, the social justice politics of those communities cross-pollinated with the corporate ethos at the heart of infrastructural expertise in ways that dissolved certain normative ideological boundaries. Put differently, efforts to decentralize energy governance can muddy the imagined political spectrum separating corporate growth from environmental justice, as traditionally oppositional agendas coalesce around a shared imperative to improve marginalized communities through EERE. While I do not wish to reify this imagined spectrum (there's no shortage of scholarship problematizing the concept of a self-evident political left and right), my point here is that the conflation of corporate growth and social equity—the win-win calculus of many decentralized energy plans—creates a sort of ideological center point for many actors in the EERE world, engendering a sense of consensus among an otherwise diverse group of liberals like Brian and the activists he marched with.[41]

Yet as decentralized energy muddies the normative political spectrum, it activates not only solidarity across ideological difference but also a multicultural facade that obfuscates the hegemony of racial capitalism. In the next section, I explore these insidious shifts, attributing them to late liberal screenwork when it interacts with the diffuse spatiality of solar and other EERE infrastructures.

FROM THE STREETS TO THE SCREEN

We can question whether a white solar salesman at an EJ protest or a white EERE VP critiquing corporate racism, classism, and sexism represents anything loosely resembling a shift in structural power. But Brian and Sean *do* point to the ways in which solar's diffuse spatiality creates openings for forms of solidarity and interpersonal care that have historically been anathema to the corporate world of electricity generation. The *in-person* acts of witnessing homes and communities that couldn't be more different than your own, getting to meet people at their kitchen tables or mosques, listening to their challenges, and standing side by side with them in the midst of a labor dispute or a protest against state technocracy are radical shifts in how white-collar energy professionals have gone about their work.[42] In the context of an industrialized energy system that has alienated most of these professionals from the marginalized spaces they directly impact, forging solidarity across difference necessarily demands *physical proximity* and *shared spaces*—a closeness at odds with the spatialized divisions that constitute differential value.

But this proximity is not characteristic of many forms of equicratic EERE work. Instead of in-person encounters, many energy equicrats attempt to forge solidarity with poor communities of color through the medium of the screen. Consider the images in figures 2.3 and 2.4, respectively: a panoptic rooftop visual with rectangular indigo strips simulating potential solar panels, and a mounted screen with a color-coded spreadsheet organizing reams of energy use data for a portfolio of housing properties. These visual interfaces represent the primary way in which many equity-minded energy professionals engage with the communities they aspire to help, designing complex rooftop solar installations from the enclosed space of their workstation or assiduously tracking energy inefficiencies with their colleagues in small conference rooms. Instead of the living rooms that Brian converses in and the boiler rooms that Sean once labored in, rooftop simulations and data management platforms offer EERE experts like Celeste remote access to communities they'll likely never set foot in. But the flattened, glossy surface of the screen necessarily facilitates surface-level apprehensions of space, "transform[ing] the representational and material world into a single, stable, universal order."[43]

A single, stable, universal order: this is an apt characterization of sprawling affordable housing complexes as they appear on the screen of Martha. A recent college graduate, a self-identified democratic socialist, and SunLight's preeminent data analyst, Martha spends her workdays glued to her computer tracking ongoing energy usage data, looking scrupulously for idiosyncrasies in the KPIs (key performance indicators) so as to improve energy efficiency in her client's properties, including those affordable housing complexes. From this perspective, one such complex, Terrence Heights—which spans four square blocks in the North Bronx—appears not as a series of austere, near-identical postwar high-rises with hundreds of working-class Black families. It instead appears as a stream of metrics in tiny white Courier font atop a black backdrop; every facet

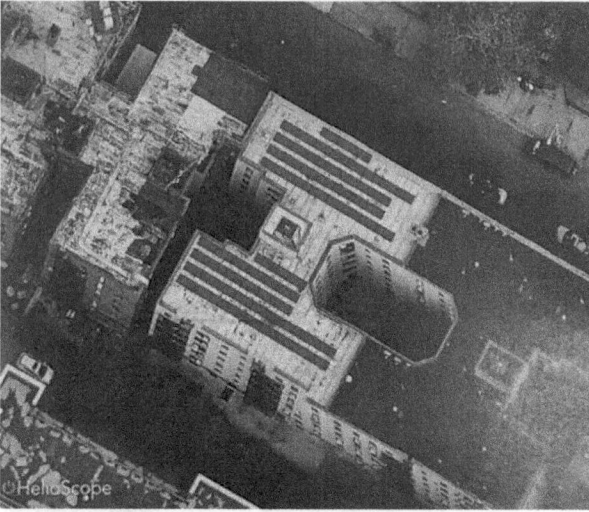

FIGURES 2.3 AND 2.4. Screenwork for solar and energy efficiency and renewable energy reduces low- and moderate-income communities of color to flattened space and datafied text that equicrats can manage at a distance. Figure 2.3 from Google Earth, generated by the author's anonymous interlocutors with HelioScope, photo by Myles Lennon. Figure 2.4 spreadsheet produced by Myles Lennon with Microsoft Excel, photo by Lennon.

of its energy infrastructure—common-area lighting, gas water heaters, elevator usage, heat consumption between 6 p.m. and midnight—is rendered in a programming language that Martha is not simply fluent in but also, on this screen, the master of (see figure 2.5). That is to say that she navigates what looks like computer gibberish to me with Zen-like attention to every character on her screen, translating these characters into infrastructure she otherwise has never encountered with a "ready-to-hand" feel.[44] She spends large chunks of her days ensconced in this practice, nibbling on her peanut butter sandwich or fiddling with the quirky toys she keeps at her desk while her eyes stay fixed on the screen as code reflects off her glasses like there's no boundary between her and the data. While she has a bubbly personality and doesn't take herself too seriously, when she sits at her workstation scanning this real-time energy usage data for hours on end, she often appears stoic and unreachable—a cog in the machine.

But that stoicism dissolves the moment that a dataset seems off—the moment, for instance, that she detects that the peak-to-average power ratio has deviated from historical data—as she is overcome with a drive to ensure that the energy inefficiencies indexed by her data don't hurt poor people of color besieged by rising housing costs. For much like Celeste, Martha views this screenwork as a "concrete" means of addressing the housing affordability crisis plaguing NYC's communities of color—a pragmatic response to a structural problem that her corporate employer nonetheless cannot solve in its entirety. As such, her formal employment is only one piece of her broader leftist praxis (as I discuss below) and is thus charged with a sense of moral urgency. Any minute outlier in the

```
<feed xmlns="http://www.x99.com/2018/Atom" xmlns:ns2="http://a9.com/-
/spec/opensearch/1.1/"
 xmlns:ns3="http://www.x99.com/2010/xhtml">
    <id>FEED_UNIQUE_ID</id>
    <updated>YYYY-MM-DDThh:mm:ss.sssZ</updated>
    <title type="text">AggregatedMetrics</title>
    <subtitle type="text">ManagedEnergySystem MANAGED_ENERGY-
SYSTEM_UUID</subtitle>
    <link rel="self"
href="https://HMC_IP_OR_HOST_NAME/rest/api/pcm/ManagedEnergySystem/
```

FIGURE 2.5. A reproduction of the code that Martha scans at her desk in search of energy performance outliers—the primary way she relates to low-income residencies that she has never encountered in person. Reproduction and image by Myles Lennon.

data calls on her to imagine a *practical* infrastructural intervention that can immediately help people struggling with their bills while reducing their fossil fuel consumption. When she detects these outliers, she scurries over to her colleague's desk and asks them, with a professional sharpness that conceals her anxiety, to consult another energy management dashboard for Terrence Heights to try to diagnose the irregularity.

I want to suggest that Martha's sober intensity represents a particular form of equicratic care that emerges from renderings of impoverished residential spaces in code. For the diffuse spatiality of EERE conjoins with an affect of datafied rigor (the *feel* of "hard" analytics that suffuses office spaces) to create a sense that one can address differential value with stoic rigor at their keyboard—the monasticism Martha displays when mining data to close raced/classed disparities in energy costs.[45] In contrast to Sean's and Brian's attempts at enacting solidarity with workers and activists, this equicratic care is a screen-based way of relating to the world—a rational gaze that transforms Terrence Heights into *a single, stable, universal order.* While the EERE goods and services that SunLight sells demand the *decentralized* provisioning of energy expertise, these goods and services also demand the panoptic perspective of *centralized* energy governance—a tension that, I will show, undercuts Martha's equicratic efforts to redress differential value.

THE DIGITAL HABITUS OF ENERGY EQUICRATS

As NYC's EERE market grew in the 2010s, SunLight slowly moved away from the "bro-y" technician culture that Sean bemoaned, progressively becoming a place where antiracism and feminism poked holes in the apolitical veneer of white male expertise. Increasingly, SunLight's employee base included more women, mostly at the company's lowest and middle levels, although there were signs that the gender divide in leadership would change soon too. While the company had a single Black employee at the time, it was slowly growing more racially diverse as well.

Some of these new employees (including Martha) wore their politics on their sleeves, and others were more circumspect, but in the course of numerous discussions with them all, they explained to me that they felt impelled to do technocratic energy work not simply because they thought it would "help" the environment or "decelerate" climate change but also because they felt it could overhaul the structures of inequality engulfing the city. Their outspoken critiques, polite suggestions to management, water cooler talk, and pointed questions in company meetings centered critiques of gentrification, their corporate clients, and the company's lack of diversity in SunLight's internal discourse.

These shifts mirrored NYC's broader political shifts at the time. The city had grown more politically progressive, with an electorate that was increasingly critical of anti-Black policing, the lack of affordable housing, corporate real estate development, and climate change's impacts on marginalized communities (which Hurricane Sandy chillingly highlighted)—and the city's new elected officials shared these concerns.[46] Furthermore, with the rise of Occupy Wall Street and the Black Lives Matter (BLM) movement, activists had shifted the terms of the city's mainstream political discourse, normalizing antiracist and anticorporate ideas that had been considered more radical during the previous decades under Republican mayors.

Still, it was not a given that these political shifts would influence SunLight's employee base. The EERE and electricity generation industries have historically been overwhelmingly run by elite white men and governed by a white-collar hierarchy—a structure that has largely proved impervious to social change. The New York State Energy Research and Development Authority (NYSERDA), which oversaw most of the state's EERE programs, showed no signs of demographic or political shifts at this time, as its predominantly elite white male leadership and bureaucracy prioritized neoliberal strategies and employed a color-blind market-based analysis that largely ignored EJ activists' efforts to foreground equity in those programs. Indeed, as the state couched its energy policies in environmentalist rhetoric, it made little effort to connect climate action with the progressive politics championed by the city's environmentalist organizations. Similarly, the city's commercial real estate industry, which included many of SunLight's biggest clients and worked with government to design EERE programs to its own benefit, had often opposed the burgeoning activism of the moment, championing the luxury housing rezonings and "quality-of-life" policing that had decimated the city's marginalized communities. Finally, as we've discussed, energy experts have long employed discourses that render energy as a technical matter that is objectively foundational to life, not as a political matter implicated in intersectional power structures. So as NYC grew more politically progressive, it was not a given that SunLight would follow suit.

Instead, SunLight's political character was transformed by the respatialization of their portfolio of EERE projects. For when the expansion of the city's solar market and the emergence of solar incentives for LMI residents brought the company into marginalized communities that they had almost never worked in before, this spatial shift endowed a sizable portion of their work with the moral gravitas of climate justice: this sense that SunLight was supporting those most vulnerable to climate change by reducing their energy bills, lowering their carbon footprint, and helping them become more resilient in the face of outages (through solar battery storage). These ethical dimensions had attracted many of SunLight's new employees (most of whom were under twenty-eight), as they expressed a strong affinity for meaningful work in places in need.

But it would miss the mark to suggest that these employees simply came to SunLight because they believed they were working for climate justice in communities of color. Instead, these employees were more specifically drawn to impactful *screenwork*—the particular form of social justice labor conditioned by "a digital habitus" that conditions life as a young NYC professional.[47] In this context, the *process* of playing with data on a screen is often edifying irrespective of its specified goals in social worlds organized around the primacy of information. Put differently, the political *ends* of equicratic work are arguably as affective as the cloud-based *means* of such work. I first noted this when Martha and other equicrats at SunLight would display affection for the everyday practice of information mastery, geeking out over data visualizations or underutilized tools on their platforms, gossiping about one another's proficiencies with code or visual satellite images. This affective orientation to screenwork was conditioned by workspaces rooted in the cloud. Indeed, the offices of tech companies like SunLight are a spatial condensation of a broader milieu of data-driven governmentality that affectively animates screenwork, endowing it with emotional depth beyond its official purpose. This data-driven governmentality is characterized by the proliferation of digital interfaces, the visualization of data, and a culture of elite "smartness" centered on "the will to improve"—a point I now expand on before connecting it to equicratic practice.[48]

As many scholars have shown, the ubiquity of smartphones, apps, and social media enmeshes modern subjects in virtual publics and digital spaces that constitute many of our relations.[49] This ever-expanding mediascape coproduces visual forms of quantitative data that economize life under neoliberalism, generating diagrams, charts, plots, tables, dashboards, and digital profiles that not only facilitate technocratic governance but also aestheticize information—bringing it into the visual texture of everyday life.[50] These visual projects, then, transform numbers into what Michelle Murphy calls *phantasmagrams*: "quantitative practices that are enriched with affect, propagate imaginaries, lure feeling, and hence have supernatural effects in surplus of their rational precepts."[51] At tech companies that purport to improve society through their corporate innovations, phantasmagrams are central to *smartness*—Karen Ho's term for the performative rendering of intelligence as cultural capital.[52] This is to say that in the SunLight office, an employee's adeptness with various quantitative practices is a mark of an enlightened social status, generating respect, pride, and camaraderie because it points to one's ability to use one's smarts to advance noble objectives like climate mitigation. For instance, in the after-work keg ritual described above—a *social* event centered on cheap beer and bro-y fraternizing—SunLight engineers would present detailed, metric-filled PowerPoints on improving solar infrastructure design, accruing their colleagues' respect by demonstrating a mastery of technical knowledge with real-world application. These PowerPoints were foundational to the social bonds of SunLight's engineering division, as they inspired awe, envy, friendship, nerdy puns, and inside jokes until the keg was empty. This weekly ritual thus demonstrates the affective intersections of smartness and phantasmagrams/visualized data in the tech corporation, as one's proficiency with information was a source of social capital and identity, shaping how employees understood themselves as valuable members of a team and instilling a sense of belonging. Of course, these data-driven relations are embedded in the broader mediascape of twenty-first-century urbanity, as smartphones, touchscreen kiosks, mounted TVs, tablets, and computers saturate every physical space of contemporary NYC. This mediascape enmeshes young urban professionals in unending screen time, linking their leisure lives with the professional spaces where they spend most of their days.

In this context, the panoptic power to situate solar panels in a poor Black community via Google Earth or the computational power to identify problematic outliers in low-income residents' electricity consumption data is enlivening irrespective of its equity-oriented objectives. Furthermore, these panoptic and computational powers coalesce with the diffuse spatiality of EERE to generate equicratic screenwork that is as affectively animated by a cult of smartness as it is by its equitable objectives. In other words, both the ability to pragmatically intervene in those spaces ravaged by racial capitalism *and* the feeling of using one's technical skills in ways that generate social capital collectively endow cloud-based energy governance with a sense of purpose among young professionals, instilling in them the sense that they can "do good by doing well" at their workstation.[53]

And it is this *particular* feeling of cleantech screenwork attuned to LMI spaces that gives affective form to the late liberal sensibility that EERE technologies like solar can redress differential value. To illustrate this point, I now return to Martha, further exploring the affective intersections of her screenwork and EERE's diffuse spatiality.

CORPORATE CARE

Martha might sound like the last person you'd expect to be emotionally invested in corporate screenwork. As a democratic socialist, she's intensely critical of her employer and what she calls *neoliberalism*—which she uses synonymously with *capitalism* and *corporate power*. When she's not in the office, she works on grassroots campaigns for racial and economic justice that lambaste corporations. And when she *is* in the office, she often espouses unapologetically leftist politics, offering her unsolicited two cents on police brutality, gentrification, financialization, and animal rights. Furthermore, while Martha isn't poorly paid, she knows she could earn a comparable salary in a senior position in the nonprofit sector doing work that's more aligned with her political ideals; SunLight's pay is not a draw. Why, then, would such a leftist take a job where she spends her days behind a screen reviewing data for a corporation that, for instance, provides energy services to many of the big banks she disdains—a corporation that actually markets its services as a way for other corporations to improve their margins?

The answer lies in a corporate form of care that links screenwork with EERE's diffuse spatiality, positioning the corporation as compatible with Martha's convictions. This care takes two forms: (1) care for the craft of data and (2) care for the built environment. The first—care for the craft of data—refers to the affective texture of the screenwork described above: the smartness of mastering phantasmagrams, irrespective of their specific ends. While Martha is responsible for solving "real-world" problems, at SunLight real-world problem-solving is, as stated, a reflexively *analytical* practice that is not simply about tangible changes but also *intellectual challenges*, fostering her affection for the *process* of scrutinizing data and identifying outliers. She finds screenwork meditative—a daily ritual with coworkers that conjures the collective feeling of purpose unbeholden to formal function. In this way, *solving problems through screenwork is an end in itself*—and this end, coupled with the company's moral objectives, ultimately attracted her to a job that was at once compatible and discordant with her politics. She would never work for a bank doing comparable analytical work—SunLight's environmental impact on marginalized communities undergirds her employment—but she is willing to forgo work that better suits her political aspirations because of the "feeling of doing something pragmatic that taps into my skill set"—a care for her craft. While she was deeply committed to her activism, too often its vision for justice did not lead to things "that you could [actually] see" on the ground—concrete change that marginalized communities directly experience. Through EERE she sought to make a tangible difference in the here and now—to lower energy costs, increase physical comfort, and reduce emissions in communities burdened by environmental vulnerability—while putting her smarts to better use by eschewing the organizational messiness of her activist work. In this way, such activist work was an entrée to her position at SunLight, as the challenges, abstractions, and drawn-out temporality of social justice campaigns made her crave a job in which her analytical craft supported justice-oriented work with fewer complications and headaches.

At SunLight, this care for the craft of data went hand in hand with a care for the built environment: an affection for the boiler rooms, building envelopes, and rooftops of the buildings they worked on. Focused on *individual* properties and *specific* real estate portfolios, this care is necessarily missing from centralized infrastructural governance, complementing the company's "care-full" screenwork in ways that made SunLight amenable to Martha's politics.[54] Specifically, SunLight's efforts to develop and implement *site-specific* plans to improve energy management and increase EERE uptake require in-depth analyses of clients' buildings, detail-oriented engineering and design practices, close relationships with property managers, and intimate familiarity with a property's logistical and financial operations—a level of long-term, hands-on interpersonal attention that is simply impossible in even the most thoughtful centralized energy infrastructure governance. Such care animates a corporate habitus dominated by technical perfectionism—a neurotic obsession with optimization in excess of the company's value proposition. For instance, SunLight's solar engineers will go through the trouble of redesigning blueprints for solar installations when they learn of an overlooked rooftop obstruction or other minor structural idiosyncrasies that could affect the placement of only a few solar panels. By SunLight's own admission, this sort of perfectionism does not help their bottom line or strengthen their brand. In fact, it often lengthens the time it takes to complete installations—delays that no company wants on their record. Yet several engineers explained to me that they engaged in this perfectionism because they wanted to look back on their projects and take pride in their work, doing everything possible to optimize their clients' properties. Similarly, their solar installation crew will spend upward of two days reconfiguring an elevated solar system's racking structure if it doesn't *look* right even if it's perfectly functional; these reconfigurations are often purely aesthetic efforts to demonstrate to clients the skill, detail, and attention SunLight puts into every solar installation. This care for the built environment permeates the company's social environment, coalescing with the care for the craft of screenwork to create a workspace that is obsessed with the continuous improvement of client services. Centralized energy governance does not create the affective space for this sort of care, as it's focused on a sprawling system of grids, pipelines, power plants, and transmission lines—a scale incompatible with SunLight's technocratic intimacy. Centralized energy governance certainly demands rigor, expertise, and attention to detail but not the granularity, customer service, and obsessive craftsmanship that individual building systems conjure at SunLight. The LMI respatialization of SunLight's project portfolio only magnified the moral stakes of this care for the built environment—endowing it with the climate justice gravitas that I discussed above.

It is in this care-full context that Martha pursued a career at SunLight focused on social equity. She was intrigued by not simply the company's work to reduce LMI housing costs but also the intimacy of client partnerships and the collective effervescence of analytics. A symbiotic relationship between buildings in need of upgrades and the screenwork she is proficient in inspired a care for her job that complements her politics. For SunLight's *decentralized* patchwork of LMI energy upgrades created an affective terrain through which its young, equity-minded employees like Martha could indulge in the analytical wizardry of screenwork, at once providing specialized care to people toward social justice ends and caring for data as an end in itself.

By infusing SunLight with such an equicratic ethos, these employees fostered a politicized company environment—a space in which almost everyone discussed and contemplated their work's impacts on society—

impacts that couldn't be reduced to gigajoules or greenhouse gases. For instance, employees frequently debated the ethics of providing EERE services to properties owned by big banks. Did emissions reductions really justify working with an "evil" corporation like JPMorgan Chase? For many, the answer was no, while others felt uncomfortable but ambivalent about this sort of client relationship. Similarly, the company's young equicrats grew increasingly vocal about SunLight's lack of employee diversity, contending that its nepotism constrained its ability to provide EERE services to low-income communities struggling with the city's affordable housing crisis.

SunLight's culture of caring for a decentralized patchwork of properties provoked a reflexive line of inquiry that's historically been absent in the self-assured spaces of centralized energy governance: Who, *specifically*, are we providing care to, and toward what ends? The implication here was that under the right conditions, EERE screenwork could pragmatically redress the differential value of fossil energy. Yet, as we will see, the panoptic gaze of energy expertise and the digital habitus it is enmeshed in would prove incapable of disentangling the city's built environment from the divisions of racial capitalism.

BLACK LIVES MATTER IN THE EQUICRATIC CORPORATION

On July 5, 2016, Alton Sterling, a thirty-seven-year-old Black man, was killed by two white cops in Louisiana for no justifiable reason. A day later, Philando Castile, a thirty-two-year-old Black man, was killed by a white cop in Minnesota for no justifiable reason. Both of these murders would likely resemble the countless acts of anti-Black police brutality that have been forgotten, downplayed, or disingenuously repudiated were it not for cell phone videos documenting them. Castile's murder, in particular, exposed the depths of anti-Black police brutality because its immediate aftermath was captured on a Facebook livestream by Castile's girlfriend, Diamond Reynolds, in the vehicle in which he was shot. This harrowing piece of footage went viral in real time, as Reynolds struggled to make sense of the deadly violence she had just encountered while calmly responding to the crazed recriminations of her boyfriend's murderer. Products of the American mediascape, these videos quickly made the rounds on social media before bleeding into the mainstream news, as the raw horror, visceral pain, and utter senselessness of Sterling's and Castile's back-to-back deaths spread from smartphones to laptops to TV with an efficiency that transformed them into household names. The spectacle of anti-Blackness has always intrigued American audiences, but as these videos played on loop for a week, they didn't simply captivate the public.[55] They more specifically inflamed a sense of outrage that had already grown more potent with the circulation of other viral snapshots of anti-Black violence in previous years. Sparking civil unrest, Sterling's and Castile's murders were arguably as symptomatic of the affective power of screen culture as they were indicative of anti-Black police brutality.

This power was palpable in SunLight's boardroom at the company's weekly leadership meeting seven days after Sterling was slain. A sterile space on the upper floors of a corporate skyscraper overlooking the Financial District, this room was certainly not where I expected to encounter the outrage that had engulfed US streets after the infamous videos went viral. And for the meeting's first forty minutes, the discussion was in no way extraordinary. Several dozen engineers, analysts, procurement specialists, and executives—the majority of whom were white and male; mine was the only Black face in the room—discussed predictable topics: energy monitoring software, fourth-quarter profits, large-scale solar projects, HR concerns. But while the boardroom is designed for this sort of meditation on productivity, it is also nested in the broader corporate office: an epicenter of screenwork that is necessarily attuned to the digital pulse of the world beyond its walls—the visual texture of virality. And it was this digital pulse and visual texture that impelled John to announce that before they adjourned he wanted to discuss a topic that had never been discussed in their boardroom before: the matter of Black lives.

As a workaholic who spent most of his days in SunLight's office and who lived in a wealthy white neighborhood, John had limited access to the world that Sterling and Castile lived and died in. By extension, the physical spaces and lived experiences of the city's poor Black communities couldn't be further from the professional bubble that comprised his day-to-day. While John was proud of his company's record in LMI communities, the diffuse spatiality of EERE could not bridge the ontological gap between him and the precarious Black and brown beneficiaries of SunLight's work. Despite his familiarity with white privilege as a concept, it wasn't until Sterling and Castile entered his life via two videos that this abstract lacuna felt urgently real.

I was not with him when he watched the videos, but I can reasonably speculate that it occurred in his spacious (but by no means ostentatious) executive office—perhaps during a quick Twitter scroll while sipping his morning coffee, or during a quick glance at the *New York Times* website in the waning hours of the already-extended tech workday before returning to several more hours of screenwork. Like all of his white-collar employees, John was squarely focused on his various screens when he wasn't in meetings—preoccupied not just with work but also intermittent respite: a minute here or there to consume headline news and viral social media. For John, the screen was a lifeline to the world beyond the office precisely because it was omnipresent *in* the office. So it was likely there that he watched the grisly murders—their egregious inhumanity contrasting with the rational inhumanity of the endless workday. And regardless of where he watched them, the murders surely infiltrated his office when his phone would buzz with an alert regarding their aftermath—protests, politicians' comments, the repercussions (or lack thereof) for the perpetrators—at all times of the day, day after day, momentarily stealing his focus, reminding him, constantly, of the violent inequality, injustice, and unrest engulfing the country. And this muted cacophony of reminders, this harrowing screen-based ritual, was, John explained, inextricable from the political grievances of his new leftist employees, the equicratic discourse percolating in the office, as the infusion of media on Sterling and Castile coalesced in his mind with his employees' opinions about SunLight's whiteness, maleness, disreputable clients, and culpability in racial capitalism. For him, the company's internal discontent was emblematic of the broader societal discontent that he encountered on social media in his office. For there, in the rational arena of screenwork, critiques related in the broadest sense to race, class, and gender were less like the visceral shrieks and tears at BLM protests in the days after the murders and more like a single blur of data for him to constantly process.

So while most of these critiques centered not on individual injustices (e.g., two horrific deaths) but on pervasive structures of power (e.g., centuries of anti-Blackness), in the blur of data, amid the calm of screenwork, John interpreted these critiques more as a call to act in the most generic sense—to do something, *anything, now*—than as a more specific call to intervene structurally. I detected this more generic imperative as he stood at the helm of the boardroom a week after Sterling's death and expressed support for the BLM protests while calling on SunLight to *generally* address inequality giving little consideration to the protests' specific demands:

There's a lot of inequality in this country, and the way people get treated based on the color of their skin or their gender or their sexual orientation is just wrong, and the stuff that's happening with the police is more than wrong, it's . . . horrifying. . . . For those of us who have experiences and identities of power and privilege—whether that's through race or gender or class or education—we need to create spaces and opportunities for us to come together and have conversation, and talk about our stories and share our thoughts, and our hopes, and fears . . . we need to think about how our actions and our words might be hurting certain communities. . . . If our company's in the business of making the world a better place, then we can't ignore these issues.

The business of making the world a better place: John here was clearly referring to the Black and brown communities that SunLight worked in—a spatiality that situated the company in a racialized political context that demanded corporate action in light of the "horrifying" "stuff that's happening with the police." His employees nodded approvingly, finally hearing from their leader the political urgency that many of them had long felt. As a white corporate CEO and a sustainable energy wonk with a passion for "the environment," John, by his own admission, was not one to spend valuable company time discussing police brutality, anti-Black racism, and white privilege. Prior to that auspicious company meeting, he went to great lengths to focus on what he called "environmental issues" as opposed to "social issues" (when, of course, he wasn't focused on the company's bottom line). It was telling, though, that this shift was triggered not by what was happening on the ground in the communities where SunLight was "making the world a better place" but, instead, by a viral Facebook livestream—by the digital habitus that enabled John to feel "horrif[ied]" about violence that his corporate confines hermetically sealed him off from. In light of this provincial virality, John spoke with a genericness that would prove troubling.

As the meeting neared its end, he announced that HR would facilitate another company meeting later that week to discuss racism, inequality, and SunLight's role in addressing these complex issues. "We're living in a time unlike any other in my lifetime," he explained to me afterward. "I don't want us sitting on the sidelines."

How can a company affirm BLM if it only has one Black office worker? Many of John's equicratic employees raised this question after the meeting, as it hovered uncomfortably in the backdrop of John's company-wide response to anti-Black racism. These employees understood that the physical violence that befell Castile and Sterling was inseparable from the structural violence of the corporate world—that which the homogeneity of SunLight's office made visible. This systemic understanding of white supremacy resonated with John; when I asked him why he decided to discuss anti-Black racism at a company meeting, he linked it to the company's dearth of racial diversity:

> You came here [to SunLight] at this crazy time in our country, and I've never felt a need to talk about things like this with the company before. I mean, the race stuff is an issue here; our management team is all white men . . . and that wasn't by design. That's just how things happened. Some women in the company brought it to our attention. I would like more diversity in the senior team. It's more important now than ever that we grow more diverse. . . . It's more a matter of right or wrong; our society fails massive segments of it, and I don't think it's good for anyone.

When John links the country's racial turmoil with his company's homogeneous demographics, problematizes all-white male leadership, expresses deference to his women employees, and asserts that a white-collar energy company has a moral obligation to address inequality, he affirms a progressive politics that is usually absent in the world of centralized energy governance. But this equicratic shift is ideologically muddied, as John conflates social equity and corporate inclusion. When asked why he addressed the Castile and Sterling murders, John's mind immediately went to diversity, corporate social responsibility, and his own company—not to a much broader structure of systemic racism, not to his own culpability (consider his claim that "that's just how things happened"), not to anything beyond who's on his payroll. Make no mistake: John's diversity concerns were sincere; he was troubled by SunLight's skewed hiring practices, knowing they reproduced inequality. Furthermore, in the often nepotistic world of energy expertise, John's stated commitment to hiring women and people of color represented an important departure from the status quo. But this does not detract from the fact that John unthinkingly reduced social unrest against anti-Black violence to a matter of racial and gender diversity, rendering differential value as a lack of corporate representation (i.e., "society fails massive segments of it," so "it's more important now than ever that we grow more diverse").

Perhaps the most telling component of this disclosure was John's focus on "diversity in the senior team." This was a notable specification because it affirmed the importance of making space for not just any old brown face but specifically ones in leadership positions. Instead of citing reductive diversity statistics that misrepresent deep-seated racial inequalities, John strove for representation that pointed to structural change. But in focusing on the face of leadership, he overlooked that the company was, in reality, growing more diverse—and in somewhat troubling ways. As SunLight expanded their portfolio of solar installation projects, they rapidly hired more blue-collar solar installers, many of whom were Black and brown. After a few months, these workers began to express numerous underacknowledged grievances with the company regarding low pay, lack of jobsite safety and protection, and problems ordering jobsite materials. In myopically focusing on corporate diversity, John's statement on equity elided the racial inequality that his company was directly responsible for through its abysmal blue-collar labor practices.

Insofar as he saw C-suite diversity as central to SunLight's success and ignored his blue-collar workers' well-being, John's antiracist concerns point to the ways in which progressive politics operates through the hegemonic structures of racial capitalism. Conceptualizing this phenomenon as "progressive dystopia," Savannah Shange contends that in the absence of abolitionist commitments, antiracist ideology can be mobilized to foment de facto anti-Black institutions.[56] John's call to diversify SunLight's upper ranks is exemplary of progressive dystopia, bespeaking a commitment to an *appearance* of inclusivity that masks the differential value of the labor market. Indeed, hiring Black and brown executives without efforts to address the grievances of Black and brown waged workers engenders a spectacle of equity that maintains racial capitalism. Following media scholars Marita Sturken and Lisa Cartwright, *spectacle* here "generally refers to something that is striking or impressive in its visual display, if not awe-inspiring."[57] *Spectacle of equity*, then, refers to a striking visual display of progressive political values: racial justice or gender equality choreographed in a way that one can see. The word *see* here literally refers to optical apprehension; the spectacle is thus a sight before your eyes. A spectacle of equity could refer to the somewhat jarring sight of Black women in SunLight's executive office. While we "see" people in different ways that cannot be reduced to optical apprehension, I narrowly characterize the spectacle as a visual appearance to theorize how equity becomes an object of spectatorship in the office's digital habitus.[58] For when office workers use cloud-based platforms to strengthen neglected spaces they have never visited and support marginalized people they will never know, equity (that is, justice for those marginalized people living in those neglected spaces) is experienced not through interpersonal relations but through data that you can see on a screen—"concrete" work you can do in the office. More specifically, in John's world, social and environmental progress is experienced at his desk and in the boardroom vis-à-vis data on carbon emissions reductions, utility bill savings, and client demographics in the course of his 9 to 5 (or, rather, his 9 to 7:25)—and this mode of inhabiting space and encountering the social animates his approach to equity. So when faced with social unrest on his screen—Diamond Reynolds's wrenching Facebook livestream and its angry aftermath—John retreats to this familiar space, imagining equity as a spectacle he can choreograph from his desk, as a matter of transforming the *look* of leadership, as an empirically verifiable appearance. This imaginary overlooks that which EERE data neglects, that which is not readily visible in the boardroom: the grievances of his Black and brown workers on scorching rooftops, the challenges those workers have in making ends meet, the injuries caused by installing solar on skyscrapers without proper safety protocols in place, the impossibility of doing your job on a roof in the rain when someone in the office fails to order the right materials. These realities can't be seen on John's screen and are therefore provincial to corporate DEI. The decentralization of energy vis-à-vis the panoptic gaze of corporate expertise creates an affective space in which the office becomes the locus of "making the world a better place," endowing it with a sense of importance that obfuscates the on-the-ground realities in those spaces that its screenwork is allegedly improving.

After spending years with John in his beloved workplace, I truly believe he was simultaneously moved by the spectacle of equity—by the thought of seeing nonwhite faces in the C-suite—and somewhat incapable of understanding the exploitation of the Black and brown (and also white) solar installers on his payroll. I offer this assessment not to minimize his culpability but rather to highlight the affective power of screenwork—a late liberal visual economy in which making the world a better place through cloud-based platforms delimits how he encounters the very world he seeks to improve.

The limits of this narrow focus on looks came to the fore when the company convened to discuss the generic equity concerns that John had raised after the Sterling and Castile murders. Martha opened the discussion with a stinging reflexive critique: "It's very easy for us to criticize the cops, but systems of oppression work together, and the same stuff that's leading to horrific police killings of innocent Black men and women is the same system of oppression that's causing gentrification, yet we don't acknowledge that we have a part to play in gentrification by installing solar for a lot of nasty building owners." While this analysis elicited nods of approval from SunLight's equicratic employees, instead of following Martha's lead and delving into the complexities of racial capitalism, John quickly shifted the discussion to the reductive matter of corporate diversity: *How can we do more to recruit diverse talent and make SunLight more reflective of NYC's demographics?*

Later, Martin, the lone Black employee in the office at the time—in a low-level administrative position—courageously spoke at length about his alienating experiences navigating the workplace as "the only one," highlighting the intersectional precarity of being a Black, queer, and low-paid person at the bottommost rung of SunLight's hierarchy. Martin's words focused on the everyday *feel* of structural racism and heteronormativity in an ostensibly liberal space like SunLight—that which can be neither understood through diversity data nor eradicated through more Black faces in the C-suite. This soliloquy brought sympathetic avowals, but instead of any substantive discussion on the affective texture of intersectional oppression, the discussion quickly steered back to diversity optics—the alleged difficulty of recruiting employees from diverse backgrounds. In this way, John and a few other white male employees focused more on making SunLight *look* like NYC's melting pot than on the structural problems that had seemingly inspired this convening in the first place—a spectacle of equity that is inextricable from the company's digital habitus.

As far as I could tell, they truly believed that the more diverse the company looked, the more inclusive it would feel—a well-supported theory.[59] Yet this focus on a diverse appearance addressed neither Martin's lived experience with the insidious dimensions of white supremacy in a progressive dystopia nor the company's more legible manifestation of raced and classed oppression: the exploitation of SunLight's blue-collar installers when they work in the very communities that the company seeks to support. While this selective social justice politics is not particular to EERE companies, the feeling of "making the world a better place" from one's workstation is particularly acute in a space where the panoptic gaze of screenwork can catalyze decentralized energy infrastructures in marginalized communities, giving a *specific* affective and spatial structure to a broader genre of late liberal hypocrisy. This genre became ubiquitous four years after Sterling's and Castile's murders when another video documenting cops killing an innocent Black man—George Floyd—captivated the corporate imaginary, inspiring a particularly capacious spectacle of equity in the form of commercials, social media, and PR campaigns centered on workforce diversity. While I cannot ethnographically speak to the late liberal hypocrisy of multinational corporations spearheading diversity initiatives in response to Floyd's murder, I speculate that many of these corporations' CEOs did not have the sort of emotional proximity to equity that John did because screenwork is affective in particular ways when it simulates "a just transition" that is more concrete in its datafied facticity than the alleged positive impacts of most corporate goods and services. In other words, I am speculating that the spectacle of equity is particularly affective in a digital habitus in which the act of looking at your computer is enriched with the feeling that you are *directly* supporting marginalized communities. This corporate space is not actually more or less attuned to the nonvirtual world beyond the office than its less wholesome corporate counterparts, but it has a heightened sense of importance rooted in EERE digital platforms that feel morally valenced in ways that most white-collar screenwork doesn't.

That said, SunLight's problematic diversity aspirations offer only a snapshot of the broader moral conundrums of decentralized energy's digital habitus. In addition to choreographing "concrete" change that obfuscates differential value, EERE screenwork can also perpetuate some of the most extractive modes of racial capitalism—work to which I'll now turn.

EQUICRATIC SIMULACRA

BlocPower, a "climate technology company that analyzes, finances and upgrades homes and buildings," demonstrates how the equicratic screenwork of decentralized energy governance can reproduce the most extractive facets of racial capitalism under a progressive veneer.[60] Like SunLight, BlocPower uses, in their words, "proprietary technology" to oversee the preliminary analysis, project scoping, financing, and contracting phases of residential EERE/rooftop solar projects in LMI communities. The company is headquartered in NYC, which is where many of their projects are located, although they also partner with municipalities and organizations throughout the United States. In contrast to my work at SunLight, New Grid Power, and Celeste's company, I did not conduct ethnographic research at BlocPower (which is not a pseudonym). Instead, I interviewed four former employees and archived BlocPower's web-based materials and social media while following the company online for years. I also worked with one of the company's leaders in a nonresearch capacity prior to their work at the company.

Paradigmatically equicratic, BlocPower was founded on the premise that information technology can optimize residential EERE upgrades and that this technology is therefore essential to fighting climate change. As such, BlocPower largely focuses on analyzing and organizing data in office spaces and software platforms that are removed from the physical infrastructure and manual labor at the core of EERE work, positing data-driven "intelligence" as key to a just transition. Similar to SunLight, BlocPower's equicratic approach to addressing climate change attracted technology experts with no prior experience in the energy field who wanted to put their skills to use doing something meaningful—professionals who saw in BlocPower's mission an opportunity to address long-standing social inequalities and take concrete steps toward lofty climate goals. But BlocPower eschews the more hands-on, intimate care for buildings that SunLight prides itself on, focusing primarily on technological innovations that could bring projects to scale.

For example, BlocPower deployed a technology that conducted thermal analyses of buildings from moving cars, used drones to map energy inefficiencies, used Intel Xeon Processers to run energy simulations of multiple buildings simultaneously, and built "a powerful, cost-effective data processing pipeline . . . in the cloud . . . that provides actionable insights for decarbonizing buildings to property owners, utility companies, municipalities, states, and other groups."[61] In developing and/or adopting these technologies, they specifically aimed to minimize the time and labor of in-depth energy analyses of individual buildings, attenuating the in-person communication that equicrats like Brian embrace. In this way, BlocPower puts a particularly high premium on office-based digital renderings of infrastructure, apprehending EERE mostly through screenwork. Put differently, their technology aims to enhance the power of EERE professionals to panoptically assess a decentralized landscape of infrastructure without leaving their desks.

I want to suggest that BlocPower's panopticism is embedded in an extractive EERE praxis that employs equicratic discourse that obfuscates the company's entanglements in some of the worst machinations of racial capitalism. For example, BlocPower briefly worked with a hydro-electric company in the Democratic Republic of the Congo (DRC) on a project to help poor Africans access renewable electricity through micro-loans. As a NYC-based startup, BlocPower had no connections to the DRC or even sub-Saharan Africa more generally. But in a bid to generate startup investments, the company's crafty CEO allegedly sweet-talked a philanthropic tech entrepreneur whose charitable pet projects are centered on (1) providing microloans in the Global South and (2) protecting the Virunga National Park in the DRC from extractive industries. According to former employees, the CEO convinced the entrepreneur that developing microloan infrastructure is tantamount to developing the infrastructure for on-bill recovery financing—a loan mechanism enabling LMI people to pay for EERE upgrades through reduced utility bills. Given BlocPower's (limited) experience with on-bill recovery financing in low-income communities in the United States, the CEO claimed that the company was qualified to develop a similar mechanism using smart meters to connect poor Congolese people to hydropower. When the project launched, BlocPower quickly hit a huge roadblock: there was practically no reliable internet in the area of the DRC that they were targeting. As such, they couldn't create the advanced smart metering system that they proposed, as their EERE expertise is centered on screenwork in corporate spaces nowhere near the communities they aim to empower. While they were genuinely surprised by the dearth of internet access, such ignorance does not excuse the fact that the CEO leveraged his company's equicratic brand to drive investment in a for-profit humanitarian project in an impoverished Black country ravished by racial capitalism—a project that could never have plausibly succeeded. In this way, the infrastructural precarity of a Black former colony served as a pretext for a technological solution that only padded BlocPower's profits.

This racial capitalist ruse appears all the more fraught when we situate it in the broader transnational economy of renewable energy production. As discussed in chapter 1, solar energy storage technology is produced with cobalt that is mined by essentially enslaved Congolese people—precarious workers (many of whom are children) whose labor generates the advanced energy infrastructure of the Global North, even as they lack access to basic energy infrastructure themselves. So when BlocPower exploits this energy insecurity in the name of an inclusive sustainable future, they give a disingenuously equicratic face to a transnational renewable energy economy that foments the differential value of anti-Black colonialism. Ironically, it is precisely this energy insecurity that doomed BlocPower's hydropower project, as the DRC's internet lacuna is inseparable from the country's minimal electrical infrastructure. This project thus demonstrates the insidious power of screenwork when it coalesces with the diffuse spatiality of EERE, as efforts to connect laypeople's personal spaces with renewable energy through screenwork lack community accountability when corporations only relate to those personal spaces through the cloud.

According to former employees, BlocPower's labor practices also egregiously deviated from their equicratic brand. While I cannot disclose the details of my interlocutors' accounts so as to protect their anonymity, I can say that multiple former employees confirmed that they were harmed by transparent violations of basic labor laws. While labor exploitation takes many forms and is rampant throughout the corporate world, if my interlocutors' accounts are true, BlocPower is guilty of pushing the boundaries of professional workplace practices in a way that suggests a detachment from reality—as though they are unmoored from the basic organizational norms that even many of the most unscrupulous companies abide by, especially in the technology sector. I am not suggesting that they are the worst of the worst in the corporate world, simply that their particular patterns of alleged labor exploitation bespeak a form of alienated fantasy—the out-of-touch audacity at the root of their DRC project.

I speculate that this phantasmatic orientation to the world derives from a digital habitus in which reality lives in the cloud, reducing the decentralized materiality of EERE infrastructure to pixilated data on a screen while detaching the company from the communities they wish to shape. To corroborate this bold claim, I want to call attention to a Glassdoor review of BlocPower that links their alienation from on-the-ground infrastructure to their unscrupulous business practices. I'm specifically singling out this particular anonymous assessment of BlocPower because it echoes many of the more specific critiques that my interlocutors offered, giving voice to sentiments that I heard in words that I cannot directly quote:

I wanted to work with BlocPower because of their incredible mission statement: accelerating the transition to clean electric heating/cooling in an ethical way. I was excited to see how the company used AI to identify the right HVAC solutions for a client's budget. . . . I discovered that BlocPower is no more than a finance company, run by used car salesmen who know nothing about how a car gets built (or how construction works) and are only focused on upselling scope to the benefit of themselves and their investors, but not their customers. The level of "analysis" completed for each project is laughable. Don't be fooled; this is not a . . . mission-driven company upholding the values of its CEO.[62]

In this testimony, BlocPower's unscrupulous business practices are directly linked to its leadership's lack of hands-on experience with energy infrastructure, as their focus on "analysis," that is, screenwork, not only prioritizes profit over quality but is also shaped by an ignorance of on-the-ground infrastructural development. Such detached screenwork generates a fantasyland in which they are experts of infrastructure that they know very little about—deception that is contiguous with the shady labor practices through which they allegedly cheated their employees. In other words, their enmeshment in cloud-based platforms authorizes an orientation to the world that is unmoored from reality, undermining their equicratic platitudes.

But BlocPower's cloud-based energy governance more directly reproduces racial capitalism when it explicitly reifies neoliberal logics. Consider figure 2.6, a screenshot from their website. Like most of their website's images, this one features a Black person; BlocPower curates a brand identity that centers Blackness in visualizing an equitable clean energy future—an equicratic aesthetic. But this image is notable not because it features a Black person but rather because it interpellates this Black person—a man dressed in a blue-collar uniform—as a "marketplace." *Marketplace*, here, is short for "BlocPower Marketplace"—a virtual platform in which everyday people can invest in EERE projects in low-income communities (see figure 2.7). By branding a Black working-class man as a marketplace, the image suggests that BlocPower's virtual platform is by, of, and for everyday people—a populist sphere of commerce centered on social justice. This sentiment is spelled out in the caption's final clause—"help lower building costs, reduce global warming, and create jobs"—as the "marketplace" is allegedly a tool for optimiz-

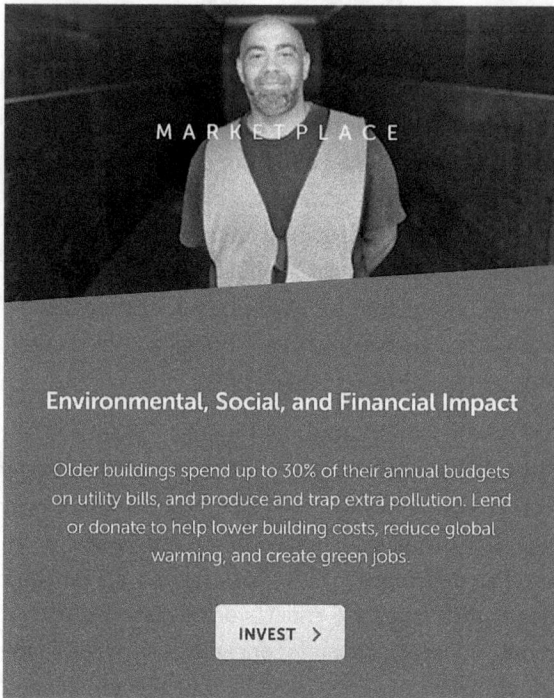

FIGURE 2.6. BlocPower's website interpellates everyday Black people as a "marketplace" as part of a late-liberal branding strategy in which the market overcomes the differential value of racial capitalism vis-à-vis energy efficiency and renewable energy. BlocPower, "The BlocPower Smart Cities Platform."

ing society, not generating profits. This "marketplace," then, is decisively neoliberal in its conception of everyday people as market actors focused on not surplus value but rather sustainable livelihood—decentralizing energy governance by subsuming life under commercial logics. In elevating a working-class Black man to the status of market totem, the image obfuscates differential value, suggesting that marginalized people can empower themselves through rational participation in the market. This virtual "marketplace," then, is paradigmatic late liberal screenwork—a progressive veneer for racial capitalism—demonstrating how the diffuse spatiality of EERE instigates inclusive energy governance imagery that suggests that commerce can foster social equity. But as BlocPower's unscrupulous labor practices and entrepreneurial audacity suggest, even well-intentioned EERE corporations depend at least in part on an extractive, exploitative economy.

As this chapter has shown, it is easy to lose sight of this violent system in the equicratic world of sustainable energy because the diffuse spatiality of EERE works in tandem with the digital habitus of contemporary corporations to position white-collar workers as environmental stewards

Discover Projects
Invest in your community and the planet

Click on any of our fully vetted, "shovel ready" retrofit projects, all of which have received thorough analysis, and learn more about their financial, environmental, and social benefits. You can lend or donate money on each project page. Thank you for investing in your community, creating green jobs and combating climate change.

CASE STUDY: New Mt. Zion Baptist Church – Completed Project

BlocPower replaced the boiler and installed modern HVAC controls, saving almost $500 per month in utilities and 1 metric ton per month in carbon emissions.

100% $47,100 Pledged   0 Days to go

Reduce Asthma and Fossil Fuel Waste In The Bronx

BlocPower has a plan to reduce asthma and fossil fuel use in the Bronx. The borough, known as the "Asthma Alley," has three times the national average for asthma-related deaths. One in five children in The Bronx has asthma. Help us use machine learning, a Wall Street financial product, and high tech sensors to fight pollution and fossil fuel waste in the Bronx.

2% $565 Pledged   40 Days to go

Unitarian Church of Staten Island

The Unitarian Church of Staten Island (UCSI), located in New Brighton, Staten Island, is an inviting spiritual community that is over 150 years old. With your contribution, our church will convert from oil to gas fuel, and upgrade our heating system and lights. These changes will reduce our CO2 emission by 41,600 pounds each year.

55% $50,090 Pledged   0 Days to go

Fordham United Methodist Church COMPLETED!

The Fordham United Methodist Church has been a cornerstone of the Bronx community for over 150 years. With your loan or donation, our church will upgrade our lighting and heating systems. We will save $24,000 per year on utility bills and avoid 33,000 pounds of CO2 emissions each year.

55% $51,160 Pledged   0 Days to go

FIGURE 2.7. A neoliberal / late liberal visual orientation to energy, BlocPower's online "marketplace" asks everyday people to redress differential value by browsing projects on their screens. BlocPower, "Discover Projects: Invest in Your Community and the Planet."

in marginalized communities. Put differently, when screenwork in professional spaces and solar panels in poor spaces coalesce, they produce a spectacle of equity that obfuscates the differential value separating those different spaces, affectively animating the corporate energy equicrat in the process. This coalescence, then, demonstrates the affective power of EERE technologies, as their diffuse spatiality positions cloud-based platforms as tools of equitable energy governance, shaping the sensibilities of white-collar professionals who aspire to change the world through their work. In other words, decentralized energy and its requisite forms of screenwork collectively create affective space for equicratic impulses, endowing the corporate office with a feel of inclusive environmental stewardship.

Very often, this equicratic work in NYC substantively addresses social inequalities, enabling leftists who are critical of racial capitalism to offer pragmatic care to marginalized communities. At the same time, the equicratic shift in energy expertise cloaks racial capitalism in a progressive veneer, as corporations that are committed to transitioning from fossil fuels reproduce the differential value at the core of the fossil-fueled order.

But the corporation is just one sphere of this equicratic shift; the diffuse spatiality of EERE has created new managerial and entrepreneurial roles for not only the corporate professionals that I discussed in this chapter but also activists from marginalized communities who have never before played a role in electricity generation. Chapter 3 explores these community-based equicrats, focusing on the ways in which solar affectively animates grassroots activism that resembles technocratic practices that have traditionally been antithetical to people power.

3

In contrast to fossil energy, solar electricity is often generated atop every-day people's homes and personal spaces. As such, solar inspires imaginaries of an energy democracy in which so-called communities—an elusive referent, as we will see—own and control the energy that their lives depend on, freeing themselves from fossil capital.

This imaginary inspired figure 3.1: a digital recreation of a photo that EQUAL (Environmental Quality for All, the environmental justice non-profit discussed in the introduction) featured on the website for their community-based solar campaign, Solarize Our Lives (SOL). The photo features me and a Latina student recruiting a low-income Black resident for SOL; both of us are wearing T-shirts with EQUAL's logo. It was paired with text that couched this community outreach as a matter of climate justice: "Frontline communities [i.e., marginalized communities that are vulnerable to climate change] . . . stand to benefit the most from solar, [but] have the least access to it. We want to change that." When viewed with this text, the image simulates grassroots power, creating the appearance of community organizers working toward a world in which poor people of color can own and control energy infrastructure from the comfort of their homes. Crucially, though, this appearance was patently disingenuous—a grassroots simulacrum—but I'll discuss this later.

Putting aside the matter of authenticity, I want to suggest that this image's populist air obfuscates the technocratic rigor and professional proficiencies that solar technology demands. From building codes to elec-tricity metering to tax rebates, a range of technical matters are essential to the development of solar infrastructure—matters that require creden-tialed expertise, not the "power of the people."[1] In this chapter I suggest

FIGURE 3.1. A simulacrum of grassroots activism promoting the Solarize Our Lives (SOL) campaign. AI generation, prompted by Christine Riggio.

that a mismatch between these technical matters and the "energy democracy" sentiments visualized in figure 3.1 has given form to a spectacle of equicratic activism: the shallow *appearance* of on-the-ground environmental justice (EJ) organizing with technocratic imperatives.

In elucidating this spectacle, I theorize the ways in which solar infrastructure affectively incubates energy equicrats in grassroots spaces: EJ activists who fight for social justice by counterintuitively adopting the managerial audit culture of white-collar energy experts.[2] I partly attribute this shift to the *modularity* of solar technologies and the *quantifiability* of solar electricity. These two material properties position solar as an ideological glue between what is often called *philanthrocapitalism*—a particular form of neoliberalism—and EJ activism, transforming anticapitalists into unwitting champions of market-based governance. Specifically, I first explore how the modularity of solar technology—the ways in which it can be rearranged and detached from other infrastructure—aligns solar with both the populist spirit of energy democracy and the calculative sensibilities of philanthrocapitalism, giving form to the grassroots energy equicrat. Focusing on EQUAL's efforts to develop a solar microgrid in their community, I suggest that this equicratic shift undermines traditional bodies-in-the-streets community organizing by amplifying the philanthrocapitalist call to achieve impact at scale. This, in turn, creates a spectacle of equity, as EQUAL choreographs visuals like that in figure 3.1, simulating on-the-ground activism that appears consistent with their populist politics while focusing on large-scale work that overlooks how their community members experience environmental injustice.

The quantifiability of solar electricity—the physical imperative to assign numerical values to electrical currents—further enmeshes EQUAL's activists in market logics, demanding technical expertise that moves grassroots organizing from bodies in the streets to disembodied spreadsheets. Specifically, as these activists fixate on the price per kilowatt-hour and the kilowatt production potential of community-based solar projects, they focus less on Black and brown people's bodily encounters with environmental injustices and more on technocratic calculations that indirectly address those encounters. In the process, EJ activists foreground a scopic identity politics that values the curated *sight* of Black and brown skin in campaign images like figure 3.1 instead of their everyday realities outside of these images. These images also complement spreadsheets, smart boards, PowerPoints, and metric-based infographics—visual performances of quantitative precision—that simulate technocratic rigor in EJ spaces known more for community organizing. I contend, then, that the quantifiability of electricity situates solar in late liberal screenwork that couples these simulations of technocratic rigor with photographic renderings of grassroots activism, producing an equicratic spectacle.

In sum, solar's *quantifiability* and *modularity* (through which it's brought to *scale*) affectively animate equicratic EJ activism that supplants what I will characterize as a Black feminist praxis with neoliberal managerialism. Thus, throughout this chapter, I show how the cross-pollination of community organizing and numerical technoscience at once empowers marginalized communities to take part in electricity generation *and* erases the grievances of the very subjects that this equicratic convergence is supposed to support. Locating this erasure in late liberal screenwork, I call on the EJ movement to shift from the spectacle of equity toward a politics of deep listening attuned to the core principles of recognition justice.

The power plants, electricity substations, transmission lines, natural gas pipelines, electricity grid, and electrical rooms of NYC were conceived, designed, (partly) fabricated, and financed in laboratories, engineering offices, factories, legal boardrooms, and political backrooms, out of sight and out of mind for the millions of New Yorkers who, often unthinkingly, use electricity, heat, and fuels every moment of their lives. So a community room with stained carpeting and broken windows in a run-down Black church in a poor (albeit gentrifying) residential neighborhood is not exactly the first place you'd expect to find people planning the electricity infrastructure of the future. Yet it's in this humble setting that I took part in an EJ effort toward that end: SOL, a local campaign to develop a community-owned, solar-powered microgrid in the middle of the 'hood. The campaign was supposed to continue EQUAL's decades-long history organizing low-income residents of color against the corrosive infrastructure that disproportionately affects them.[3] At SOL's kickoff meeting in the run-down church, most of the thirty participants were Black and Latino, a plurality of whom were Black women in their fifties and sixties who live in the community's housing projects.

This church's transformation into a site of inclusive energy governance evinces solar's affective power: its capacity to, for instance, enable those low-income Black women to feel they could participate in a system of industrial production traditionally dominated by elite white men. For as I discussed in chapter 2, the *diffuse spatiality* of solar welcomed new people into the technocratic realm of electricity generation, emboldening EJ activists to imagine the homes and everyday spaces of low-income communities as nodes of energy governance. This diffuse spatiality was at the heart of EQUAL's vision for SOL, as the campaign aimed to empower local residents to develop, own, and operate an energy infrastructure on their buildings' rooftops, attenuating their dependence on electricity generated by unaccountable investor-owned corporations.

Yet SOL's political vision was animated not only by this decentralized topography but also by another material property of solar: its *modularity*. As sustainability scholar Ali Kharrazi explains, "Modularity is a system property which measures the degree to which densely connected compartments within a system can be decoupled into *separate communities* or clusters which interact more among themselves rather than other communities" (emphasis added).[4] Solar panels are modular because they can be arranged into different-shaped clusters of variable sizes in most out-side spaces with sun exposure, and such arrangements allow the panels to work in tandem as *a system*—a *discrete site* of electricity generation.[5]

This modularity is especially pronounced when solar panels are incor-porated into a microgrid, for several reasons. First, unlike most tradi-tional rooftop solar electricity systems, solar microgrids include batteries that store energy for use at night and on cloudy days. As such, buildings that are partly powered by solar microgrids are less dependent on the central grid than buildings that are powered by traditional rooftop solar. Microgrids therefore operate with a greater level of autonomy—like dis-crete systems—than most clusters of solar panels. Moreover, solar micro-grids often include more than one low-carbon, on-site energy production technology, such as fuel cells and CHP (combined heat and power), which further minimizes microgrids' dependence on the grid, enhancing their autonomy. Crucially, the combination of batteries and on-site energy production allows microgrid-powered structures to "island off the grid" when centralized power is disrupted (due to storms or blackouts).[6] In other words, microgrids can enable a cluster of buildings to operate inde-pendently of the grid for extended periods of time. Finally, as an infra-structure that usually powers multiple contiguous buildings—not simply a single home—solar microgrids connect different people's dwellings to form a unified site of electricity generation, leveraging solar's modular-ity to create unique spatial configurations of power. Put differently, a solar microgrid transforms several rooftops into a *united confederation* of energy production, configuring panels in a modular arrangement that exceeds a single building.

As a result, these modular clusters can create new opportunities for community cohesion—for local collaboration. This modularity was there-fore appealing to EQUAL activists, who were more interested in orga-nizing local residents for broad-based (infra)structural change than in simply installing solar on individual buildings. They recognized that their organizing imperatives were affectively aligned with a community-owned electricity system that clustered homes into a semiautonomous zone; the modular ability to "island off" the grid and link together other-wise discrete buildings could generate a sense of community solidarity, magnifying their collective power.

Samir, an EQUAL community organizer, testified to this affective power at the SOL kickoff meeting. He stood at the front of the church's commu-nity room and explained in impassioned detail how the (still hypothetical) geographic boundaries of the proposed microgrid—the *contained space* of several neighborhood blocks—could disrupt the power that the city's investor-owned utility held over the community, liberating residents from the utility's grip. The proposed boundaries would, in Samir's telling, delineate a sphere of *cooperative* power that local EQUAL activists would communally govern. More specifically, Samir suggested that a "worker-owned co-op" would emerge in tandem with the development of the microgrid by virtue of its spatialized autonomy—as if the imagined contours of community-based electricity infrastructure were naturally coterminous with community-based labor governance. With a visionary fervor, he spoke of community members securing jobs designing, manufacturing, installing, and maintaining the microgrid, as if its detachability from the grid—its local character—was inherently compatible with local employment opportunities and local worker organizing. Indeed, the proposed microgrid's modularity—its capacity to operate as a "separate community," to use Kharrazi's term— anchored Samir's hyperbolic pitch, as he zealously recited the words *empowerment* and *democracy* multiple times when attesting to the radical potential of this modularity.

When I pressed him on the *specifics* of his proposal after the meeting, Samir seemed unconcerned with the questionable veracity of his suggestion that a microgrid was naturally compatible with a worker cooperative that disrupted the local utility's omnipotence. And he didn't once address how on earth he would generate the funds, expertise, political support, and, crucially, physical engineering conditions necessary to build an extraordinarily complex infrastructure like a solar microgrid in the middle of the 'hood. His utopic leaps in logic were attributable to the inspiring affordances of an imagined microgrid—the power of an *idea* of a detachable, semiautonomous zone of energy production. For even a hypothetical microgrid can be affective, as its modular properties—its localized contours, battery-powered detachability, and building-to-building linkages—make the lofty ideals of people power and worker power visualizable. As Samir explained to me, "Our job is to educate [the community] about what's possible, to *open their eyes*. Get the community inspired to take action so that ultimately the microgrid will be whatever the community wants it to be" (emphasis added). When Samir finished his spiel, two of the older Black women at the SOL meeting further testified to the affective power of solar's modularity. They explained that they were interested in SOL because they were longtime friends who did not live in the same building and wanted to *connect* their respective buildings—which they had lived in for most of their lives—to do something "positive for the community." Solar's modularity, in the form of a proposed microgrid, allowed them to envision multiple dwellings as a *united* force for change.

Thus, both the diffuse spatiality and modularity of solar created the affective conditions for transforming a church's community room into a space for planning a complex energy infrastructure. On the surface, this is a remarkable development. The Black church has historically been a site of political organizing and mutual aid in some of the most marginalized communities in the United States, serving as a central space for grassroots activism in the EJ movement's formative years.[7] That such a space could be a site for not simply addressing energy production (which the EJ movement has long done through antipollution campaigns) but more specifically for developing electricity infrastructure—work traditionally dominated by white men in white-collar spaces—gestures to the transformative intersectional power of renewable energy.

Yet when we think outside these optics of representation—the *appearance* of Black women conspiring in a Black church to develop electricity infrastructure—this inspiring scene shows hints of something more insidious. Beneath the communal surface, I sensed an equicratic ideology that seemed to put greater emphasis on technical managerialism than grassroots power. Consider, for instance, these excerpts of Samir's remarks at the SOL kickoff meeting: "*Energy democracy* . . . in this community includes develop[ing solar] in a way that is cooperatively owned and managed by *stakeholders*. . . . We need renewable energy but in a way that empowers *a certain demographic*. . . . We're gonna have to have renewable energy spread all across the city in our low-income communities if we're gonna meet [the city's] *carbon dioxide emissions reductions goals*" (emphasis added). While this endorsement of a democratic energy transition sounds mundane, in both words and tone something felt off about it. For this statement conjoined the activist language of *energy democracy* with technocratic corporate-speak, evoking an ideological muddiness that is at the heart of grassroots equicratic governance (as this chapter shows). Specifically, energy democracy, as a concept, refers to an equitable renewable energy transition that "prioritizes the needs and concerns of working families, Indigenous communities, and communities of color in the struggle to define a new energy future."[8] Well aligned with the populist affect of the Black church, the concept was broadcasted by the EJ and labor movements over the past decade (which often use it interchangeably with the phrase *just transition*). Yet when Samir directly addressed the marginalized people whom the term *energy democracy* evokes, instead of speaking to their struggles or inquiring about their experiences—instead of cultivating the nonhierarchical spirit of democracy—he oddly rendered them as "stakeholders" and a "certain demographic," two technocratic terms that have historically worked in service of hierarchical corporate governance.[9] For instance, private industry pioneered *stakeholders* to create a veneer of inclusiveness that managed anti-neoliberal dissent.[10] While these terms have migrated from financial and statist contexts to contemporary civic discourse, they still often frame political actors as subjects to be classified, tabulated, and governed as pieces in a managerial puzzle. These terms, then, are often deployed to counter protest, placate publics, or maintain the status quo—not to describe political agitators—undermining the EJ movement's subversive politics.

Crucially, the undemocratic sentiments that have traditionally animated *stakeholder* discourses seemed present in Samir's heavy-handed facilitation of the meeting. Samir did most of the talking, lecturing about the virtues of energy democracy and the proposed microgrid, and when he opened up the space for commentary and questions, he severely limited the time that people could speak, rebuked people for not discussing the proposed microgrid, and perfunctorily acknowledged grievances about public housing problems raised by the older women without committing to deepening this discussion in any meaningful way.

In numerous conversations with Samir before and after the meeting, he affirmed his fervent belief in democracy—in empowering "the people" to collectively fight for their livelihoods—but this sentiment seemed less like the foundation of a radical praxis and more like an abstract idea divorced from his interpersonal relations. This hunch, though, was incongruent with the person I knew Samir to be. As a Muslim American son of immigrants, with a master's degree in public policy, he had committed his life to EJ out of a conviction that the neoliberal policymakers that ran NYC would never work in the interests of poor people of color—that real change happens at the grassroots. Frustrated with the political incrementalism of the city's liberal elites, he championed an ecosocialist agenda centered on large-scale economic transformation driven by marginalized communities. The emergence of solar in the city (described in chapter 2) tapped into these ecosocialist sentiments, as its modularity created an opportunity for communities to seize the means of (energy) production. But infrastructural change is no proxy for structural change, and it wasn't clear that Samir was truly trying to tap this opportunity. While solar's modularity inspired EQUAL's microgrid vision, I wondered if this vision, in turn, was eclipsing the populist spirit of the EJ movement, eliciting technological imaginaries untethered to community organizing. As Samir spoke at length about this vision, the people before him—the silent "stakeholders"—appeared as extras in the room: bodies whose Black and brown skin made legible his generic script of community empowerment, not complex individuals with their own political agency.

In this vein, his suggestion that solar needed to "spread" to "low-income communities if we're gonna meet [the city's] carbon dioxide emissions reductions goals," rendered these "stakeholders" as pieces in a decarbonization plan untethered to a local vision. In this schema, community-based solar mattered insofar as it worked in service of a climate agenda *beyond* the community. Samir, then, was suggesting that solar was a conduit between local EJ priorities and mainstream environmental politics. I wondered if the modularity of his proposed microgrid—the capacity to not only detach from the grid but also link his community with the macrorealm of electricity governance—had infused nonlocal sentiments into Samir's local organizing, rendering EQUAL's community members as a mere demographic who mattered insofar as they were part of a greater cause. In other words, *could solar's modular properties shift the movement's local politics to a scale beyond the community?*

In Samir's managerial terminology and focus on scale, I sensed a different sort of equicrat than the one discussed in chapter 2: a grassroots activist who apprehends social equity through a technocratic lens, privileging infrastructural change over on-the-ground agitation. But perhaps I was being too harsh. Samir's equicratic impulses did not necessarily preclude his radical aspirations. As an anticapitalist EJ activist, Samir was well aware of the tensions between technocratic energy governance and community organizing; perhaps this consciousness would become legible as SOL evolved. Among the mostly mum locals in the room, I reminded myself that this was just the first meeting, hopeful that EQUAL would eventually amplify their voices.

And even through most people didn't speak at length, the room was heavy with excitement—a sense of opportunity activated by the idea of transforming local rooftops into a united front of renewable energy. "In my sixty-three years, I never thought much about energy, but this here is some exciting stuff, something good to get involved with," exclaimed one of the older women, encapsulating an affect animated by Samir's vision of modular power. I left the meeting hopeful that SOL would become the sort of inclusive campaign that EQUAL was known for. But I underestimated the power of modular technology to shift EJ activism beyond the grassroots. This shift, I would learn, was consistent with the broader political winds moving EQUAL's work from local bodily problems to global climate change.

EJ'S BODY POLITIC(S) AND PHILANTHROCAPITALISM

Prior to SOL's launch, EQUAL did not work on renewable energy. Instead of developing sophisticated infrastructure, the organization focused on their community members' ability to breathe, dwell, and move in their everyday spaces.

EQUAL was founded by Black women in the early days of the EJ movement as part of a community struggle against a toxic facility that had made it difficult for locals to breathe. When the state deemed the facility safe, the community members knew better due to their bodily encounters with the facility's pollution, so they took to the streets to protest. Eventually, they won a legal settlement and structural repairs to the facility. This infrastructural upgrade was directly attributable to a mode of activism centered on the bodily well-being of marginalized people. In the decades after this foundational struggle, EQUAL continued to highlight their community members' exposure to toxins and contaminants—from asbestos to particulate matter to toxic products—organizing locals to challenge the official word of experts who lacked a corporeal relationship with the pollutants that the community lived with every day. At the same time, EQUAL evolved into an incorporated nonprofit that employed their own credentialed experts (lawyers, urban planners, public health specialists, lobbyists) to redress the community's bodily precarity through data, modeling, and white-collar professionalism. As such, EQUAL was equicratic far before they embraced solar, employing both community organizing and technical expertise with the ultimate goal of protecting the bodily well-being of local residents.

I want to suggest, though, that their shift to solar not only deepened this equicratic practice but also disrupted it, minimizing the role of bodily well-being in their work. Prior to EQUAL's solar efforts, the organization balanced what I conceptualize as a Black feminist ontology of the body with a datafied ontology of the body—an equilibrium through which they addressed both the physical impacts and the affective texture of environmental injustice locally. A philanthrocapitalist fixation with solar upset this equilibrium, shifting EQUAL's work from structures of intersectional power to infrastructures of electrical power.

To unpack this shift, let me first elucidate its starting point: a Black feminist ontology of the body. Black feminism conceptualizes Black bodies as assemblages of biological material and structural violence, physical organs and historical origins, human corporeality and less-than-human mythology.[11] Specifically, many Black feminist scholars highlight the ways in which symbolic structures of power such as race, gender, and class always operate through the physical forms called *bodies*.[12] They use this term—*bodies*—to illuminate the material-discursive character of Black existence: the ways in which the corporeal features of racial others (e.g., phenotype, hair texture, respiratory system) become legible through discourses and affects ranging from overt white supremacist logics to an inchoate sense of intersubjective difference, from the codified color line to the front lines of antiracist struggle.[13] In this spirit, we can think of Black bodies as not the organs or flesh of people marked as Black but, instead, a way of conceptualizing how these people's corporeality takes form through, on the one hand, the optics and visual code of alterity and, on the other, the lived experience and interpersonal feeling of racialized marginalization—ocular apprehension *and* affective "frequencies."[14] Thus, the body, as it's been conceptualized by Black feminist scholars, is neither an autonomous entity nor a strictly physical form. In this sense, Black bodies were at the heart of EQUAL's work, as the physical sensations of inhaling toxins, dwelling by bus depots, struggling to breathe, living with lead in your home, and lacking access to fresh food were, in EQUAL's analysis, inseparable from the psychic sensation of marginalization, racialized existence, and social positioning. In other words, EQUAL conceptualized bodily health as inextricable from the feeling of being heard, the experience of collective power, the knowledge of ancestral lineages. In this spirit, EQUAL's activism focused on the *affects* of "getting bodies into the streets": the catharsis of occupying public space with folks who look and feel like you; the sight of dark skin paired with the deafening sound of defying silence; the representational power of corporeality in its most everyday form. In the EJ movement's insistence on marching in public spaces, blocking roads, and staging die-ins was a recognition that bodies are not just repositories of structural violence but also mechanisms of change—a material terrain through which we exercise power.

Yet when EQUAL's professional staff met with lawmakers, proposed policy, and wrote reports, they focused on stylizing epidemiological data and state statistics, rendering the Black body as a physical form—not a material-discursive assemblage—through late liberal screenwork. In mapping the differential value of air quality, life expectancy, and asthma hospitalizations through colorful graphics, these experts datafied the body, eschewing the cathartic feel of EQUAL's bodies-in-the-streets approach. For decades, EQUAL balanced these two ontologies of the body, framing their equicratic screenwork as complementary to community organizing focused on the lived experience of injustice and the affective power of demonstrations.

But in the decade before they launched SOL, EQUAL gradually focused less on getting bodies into the streets for protests and more on getting biostats under the gaze of policymakers; the organization's datafied apprehension of bodily pain began to eclipse their focus on locals' lived experience. The reasons for this shift are multiple, but for the purpose of contextualizing SOL, I want to highlight one in particular: the neoliberalization of social justice—a shift that oriented activism around the concepts of *scale* and *quantifiability*. When neoliberal policies privatized the social safety net in the final decades of the twentieth century, corporations, policymakers, and philanthropists looked to nonprofits to provide social services and address social problems exacerbated by free market reforms.[15] In other words, an alliance of economic and political elites regarded nonprofits as essential to the societal shift to free market governance, ushering in a post-Keynesian form of racial capitalism. These elites believed that because nonprofits lacked the rules, structure, and labor restrictions of state bureaucracies, they could deliver services and solve problems more efficiently than public institutions, protecting the welfare of the most vulnerable while the unleashed entrepreneurialism of a deregulated private sector enhanced the rest of society. This logic catalyzed a nonprofit industrial complex fueled by a neoliberal form of charity often called *philanthrocapitalism*.[16] Guided by free market principles, philanthrocapitalism is an approach to funding nonprofit work focused on maximizing returns on private investment in "social" "goods and services," rendering social change as a commodity.[17] Philanthrocapitalism is thus uninterested in contesting differential value; it invests in social change through flows of capital that paradoxically keep intact the raced and classed stratifications it is intended to attenuate.

At the same time, neoliberal policies decimated the social infrastructure and communal ties that had upheld grassroots activism throughout the twentieth century.[18] Prior to the ascent of market-based governance, social justice work largely operated through mutual aid, sweat equity, volunteerism, and direct action, as community organizing lacked professional hierarchies and institutional credentials. But widening socioeconomic inequality hampered these support systems, impelling social justice organizations to depend more on private foundations for funding. In this context, the nonprofit industrial complex professionalized grassroots activism, aligning community organizing with neoliberal principles.[19] Specifically, private foundations demanded that community-based organizations maximize their social impact by bringing their work to scale or identifying replicable models. Foundations also demanded that these organizations quantify their impact to "objectively" evaluate their efficacy. This focus on scale and quantifiability shifted many activists away from what Martin Luther King called "the beautiful struggle" and what Cedric Robinson called the "Black radical tradition"—the sense that justice is a lifelong praxis (not a quantifiable objective), that the struggle never ends—and toward teleological efforts to efficiently maximize impact.[20] As neoliberal policies constrained laypeople's capacity to organize through nonmonetized labor, grassroots activism was increasingly beholden to the nonprofit industrial complex (NPIC)—institutionalized according to "maximize and measure" dictates.[21]

As a result, EQUAL focused less on the Black feminist ontology of the body and more on its datafied correlate, subordinating their community members' lived experiences to the biopolitical screenwork of their professional staff. I'm not suggesting that EQUAL's commitment to data was solely attributable to the NPIC; technical expertise has long been central to community-based EJ organizing.[22] Instead, I'm suggesting that EQUAL's screenwork *gradually eclipsed* this organizing, as the NPIC's neoliberal demands shifted EJ organizations away from their two-pronged approach—bodies in the streets *and* disembodied spreadsheets—and toward the latter *predominantly*: screenwork that could be quantified and scaled.

In the 2010s, modular solar infrastructure offered EQUAL a new way of scaling and quantifying their impacts, further shifting their activism away from the Black feminist ontology of the body and toward the data-fied ontology. This is not to say that EQUAL embraced solar solely due to philanthrocapitalism. Instead, climate change was the main catalyst for EQUAL's pivot to solar. But as I now discuss, climate change enabled EQUAL to maximize and measure their work in ways that aligned with the NPIC's priorities, further elevating biopolitical screenwork over bodies in the streets.

CLIMATE CHANGE AND THE NPIC

The EJ movement has long regarded climate change as an environmental justice problem, demanding efforts to protect low-income communities of color who are disproportionately vulnerable to sea-level rise, extreme weather, droughts, and other climatic shifts.[23] This climate justice ethos is rooted in the movement's focus on bodily care, as EJ activists recognize that climate change poses multiple threats to their community members' physical well-being. As such, EQUAL and their peers contend that a just transition is essential to protecting low-income communities of color from numerous bodily problems related to climate change, including heat exhaustion, strokes, residential flooding, and precarious food access.

But unlike the movement's traditional objects of contestation—petrochemical plants, for instance, or unfairly sited highways—climate change is not localized, lacks clear physical contours, and cannot be directly apprehended by one's bodily senses. While heatstroke can be attributed to climate change, this causal connection does not animate the bodily *experience* of dangerously elevated heat in the same way that, for instance, the causal connection between a petrochemical plant's smokestacks and one's respiratory problems shapes one's conscious apprehension of one's physical precarity. Put differently, as a causal agent of bodily problems, smokestacks are a lot more present and discernible than climate change. Thus, a person experiencing heat exhaustion might know that their condition is caused by record-breaking heat waves (a symptom of climate change), but this knowledge is different from seeing, hearing, smelling, touching, or tasting the plumes of smoke from a petrochemical plant that causes breathing problems—from *experiencing* the most proximate causal agent of physical pain. You cannot see, smell, hear, touch, or taste climate change—it is a concept with multiple definitions and contested etiological properties that render it somewhat elusive.[24]

In this vein, efforts to mitigate climate change through renewable energy do not directly relieve marginalized communities' bodily problems. While decarbonization strategies can ameliorate *long-term* physical precarity, they usually do not appreciably improve people's bodily well-being in the *here and now*. Installing a few megawatts of solar locally does not have the same positive public health impacts as a petrochemical plant closure or a highway decommissioning effort, for instance. Of course, petrochemical plant closures and highway decommissioning efforts are also decarbonization strategies, so the distinction here is not only between solar and traditional EJ interventions but also between the ways traditional EJ campaigns were framed (as a matter of bodily health) and the ways climate-oriented campaigns are framed (as a matter of reducing carbon emissions). In other words, *the EJ movement has long championed actions that reduce carbon, but it has only recently championed carbon reduction as a defined end in itself* (albeit *equitable* carbon reduction).[25] As a defined end, carbon reduction is concerned with achieving quantifiable impacts *on a national and global scale*. Even when EJ actions are focused on justice locally, if they are *framed* in terms of carbon reductions, they are symbolically linked to *global* efforts to reduce emissions. In fact, several EQUAL community members and staff contended that SOL was part of a broader global effort to reduce global emissions. This focus on scale and quantity is divorced from the locus of bodily pain that animated EQUAL's work for years; scale and quantity are not directly related to people's ability to breathe, for instance. So, while EQUAL's work *always* contributed to climate mitigation, when the organization discursively shifted to climate justice vis-à-vis solar, they moved further away from community organizing that addressed bodily precarity and toward strategies that linked that precarity to a scale beyond the community.

Climate justice, then, created conceptual space for EJ work that could be *quantified and scaled*, discursively aligning EJ with philanthrocapitalism. Although private foundations grant a tiny fraction of their environmental funding to EJ nonprofits (philanthropy has historically prioritized white-run, mainstream environmental organizations), EJ groups like EQUAL were granted greater access to philanthropic funds when they began to frame their work in terms of climate justice and emissions reductions.[26] Specifically, a small group of private foundations looked to more polished EJ organizations to link community-centered organizing with state, regional, and national climate policy. These foundations believed that their funding could transform "communities" into testing grounds for innovative climate mitigation strategies that could reduce emissions before being brought to scale through policy.

As a modular technology, solar microgrids seemed ideal for such scalable strategies; by linking community-based electricity generation with central grid management, they enable climate resilience at the local scale while also reducing emissions at broader scales. Their modularity, then, animates a vision of a just transition in which communities simultaneously strengthen their local autonomy *and* address energy problems beyond their geographic boundaries.

This vision was at the core of EQUAL's grant agreement with one of the largest environmental philanthropies—a funding pact that further steered EQUAL to equicratic screenwork. The agreement funded EQUAL to initiate a local solar microgrid project and to push for policies that could enable other communities to start their own microgrid projects—*to bring the local to scale.* In other words, the grant transformed EQUAL's general aspiration for a local microgrid into a *specific* policy agenda to develop microgrids on a scale far in excess of their community, even as their community members knew almost nothing about microgrids. This grant therefore aligned well with the foundation's focus on scale, but it committed EQUAL to ambitious priorities that were not centered on the community's stated needs, such as child care and open space. Instead of listening to locals, Samir and other EQUAL staff facilitated quasi-participatory community forums regarding the microgrid that resembled the SOL kickoff meeting: settings in which participants' views were an afterthought—solicited to rubber-stamp climate justice plans that were predetermined in grant agreements. Yet EQUAL meticulously documented these forums in photos, as all staff were tasked with capturing even the most mundane sight of a dozen community members sitting together in EQUAL's office. These images populated EQUAL's social media and website, simulating a community-based brand by suggesting that everyday people were calling for a microgrid.

In the absence of bodies-in-the-streets organizing, EQUAL leaned heavily on its datafied concept of bodily precarity, producing glossy maps and colorful infographics that rendered the community's climate vulnerability in technical terms (such as hurricane evacuation zones) while curating an allegedly community-driven vision for climate resilience locally. But these maps and infographics focused on ambitious infrastructural project proposals, notably the microgrid—projects that poor Black community members had not, in reality, championed. These community members were concerned with far less sexy matters that more directly related to their bodily well-being, such as the scaffolding that hindered movement on sidewalks outside their buildings and the chronically broken elevators in their high-rise housing projects. Ignoring these concerns, Samir zealously utilized his maps and infographics at EQUAL's community meetings to choreograph a future in which microgrids would enable the community to survive and even thrive in the face of climatic volatility. These visuals lent a sense of technical rigor to an otherwise nonspecific pipe dream promised to philanthrocapitalist donors. For the infographics suggested that innovative solar infrastructure could serve as a connective tissue of sorts, linking worker cooperatives, affordable housing, community land trusts, social hubs, places of worship, and a community bank to form a united front of local power. In this way, the modular properties of an imagined solar microgrid—its detachability and connectivity—inspired a climate justice vision untethered to local residents' bodily experiences, as the philanthrocapitalist imperative to scale aligned with a scalar technology that most laypeople know nothing about. The image in figure 3.2 reproduces a section of a climate justice infographic that Samir developed, simulating "cooperatively owned microgrids" in a utopic fashion by positioning rooftop solar panels next to both rooftop gardens *and* rooftop wind turbines—a unified front of sustainability unmoored from the engineering realities that make this sort of contiguity at best a pipe dream. Between the colorful infographics and mundane meeting photos, I detected not the EJ movement's aforementioned equicratic equilibrium but instead a simulation of both "the people" and technical rigor—an equicratic spectacle.

The spectacle here is a world in which a climate-vulnerable community cultivates sovereignty through the modular properties of a technology that in reality necessitates expertise that easily exceeds said community's capacities. In the context of the NPIC, this tension is vanished by infographics (e.g., figure 3.2) and Instagram pics that disingenuously suggest a technocratic populism: everyday people with the technical knowledge to live autonomously and collectively vis-à-vis renewable energy. Put differently, the modularity of solar works in tandem with philanthrocapitalism and late liberal screenwork to simulate a future in which a poor community of color could sustain itself in the face of climate precarity through community-controlled infrastructure (as figure 3.2 suggests). But the reality, in contrariety to that simulation, would prove a lot more complicated.

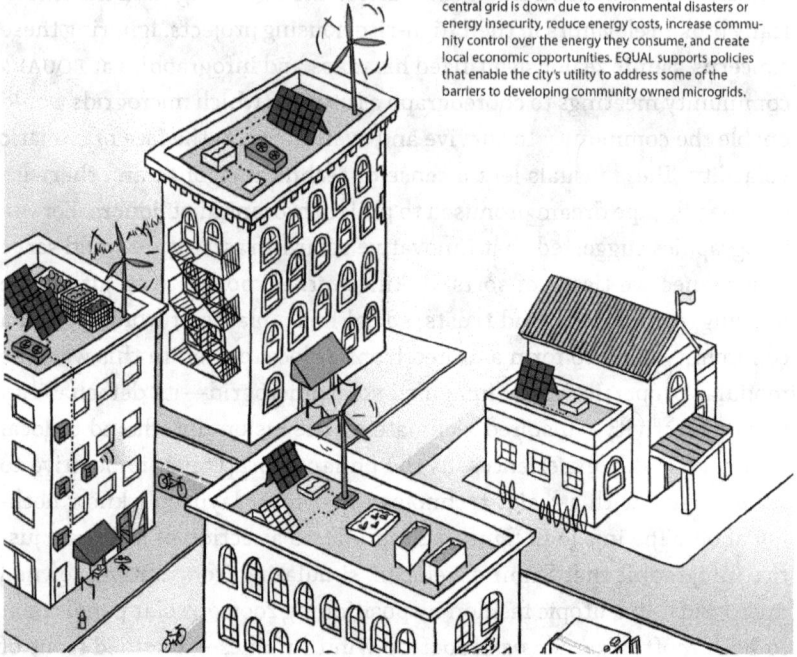

Community Microgrids

Microgrids can support climate-vulnerable areas in multiple ways. Microgrids deliver power when the central grid is down due to environmental disasters or energy insecurity, reduce energy costs, increase community control over the energy they consume, and create local economic opportunities. EQUAL supports policies that enable the city's utility to address some of the barriers to developing community owned microgrids.

FIGURE 3.2. EQUAL's utopic rendering of "cooperatively owned microgrids," which connects solar with wind turbines and gardens. Graphic by Cynthia Zhang.

ENUMERATING THE COMMUNITY SOLAR IMAGINARY

A year after SOL launched, the sense of possibility that animated the kickoff meeting had morphed into a more circumscribed sense of anticipation. As eleven of the thirty community members from that meeting sat in the same community room in the run-down church, they didn't blithely ruminate on a transformative, cooperatively owned microgrid, didn't speak in optimistic generalities. They instead soberly scanned a printed spreadsheet with electricity pricing metrics, focusing on the numerical facets of a solar campaign that was no longer the bold intervention they initially envisioned at the kickoff meeting (see figure 3.3).

While this spreadsheet appeared like a run-of-the-mill grid of numbers, it actually possessed an affective power that couldn't be detected by visually scanning its digits and boxes. Such a scan would reveal only the

Tier Pricing Proposal (Mandatory)							
Housing Type:				Multi-family affordable			
Tier Info - Aggregate Campaign Capacity			System Cost ($/W)				
Tier #	Low	High		5-15	16-30	31-50	>50
1	1	25		$4.56	$3.25	$3.13	$3.08
2	>25	50		$4.51	$3.20	$3.08	$3.03
3	>50	100		$4.46	$3.15	$3.03	$2.98
4	>100	200		$4.41	$3.10	$2.98	$2.93
5	200+			$4.36	$3.05	$2.93	$2.88

FIGURE 3.3. A snapshot of a SOL spreadsheet, which visualizes solar as a matter of quantitative data. Produced in Microsoft Excel by the author's pseudonymous interlocutors.

spreadsheet's surface: data pulled from responses to a request for proposals that EQUAL had disseminated to solar contractors two months earlier; the spreadsheet highlighted these contractors' cost estimates for bringing solar to EQUAL's community. As the community members' eyes flitted from metric to metric—the dollar per watt on a twenty-kilowatt system, fifty-kilowatt system, and hundred-kilowatt system, for instance—they expressed tempered excitement that their plan to install solar locally on low- and moderate-income (LMI) co-op buildings appeared plausible. Indeed, these metrics suggested they could realize their solar aspirations—albeit not on the scale of a climate-resilient microgrid. In place of a cathartic display of one's body in the streets, the numbers were weighted with a sense of purpose—an equicratic shift that amplified the EJ movement's predilection for data. Samir weighed in on their solar prospects with some tentativeness, mentioning potential roadblocks, yet the spreadsheet had clearly put everyone in a positive mood, corroborating their feeling that they could not so much *transform* but, rather, *improve* their local electricity generation system.

This restrained optimism reflected the pragmatic evolution of SOL. Shortly after Samir passionately proposed a community-owned microgrid at the kickoff meeting, he and other EQUAL employees switched gears, reorienting the campaign around practical, short-term milestones that could presage the ambitious, philanthrocapitalist plan for microgrids in the longer term. For all of their grant agreement's rhetoric, EQUAL knew from the outset that developing a microgrid would be difficult. It would require navigating a web of bureaucracies related to electrical metering, city codes, and solar financing to generate large-scale energy infrastructure on multiple buildings with off-grid capacity—a complicated, costly endeavor that is structurally impossible in many pockets of NYC, especially where economic resources and political power are limited. While the region's only solar microgrid was located in one of NYC's poorest, Blackest communities, extremely particular circumstances made this project viable—circumstances that are not transferrable to most other communities in the city.[27] As such, EQUAL's initial microgrid plan in their grant agreement was never a serious proposal. It was, instead, an equicratic spectacle: a utopianism intended to inspire aspirations for an equitable high-tech future.

Having succeeded in this regard, EQUAL refocused SOL on a modest, immediate-term goal of installing 150 kilowatts of solar locally—a goal that lacked the complexity of a solar microgrid and abided by the NPIC's imperative to quantify impacts. Toward this end, they adopted the "Solarize" model for developing residential solar, which had the insidious effect of injecting neoliberal dogma into a space opposed to racial capitalism (as discussed below). The Solarize model leverages community ties and economies of scale to reduce residential solar costs. Community groups that adopt this model facilitate grassroots campaigns to recruit local residents to participate in a solar purchasing pool: a coalition of neighboring homeowners who commit to purchasing solar together. Specifically, the community group—which could be a neighborhood association, a house of worship, or a nonprofit like EQUAL—selects a single contractor to install solar on all of the participating homeowners' roofs at a reduced cost for a limited period of time, making solar affordable through collaboration. The more homeowners the community group recruits to join the purchasing pool, the more leverage it has in negotiating solar prices with the selected contractor. Solarize campaigns consist of three phases. First, the community group solicits and evaluates tiered pricing proposals from contractors. These proposals indicate the prices that contractors will charge for different-sized purchasing pools; bigger pools are charged lower prices. After the community group chooses the contractor—usually the one with the lowest prices—they engage in door-knocking and grassroots outreach for a predetermined period to recruit as many homeowners as possible. On completion of recruitment, the contractor installs solar at a price pegged to the purchasing pool's size.

A technical assistance provider hired by EQUAL developed the SOL spreadsheet as part of the first phase: contractor selection. The spreadsheet compiled data from contractor proposals to enable community members to identify the contractor with the best value proposition for installing 150 kilowatts of solar. If the campaign succeeded, these 150 kilowatts would represent an infinitesimal fraction of the city's overall electricity usage—nowhere near the scale needed to appreciably reduce carbon emissions.[28]

Yet as the community members compared notes on the spreadsheet's various metrics—for example, the number of megawatts of solar installed by different contractors in the recent past—the *relative* insignificance of SOL's targets did not detract from the palpable sense in the room that they were doing something important. "We're now making things happen," exclaimed one of the community members, as though the mere sight of metrics was an end in itself. I couldn't deny that I shared this sentiment. The spreadsheet had this oddly thrilling quality; it seemed to validate our (relatively modest) plan for a community-driven energy transition. But when I scrolled back and considered the sight before me—a bunch of clean energy novices oohing and aahing at metrics on a printed grid—this ritual seemed quite strange. Why did merely *looking* at a spreadsheet's numbers feel so consequential? Why did these numbers instill so much validation, as though we had accomplished something? More crucially, why was a group of EJ activists—many of whom disdained racial capitalism—focusing their time and energy on a market-based spreadsheet and not on policies or protests aimed at empowering their community?

The answer here lies not in that particular spreadsheet but rather in the visual genre of spreadsheets more generally. Spreadsheets are a form of corporate screenwork that facilitates feelings of control, visually organizing data in rigid ways that can instill a sense of authority in the viewer—a sense they can manage the topic at hand.[29] As such, spreadsheets are a terrain of power that is compatible with late liberalism; their capacity to visualize the complexities of contemporary life and distill them into palatable cells engenders a form of economistic managerialism that is central to the governance of racial difference in a market economy. That is, spreadsheets are an effective platform for economizing difference, visually rendering social stratifications toward particular managerial ends, informing a viewer's sense that they can rationally intervene in inequality.[30]

Exemplary of this late liberal managerialism, SOL's spreadsheet endowed EQUAL's community members with the sense that they could overcome raced and classed barriers to solar. In visually plotting out a quantitative path to a local, cost-effective just transition, the spreadsheet's systematicity engendered a feeling of technocratic authority that was missing when community solar comprised a generic, embryonic aspiration without technical knowledge backing it up. For when the results of the request for proposals were cogently organized in printed reams of colorful rectilinear cells, they offered a sense of order that was undaunted by the structural problems and infrastructural limitations that differentiate access to renewable energy across demographic lines. From the shadows of the city's skyline to idiosyncratic rooftops with oddly positioned pipes to the damaged building envelopes of older dwellings, the particularities of the built environment make installing solar in NYC's marginalized communities far more difficult than it is in less dense suburban areas with more uniform buildings and often wealthier homeowners. These challenges are exacerbated by the fragmented, neoliberal solar contractor market (described in chapter 2) and the city's building and fire codes, all of which present logistical, bureaucratic, and financial barriers to adopting solar in LMI communities of color, which lack both the requisite financial *and* technocratic resources. In this context, a spreadsheet with cost estimates suggesting that installing solar throughout EQUAL's community was possible was a source of pride, demonstrating the community's ingenuity and authority. In other words, the organized array of numbers pointed to the power of laypeople to overcome the differential value that often kept renewable energy out of the 'hood, endowing the spreadsheet with an aura of opportunity that saturated the room.

But as I considered our intense focus on this spreadsheet, I wondered if we were affected not only by the transgressive possibilities that its numbers represented but also by the numbers themselves. Indeed, my comrades seemed to enjoy working with metrics, irrespective of what the metrics specifically indexed. I want to suggest that this data-philia bespeaks the affective power of the spreadsheet genre in the context of energy transitions. For as I will show, the *invisibility* of electrical currents amplifies the managerial mana of a platform that is primarily centered on *visualizing* metrics; the fact that we can't see electricity makes its metrics affective in particularly potent ways. Specifically, I argue that the quantifiability of electricity, broadly speaking, and of solar electricity, in particular, works with a late liberal mode of accounting epitomized by the spreadsheet to inspire market-oriented imaginaries and neoliberal dispositions among activists who consciously oppose racial capitalism. Put differently, solar's *quantifiability* animated our excitement in the face of a prosaic spreadsheet, transforming community activists into energy equicrats.

Janice, a spunky seventy-year-old retiree who is passionate about energy democracy, exemplifies EQUAL's equicratic shift. An active volunteer at EQUAL, she stuck with the SOL campaign long after the microgrid idea had dissipated. Three days before the SOL spreadsheet meeting, I was in EQUAL's office when Janice started an impassioned conversation with the organization's interns on why "capitalism and neoliberalism are to blame for climate change." She repeatedly called for a just transition that dismantled racial capitalism. These stated political commitments were consistent with her lifestyle and habitus. From community gardening to organizing protests for affordable housing on her block, Janice reveled in collaborative local spaces where she could "be the change" she wished to see. A thoughtful listener, she rejected the extractivism of social media and characterized communing as an end in itself. But when the SOL planning meeting began, her antiestablishment spirit and critique of racial capitalism were nowhere to be found.

Instead, as community members scanned the spreadsheet, Janice practiced a technocratic managerialism that lacked her usual political fervor. Specifically, she instructed everyone to "look carefully" at the data from SOL's six prospective contractors before determining which was best. She then implored her comrades to be as "objective" as possible when they rated the contractors on a scale from 1 to 5 with regard to prices, past performance, and proposal scope. The solar technical assistance provider hired by EQUAL echoed Janice, underscoring the importance of reviewing *all* of the data to ensure "objectivity." Janice's comrades diligently followed these instructions but raised concerns that they wouldn't be as "objective" as possible because they were "not experts" and because not all contractors supplied data in comparable ways, making it difficult to compare data "objectively." The technical assistance provider proceeded to offer instructions to help community members get over these uncertainties so they could "objectively" assess the data.

I found this dogmatic focus on objectivity bizarre. As a mythic ideal, objectivity gestures to an assessment of "the cold hard facts," so to speak, and it therefore doesn't involve personal beliefs or aspirations, so, in theory, *there's no need to ask everyday people who don't fully understand those cold hard facts to do such an assessment.*[31] Put differently, the whole point of centering laypeople in any grassroots campaign is to engage with their personal beliefs—their *subjective* views. Why ask laypeople to evaluate something if you are disinterested in their unique perspective? In theory, neutral analyses are the domain of credentialed experts. If SOL was a community-driven campaign responsive to local needs and perspectives, then, I figured, community members should be empowered to express their personal views and evaluate contractors based on the information that matters to them, without any pretenses of objectivity. As the technical assistance provider continued to explain how to conduct objective assessments, I couldn't help but articulate these thoughts.

But my objection to objectivity pissed Janice off. She indignantly explained that community members needed to have a technical understanding of solar data so they could sell solar to local co-op boards as part of their effort to form SOL's purchasing pool. In other words, community members needed to have an objective analysis to market SOL's contracting services to the community. In this schema, to be objective is to be a rational market actor who sorts through information and chooses a service based on transparent reasons. Janice, then, was suggesting that only rational market actors could recruit local building owners for a Solarize campaign. Despite her trenchant critiques of neoliberalism, she essentially endorsed a neoliberal form of environmentality, that is, environmental governance vis-à-vis everyday people in the idealized space of the free market.[32] In this way, the community-driven marketing of Solarize campaigns gives a grassroots appearance to neoliberal energy transitions. Put differently, under the Solarize model, everyday people are rendered as rational market actors, incentivized to offer unpaid promotional services to solar corporations by leveraging their communal ties in the interest of environmental protection.

While Solarize is heralded as a way of utilizing neighborly bonds and local institutions to empower communities through electricity generation, this progressive reputation belies how it (1) transforms environmentally minded citizens into a free marketing service for private companies and (2) focuses community-based action not on public investment but on purchasing power of small groups of consumers. When the market is the primary domain of social change, community empowerment is coterminous with rationally navigating the market. Thus, for all her anticapitalist politics, Janice called on her comrades to become a sort of *homo economicus*: a consumer who can maximize their own good through an objective analysis while selling the gospel of rational choice to others—a contradiction I now explore.[33]

Why would Janice, a critic of neoliberalism and its utilitarian habitus, embrace the hyperrational logics of the so-called free market? While her ideological contradictions cannot be reduced to a single cause, we can answer this question by considering the commodity animating her equicratic imagination: electricity.

I want to suggest that industrialized electricity *necessarily* operates through market logics such that even the most anticapitalist energy democracy vision cannot be extricated from capitalist values. This market enmeshment is attributable to the material-discursive nature of electricity—the particular ways it troubles a dualist metaphysics that separates mind from matter.[34] For electricity at once powers the physical terrain of the metropolis and exists in the symbolic world of metrics. Its industrialized form is an ineluctably material force but also literally nothing without abstract enumeration. Specifically, the grid demands that we quantify the electricity that we inject into it and withdraw from it to ensure that energy is rapidly consumed when it's produced because electricity cannot be stored in large quantities, unlike most other commodified goods.[35] Industrialized electricity cannot exist without being quantified, in contrast to many other commodities, such as potatoes or cars. Of course, many commodities are necessarily numericized in myriad ways; you cannot produce a car without measurements of its dimensions, for instance. But if you shut down a manufacturing plant after it produces twelve cars without determining this quantity beforehand, the cars it produced still exist; you did not need to determine this amount at the outset for the manufacturing process to occur. Conversely, you cannot produce industrialized electricity without proactively quantifying it, as failing to do so would cause a blackout. Therefore, electricity metrics, such as kilowatt-hours, trouble the specious line between material and symbolic. These metrics are no less material than light bulbs, as they are *inextricable* from the physical flow of power into and out of the grid, making modern energy as we know it.

Crucially, the materiality of these metrics is not ideologically neutral; as Canay Özden-Schilling shows, the inextricability of electricity infrastructure and quantification legitimizes and reproduces capitalist ideology.[36] She contends that the material imperative to quantify the grid's electrical inputs and outputs generates a dogmatic faith in supply and demand among white-collar electricity professionals.[37] In other words, these professionals regard the necessity of quantitatively matching inputs and outputs as evidence that the world works when supply and demand are in equilibrium. At the same time, this conception of equilibrium is foundational to the free market fantasy: the imagined harmony between producers and consumers in the space of exchange. Thus, the inextricability of electrical infrastructure and metrics legitimizes this politicized myth, such that the physical act of electricity generation can corroborate capitalist ideology.

In this vein, electricity metrics often work as *phantasmagrams*, Michelle Murphy's concept that I introduced in chapter 2, which "draws attention to the affectively charged and extraobjective relations that are part of the force of numbers. Phantasma accompany quantification as another kind of output, as the felt, aspirational, and consequential imaginaries that structure the world in profound ways, as a potent aura in surfeit of facticity."[38] By aligning supply and demand and thus naturalizing market logics, many electricity metrics operate as phantasmagrams, instantiating neoliberal imaginaries in the electricity generation industry. While Janice is not concerned with maintaining grid equilibrium, electricity metrics still operate as phantasmagrams for her for two reasons that I expand on shortly: (1) these metrics are primarily structured according to monetary exchange, and (2) they are the primary way that everyday people like her relate to charged electrons that are simultaneously ubiquitous and imperceptible. As phantasmagrams, measurements of megawatts and kilowatt-hours inculcate anticapitalist activists like Janice in free market schemes, seducing radical visionaries with fantasies of rational futures. Put differently, energy democracy advocates cannot elide neoliberal values because *those values are baked into the metrics that make industrialized energy real.* Thus, electricity phantasmagrams cultivate a counterintuitive faith in neoliberal practice among anticapitalists like Janice.

This equicratic transformation primarily occurs through the phantasmagram of price. In a capitalist context, electricity physically depends on prices, as they are the main medium through which we balance the grid's inputs and outputs. While grid equilibrium does not *theoretically* necessitate monetized electricity, in practice equilibrium is partly achieved by creating a high price per kilowatt-hour when electricity demand is high and a lower price when demand is low. Although a power plant could, in theory, generate electricity without monetary values, it is difficult to extricate socially constructed electricity prices from the material process of achieving grid equilibrium. Put differently, price is embedded in electricity generation infrastructures—part of the *physical* imperative to quantify inputs and outputs. Even the Soviet Union charged everyday people for the electricity they consumed. Electricity, then, is predisposed to the commodity form; *capital is largely inextricable from the grid.* This was the case even before neoliberal policies created competitive "markets" for purchasing electricity a few decades ago. So while Janice subscribed to utopic, anticapitalist visions of clean energy futures, those visions never addressed how we could get beyond commodified electricity. Unlike radical imaginaries of collective, nonmonetized housing or food production, leftist aspirations for community solar are necessarily beholden to the commodity form. Janice, then, had no choice but to conceptualize electricity as a market commodity, no matter her anticapitalist aspirations.

We can attribute this persistence to not only the ways the grid is managed but also how electricity prices operate in her everyday sensory landscape. Electricity prices are how most people know and engage with the industrialized energy they depend on, giving a legible form to streams of electrons that we otherwise can't detect.[39] While we can see the effects of electricity—a bright light bulb, for instance—the actual currents that constitute it remain imperceptible to the eyes, invisible to most of us. Market metrics, though, overcome this invisibility; when laypeople review, contemplate, pay, and/or contest their utility bills, prices enable them to engage with electricity generation processes that are largely hidden. Beyond the market values they symbolize, electricity prices affectively orient us to the electrified world we exist in. Thus, the commodity form is the central ontological construct through which most laypeople apprehend the ubiquitous but intangible entity we call *electricity*.

Of course, other commodities' prices are similarly affective. We largely relate to our cars, for instance, through prices. Yet electricity prices are affective in ways that most other commodities' prices could never be, because, again, they are central not merely to how we value energy economically and socially but more crucially to how we apprehend energy's very existence—how we experience it, encounter it, know it, and interact with it in contemporary Global North spaces like NYC (where disruptions in energy generation are very rare and where everyday people have almost no formal role in the maintenance of centralized electricity infrastructures, unlike, for instance in sub-Saharan Africa).[40] While all people in these spaces have felt the welcoming glow of a well-lit room, the heat of a plugged-in iron, the sound of the TV filling their living rooms, most people who don't work in the electricity generation sector have no sensorial relationship with the electric currents that power our lives and don't think much about electricity itself *except in the context of utility bills, costs, and economizing metrics.*[41] At the same time, many marginalized communities in the Global North are directly exposed to power plant pollution, and most people encounter transmission lines and other electricity infrastructure components in their everyday spaces. But the flows of charged electrons that we call *electricity* are not something that most of us can touch, hear, smell, taste, or, as stated, see. Prices render electricity knowable, giving it a visual form we can grasp.

Along these lines, most laypersons like Janice can't even grasp electricity through *nonmonetary* numeration. For example, if I say, "I have five cars," most of us can pretty effortlessly conjure a mental image based on this statement; if I say, "I have a solar electricity system that can generate five megawatts of power," few people besides solar professionals can envision what that actually looks like.[42] And while most laypersons do not associate their utility bill charges with a specific amount of electricity, those charges are ultimately how they conceptualize and orient themselves to the electricity they consume. Price—not other forms of quantification—is the primary mode through which laypersons conceive electricity as a discrete entity; dollars and cents are more central to our conscious engagement with electricity than they are to other commodities we can touch, see, and reflexively relate to in other ways. Thus, through the medium of price, electricity's quantifiability centers capital in our apprehension of our everyday realities, rendering market forces as an inherent facet of our electrified lives. This naturalization of capitalism positions electricity prices as phantasmagrams, "propagat[ing] imaginaries" in ways that attest to "supernatural effects in surplus of . . . rational precepts."[43]

Solarize campaigns epitomize this affective power. Organized around the imperative to reduce residential solar prices, campaigns like SOL offer a pragmatic structure for everyday people to intervene in an electricity generation system that they otherwise can't affect with their physical labor, everyday senses, or honed proficiencies. Solarize campaigns entice activists like Janice not only because they aim to make solar affordable for cash-strapped people but also because they operate through the one medium that most people can actually apprehend electricity with— *price.* Although Janice is dogmatic about conserving energy in her home through conscious consumption, as someone without the hands-on skills to install solar panels, SOL's price reduction approach offers her a clear path for enacting her solar aspirations in solidarity with her neighbors.

Practically speaking, the spreadsheet situates Janice on this path. For in the context of SOL, the spreadsheet is a totem of price—giving it a material form, incorporating it in the social world of community activism—enabling Janice to intervene in a realm of electrical power that she otherwise has no control over. This intervention entails comparing solar contractors' proposals through market metrics—figures that reduce contractors' capacities to small boxes on a grid—impelling the spreadsheet's viewer to apprehend energy as a matter of rational calculation focused on aggregated demand throughout the community. Thus, the spreadsheet telegraphs an edict to economize rooted in an imperative to collectivize, leveraging the populism of a communal project to generate consumer benefits. This edict is immanent in Solarize Campaigns that call on everyday people to not simply secure competitive prices for themselves but also stylize competitive prices through neighbor-to-neighbor organizing for collective climate action. The spreadsheet connects these consumer and communal objectives, valuing the collective power of clusters of neighbors who commit to purchasing solar together, enabling activists to enact their communalistic impulses through commercial transactions.

This community-oriented consumerism informed Janice's directive to perform an "objective" *homo economicus*. For the imagined subject here is less a rational market actor and more a rational market *activist*, utilizing competitive prices in ways that mobilize local people to collaboratively press for change through private contracts. By extension, the spreadsheet's competitive prices are not just indexes of market value but also phantasmagrams—affective metrics that prompt people to take environmental action, transforming them into neoliberal environmental stewards. As such, Janice's objectivity is a pragmatic device for leveraging the spreadsheet's utilitarianism toward her community organizing ends.

While many scholars have conclusively established that the rational market actor is a mythic trope, others have noted that racial capitalism partly operates through people's efforts to *perform* this trope—their *attempts* at *homo economicus*—as they embrace a subjectivity premised on "objective" market analysis.[44] Prices affectively animate this sort of identity formation; monetary accounting has historically enabled subjects to see themselves as objective decision-makers.[45] Prices have thus always been phantasmagrams, cultivating free market subjectivities partly rooted in the myth of objectivity. As such, solar's inherent compatibility with price—the material imperative to numericize electricity—muddies the ideological contours of ecosocialists who envision a just transition, obliging anticapitalists to adopt the reductive calculi they abhor. By curating prices, the spreadsheet offers one such anticapitalist, Janice, a practical tool to mobilize her community for local solar in the here and now. Her "objectivity" is therefore a sensible mode of activism that is at once produced and constrained by spreadsheet logics.

In sum, laypersons' lack of hands-on, infrastructural proficiencies and their metric-centered apprehension of electricity give form to an economizing visual medium—the Solarize spreadsheet—through which activists like Janice can manage the ubiquitous streams of charged electrons usually controlled by corporate technocrats. In this context, the spreadsheet simulates late liberal governance, positioning strategic consumerism as a means of overcoming the differential value that has traditionally prevented low-income communities of color from affording solar. In the process, the spreadsheet normalizes an "objective" calculus that only reinforces racial capitalism. For Janice's call to view contractors' proposals objectively through the spreadsheet had unintended consequences, exposing the contradictions of the EJ movement's equicratic shift, as I now discuss.

REPRODUCING RACIAL CAPITALISM

The problem with the spreadsheet's economistic collectivism became legible when the community members numerically rated prospective SOL contractors based on the spreadsheet data. They conducted this rating with a scoring rubric that the community members had allegedly developed at a previous meeting with the technical assistance provider's support. But, to my surprise, the rubric focused only on technical market matters like the price per watt, the cost of "adders," solar panel efficiency, and power purchase agreement (PPA) offerings, eschewing any mention of bodily well-being, climate change, or Samir's ecosocialist vision.[46] (The abbreviation PPA refers to a mechanism for financing rooftop solar projects in which building owners do not pay for their projects upfront.) I doubt that EQUAL considered community members' views in producing this rubric; where were the concerns for community health or good jobs that local residents often articulated? It seemed as though EQUAL wanted to choose a contractor based on commercial bottom lines, not community well-being.

Unsurprisingly, then, the contractor who won the bid to install solar on LMI residential buildings had incredibly low prices; they were the only contractor with projected costs under three dollars per watt. But their bid had one major catch: they primarily employed *volunteers* to do most of their installation work. In other words, they had extremely low prices because *they didn't pay most of the people who worked on their projects.* This was legally permissible because they are a nonprofit that bills their work as workforce training. In this way, the NPIC gives moral cover to this contractor's extractive operations. Audaciously, they market themselves as a "climate justice" organization that creates "green jobs" for low-income communities of color who are more "vulnerable to climate change." This philanthrocapitalist rationale greenwashes the exploitation of under-privileged laborers who "volunteer" in a usually futile attempt to secure full-time solar installer jobs. But this doesn't mean that this nonprofit contractor has nefarious intentions. When I spoke to several employees, they seemed very committed to the idea of a just transition that provides economic opportunity for marginalized communities. Even so, their model exploits precarious workers under the banner of social change, as most of their unpaid workforce trainees don't actually get hired by solar companies on completion of their volunteer experiences. As such, the contractor in no way adhered to Samir's ecosocialist aspirations for SOL.

When it was clear that this contractor had gotten the highest score, winning the bid to do the bulk of SOL installations, I incredulously raised the matter of jobs and labor standards, hoping it could disrupt the philanthrocapitalist focus on prices and efficiency. Everyone in the room strongly agreed that SOL needed to prioritize good jobs. But in another surprising twist, most community members said that they couldn't award the contract to a different contractor because (1) they had agreed on scoring parameters and it would be "unfair and undemocratic" to change the parameters just because they didn't like the chosen contractor, and (2) they needed to pragmatically confront the reality that low-income communities like their own need competitive solar rates because solar is so expensive.

The first point, about maintaining their scoring parameters so as to remain "democratic," betrayed a counterintuitive managerialism—a sense of professional accountability toward *private contractors*—as though the community members were bureaucrats overseeing a public bidding process and not local activists with the prerogative to work with whomever they wanted. Similarly, their second point bespoke their disinterest in fighting for a strong public sector to counteract the exploitation of private capital. Market logics dwarfed their ecosocialist aspirations, as the spreadsheet simulated a marketplace of rational consumers whose objectivity dimmed any vision for government action. Taken together, these two arguments point to the spreadsheet's affective power—its ability to orient laypeople to the economizing imperatives of philanthrocapitalism by injecting a sober utilitarianism into an ctivist space. With the spreadsheet beneath their gaze, Janice's comrades evacuated their values from their activism, orienting EQUAL's work away from public health and toward private grant deliverables, as I now show.

THE MUDDLED MESS OF EQUICRATIC IDEOLOGY

When all was said and done, a group of EQUAL staff and community members that included outspoken critics of neoliberalism and avowed socialists chose a contractor that didn't pay most of its workers. Perhaps the oddest thing about this choice is that they characterized it as a *non-choice*: they *had* to objectively follow the scores based on the objective data—it was only fair! This technocratic moralism butts heads with the neoliberal dogma of consumer choice: the idea that consumers should be free to choose whichever private entity they want to work with. It would be wrong, then, to suggest that EQUAL was under the dogmatic spell of neoliberal ideology. At the same time, the community members' *nonchoice* is reminiscent of the market fundamentalism that decimates EJ communities: the logic of polluters who locate their facilities in poor communities of color not necessarily because they harbor animosity toward these communities but because it makes economic sense to do so—as though they have *no choice* but to go where land is cheapest, where prices are most competitive.[47] In suggesting that EQUAL had no choice but to choose an exploitative contractor, Janice and crew reproduced the sort of crude cost-benefit analysis that ravages public health in communities of color, pointing further to EQUAL's shift away from the politics of bodily well-being.

Therefore, the community's fealty to the spreadsheet suggests neither a full-fledged endorsement nor a renunciation of market logics; it instead points to something more muddled. As EQUAL's equicratic balance between bodies in the streets and data on spreadsheets grew lopsided, a coherent politics didn't emerge in its place. Janice, for instance, espoused the central tenets of energy democracy, technocracy, rational choice theory, and socialism—a hodgepodge of conflicting values. Similarly, EQUAL staff regularly code-switched between ecosocialist ideals and philanthrocapitalist principles. The equicrat, then, doesn't adhere to the crude political rationality that they endorse, composed as they are of different, often oppositional forms of liberal thought.

Of course, there is no such thing as an ideologically pure subject, but the muddled political attributes of energy equicrats in this instance point not simply to the fallacy of the coherent liberal actor but also to the affective power of solar prices placed in technocratic visuals like spreadsheets. While no pure free market has ever existed, solar's enmeshment in market values vis-à-vis the spreadsheet contributes to an *appearance* of something approximating a free market in EQUAL's community—a confusing simulacrum of consumer choice—that conceals the fact that the local solar industry is more a product of technocratic and state governance than it is a locus of exchange among consumers and corporations.[48] That said, many spheres of economic activity in the contemporary world—not just those of solar—simulate specious free market discourses. Yet as I've discussed, NYC's solar "market" is notable because it does not simply interpellate laypeople as consumers but more specifically does so in a context in which they lack the basic decision-making agency that is foundational to most forms of consumer capitalism. As a result, the simulation of a market vis-à-vis a price-centered spreadsheet creates the false hope for consumer empowerment that is at the heart of Janice's muddled equicratic ideology.

To better understand this ideological muddledness, then, I now look a little closer at the ways the spreadsheet simulates a market, honing in, in particular, on the phantasmagram of price.

SIMULATING THE FREE MARKET

Figure 3.4 illustrates how late liberal screenwork simulates "the solar market," generating modes of collaborative economization that seduce even the staunchest anticapitalists into performing empowered energy consumers. This image is an extract of the "Evaluations" tab on the SOL spreadsheet, featuring Janice's evaluation of four of the contractors who bid on SOL. Each cell corresponds to a different contractor's proposal. In these spreadsheet cells, Janice performs an energy consumer who engages with not a single market price but rather multiple forms of financial valuation to overcome the differential value that makes solar inaccessible.

Pricing structure: comments	▼
Concerned about the pricing structure. Starts a lot higher than the others but drops down to lower cost if we have a lot of installations. Would be the most expensive if we have a small number. Does offer PPA	
starting price is 15c more than the lowest but goes lower than the others for larger number. Considerably lowesr cost add-on for canopy	
The least expensive if we have a small number of installations but does not provide as much of a discount for larger quantities.	
Per watt pricing is very competitive. PPA pricing seems high compared to the per watt pricing, but still competitive. Concerning that there is a 2% escalator (vs 1% with other bidders). Could be worth attempting to negotiate this term. Good PPA contract and turnkey contracts.	

FIGURE 3.4. Janice's comments on the pricing structure of contractors' proposals mapped out in the SOL spreadsheet. (PPA stands for "power purchase agreement.") Produced in Microsoft Excel by the author's pseudonymous interlocutors.

While figure 3.4 features Janice's comments, SOL's other community participants also wrote contractor assessments that the spreadsheet compiled, broadcasting their consumer preferences to establish a consensus regarding purchasing solar. In this way, the "Evaluations" tab simulated a democratic marketplace of sorts—a visual grid empowering individual consumers to use commerce toward a collective good. Without this screenwork, there's not a commercial sphere in which EQUAL's community members can collectively purchase solar, as an unsentimental spreadsheet generates a populist market in the space of their disheveled office. By performing an informed consumerism through small spreadsheet cells, Janice and her comrades create a realm of rational economic life that doesn't really exist without their virtual grid.

I want to suggest that this particular simulation of a market is rooted in competing political commitments that collectively inform the ideological incoherence of Janice's equicratic ethos. Specifically, the phantasmagrams of solar prices reinforce neoclassical economic theories and neoliberal discourses that are at odds with solar's dependence on the public sector, setting the stage for a muddled equicratic ideology. To unpack this claim, I now further explore these phantasmagrams before returning to Janice's ideological pastiche.

As discussed above, price is embedded in the production of electricity in ways that it is not with the production of other commodities, situating electricity in a supply-and-demand discourse that naturalizes the free market. In this way, electricity prices legitimize neoclassical economic ideas about "natural market forces," as electricity is thought to operate according to the idealized rules of consumer exchange.[49] *Solar electricity is especially amenable to these ideas. Because solar can offset some consumption of electricity from the grid, solar is often understood in terms of the following logic: the *supply* of more solar electricity can reduce *demand* for electricity from the grid. Specifically, many industry professionals espouse some version of this (simplified) statement: *If we can get more consumers to demand rooftop solar, we can reduce the burden on the grid.* While this statement is not false per se, it simplistically suggests that individual consumers' desires and needs (i.e., their demand) are central to transforming our energy system—a neoliberal premise. But any strategy to ramp up solar through large public sector investments and governmental mandates could also reduce burdens on the grid—irrespective of consumer demand. For instance, NYC's government could amend the NYC Façade Inspection and Safety Program to mandate that when buildings are upgraded to ensure compliance with the building code, building owners are required to install government-subsidized rooftop solar if their roofs meet the necessary specifications. Such a reform would ramp up the supply of solar electricity and thereby reduce burdens on the grid regardless of whether consumers demand solar or not. But solar professionals often don't consider aggressive government regulation as a way of optimizing the grid vis-à-vis solar, privileging the capitalist conception of supply and demand in a world in which the material imperative to match grid inputs and outputs naturalizes consumer markets.

As such, these professionals attempt to grow demand through government policies that put a market price on solar electricity. For instance, net metering policies essentially affix prices to solar electricity to enable people with solarized homes to get compensated for the electricity they add to the grid. (While New York replaced net metering with the Value of Distributed Energy Resources [VDER] policy described in chapter 2, VDER also creates prices to optimize the supply of solar electricity to the grid.) These price-based policies hinge on the idea that relieving grid demand requires that we incentivize individuals (including utility ratepayers, large investors, and private corporations) to buy and sell energy commodities. In the process, these policies interpellate everyday homeowners as rational economic actors capable of transforming society by heeding "the market's price signals." This, then, creates the *appearance* of a free market: the unquestioned sense that solar materializes when prices mediate exchange among rational actors in a private economic sphere—the *feeling* among many pro-solar subjects that the energy transition depends on consumers' desire to pay and be paid for solar goods and services.

But the city's solar industry does not at all resemble a free market. For starters, the "market" would be nothing without numerous government initiatives—from tax rebates to subsidies to state marketing services to laws governing energy monetization.[50] Although no capitalist industry is ever autonomous from the state, the city's solar industry is so enmeshed in the public sector that its existence would simply be inconceivable without numerous, ongoing governmental efforts at the city, state, and federal levels. While I spoke to many energy efficiency and renewable energy (EERE) bureaucrats who insisted that their mandate was to "strengthen consumer choice," "help move the marketplace forward," "grow the industry, "help customers make the right decisions," and "develop competition," in reality, they *were* the marketplace—more entrenched in the industry than comparable government actors in other technological "markets." Along these lines, EQUAL could *never* have launched SOL without government financing. More to the point, the process of pricing solar electricity is driven not by private corporations who are attuned to consumer demand but by the government, such that the market value of solar electricity is a public sector construct.

At the same time, the everyday NYC residents whom solar policies interpellate as consumers depend extensively on experts to navigate the solar market, undermining the idea that they respond rationally to price signals. Their lack of autonomy stems from the fact that the price of solar electricity is only one data point among many that NYC "consumers" must consider before adopting solar, including different state and federal incentives and tax credits that can offset the costs of different components of solar installations; their building's ownership structure; the size of the planned solar installation; their building's energy usage data and projected energy savings; electricity ratepayer payment structures; and their building's socioeconomic demographics. While *homo economicus*, the mythic figure, is obviously adept at rationally evaluating these data points, the reality is that even the most engaged consumers cannot wade through all this technical information without expert support. Indeed, all of the SOL campaign's participating building owners lacked the technocratic know-how to purchase solar without the close hand-holding of EQUAL's solar technical assistance provider. Price signals were therefore incapable of motivating consumer action without the sustained initiative of technocrats. Expertise independent of "producers" and "consumers" moves the market in ways that trouble any suggestion that competitive prices are the primary mechanism for reconfiguring supply and demand.

Janice, her comrades, and EQUAL staff knew well that SOL required the support of government and technocrats; no one was under the impression that the solar industry was an insulated sphere of exchange between producers and consumers. Yet they also endorsed the idea that competitive prices stimulate consumer choices that can catalyze a just transition, echoing technocrats' neoliberal language without interrogating its ideological baggage. This echo, then, was animated not by their conscious ideological commitments but rather, I contend, by the electricity prices they worked with. When the material process of electricity production is inextricable from numerical metrics that are usually coded as market values (as discussed above), it is arguably impossible to shape that production without engaging with those market values. While Janice is not, to my knowledge, aware of this inextricability, it naturalizes electricity prices, foreclosing the possibility that electricity can exist unmarked by capital, such that Janice unthinkingly deferred to "objective" forms of monetary accounting in the course of her energy activism. Thus, even as she consciously opposed capitalism, she put her faith in an energy governance approach that was unmistakably market driven. The spreadsheet merged this consciousness and faith, animating a populist consumerism in which laypeople work together to move the market—transforming EQUAL's rundown office into a commercial sphere with a democratic twist.

This is why Janice's equicratic sensibilities are so all over the place. On the one hand, she knows that solarizing her community requires government assistance (including funding for technocratic expertise), believing deeply in the importance of a strong public sector. But, on the other hand, the act of solarizing her community is enmeshed in a realm of data that is affectively animated by a sense that *the* so-called market—a mythic sphere of exchange between producers and consumers—is unavoidable. Janice acts on this sense vis-à-vis the spreadsheet's "Evaluations" tab (figure 3.4), which gives a clear visual form to the abstract concept of the market, allowing her and her comrades to perform the role of woke consumers. We can say, then, that the quantifiability of electricity intersects with late liberal screenwork to generate an uninterrogated sense that individuals can transform energy infrastructure in the mythic sphere of the market. Out of this configuration emerges an ideologically amorphous mode of consumer activism—not a coherent neoliberal philosophy but instead a mishmash of market fundamentalism, collectivist impulses, Keynesianism, and technocratic sensibilities. This ideological amorphousness obfuscates solar's dependence on the public sector, aligning anticapitalist activists who care for their community with the neoliberal status quo they usually oppose.

AN OBSESSION WITH QUANTIFYING SCALE

150. This figure glowed at the top of EQUAL's smart board, emanating a quotidian professionalism mixed with an ineffable presence—a sober magic. It hovered over a red rectangular bar, which indicated SOL's progress to date: 108!! It was over two-thirds of the way to the magic number at the top. What these two figures represented felt irrelevant when staring at the smart board; quantitative progress appeared on this giant LCD screen as an end in itself. If pressed, though, everyone in EQUAL's meeting room could tell you that 150 and 108 referred, respectively, to the number of solar kilowatts that SOL aimed to install and the number of solar kilowatts that building owners had committed to install under the campaign's auspices. I found, though, that most of these people couldn't tell you what a kilowatt is or what 150 kilowatts of solar looks like or what expertise, time, complications, negotiations, contracts, and labor that those 108 committed kilowatts demanded. In other words, EQUAL's employees understood 150 kilowatts as their goal without knowing what this metric actually meant—a goal brought to life through the smart board's shine.

Selena, EQUAL's policy coordinator, highlighted the glowing metrics as she updated her colleagues on SOL's progress—but not by talking about the campaign's on-the-ground recruitment efforts or energy democracy vision. The focus appeared solely on the 150 lighting up the smart board. This fixation with the smart board's magic metrics further revealed the ways in which solar's quantifiability interacts with late liberal screenwork to affectively situate anticapitalist EJ activists in a simulacrum of a solar consumer market. I have already shown how philanthrocapitalism worked in tandem with the quantifiable properties of industrialized electricity, the scalar properties of modular solar infrastructure, and the pricing regimes of solar "markets" to move EQUAL from bodies-in-the-streets activism to technocratic screenwork, muddling their ideological commitments. I want to suggest now that the *quantitative thresholds* of the Solarize model—the magic 150-kilowatt goal glowing on EQUAL's smart board—also catalyzed this transformation, as these thresholds worked with technocratic screenwork to further shift EQUAL's focus away from on-the-ground activism toward an in-the-office calculus. Put differently, the Solarize model affectively animated a political imaginary centered on the rather mundane matter of solar kilowatts. As a result, EQUAL primarily apprehended solar as a disembodied figure, distancing their energy democracy work from the community's lived experiences.

Of course, EQUAL adopted the Solarize model precisely because they believed it would do the opposite: connect solar to local residents' everyday lives. Solarize campaigns emphasize the importance of organizing a purchasing pool by meeting people *in-person* where they're at: talking about solar's benefits to your friend after church, or flyering about solar outside of the local elementary school during PTA night, or organizing your neighbors to canvass your block after work. Through person-to-person contact, Solarize campaigns are supposed to make real the prospect of solar on one's roof, cultivating a shared spirit of local environmental action. Phone banking or a mass email does not have the affective power of familiar faces in the flesh—the collective effervescence of working together in-person. As such, the Solarize model appears aligned with a bodies-in-the-streets ethos that speaks to laypeople's lived experiences. It is a tangible form of organizing that emphasizes physically showing up in people's intimate spaces—contiguous with the grassroots work that EQUAL had done for decades before SOL.

Yet, as discussed, this on-the-ground activism has a neoliberal underside because it transforms everyday people who care about EJ into an unpaid marketing service for private companies, focusing collective action not on advocating for public investment but on enhancing consumer purchasing power. Solarize campaigns therefore demonstrate the often overlooked synergies between grassroots activism and neoliberal governmentality.[51] When everyday people organize locally to leverage solar tax credits, they foment neoliberal policies that transform those people into servants of private capital. Even the most community-oriented Solarize campaign aims to support for-profit industry. Nonetheless, I assumed that EQUAL would focus more on bodies in the streets than on profits in contractors' pockets.

But community organizing often felt like an afterthought. For example, at no point did EQUAL engage in *any* door-knocking to recruit community members for the purchasing pool. Instead of strategizing on how best to reach people, Selena would, as stated, focus campaign check-ins on the quantitative commitments that SOL had secured toward the 150-kilowatt goal. As she pointed to the smart board's red progress bar, she would update us on the kilowatt contributions of potential projects: "The

Benson Homes are close to committing, which would give us another 23 kilowatts" or "We're in talks with the church on Jefferson and . . . that would net us another 12 [kilowatts]." Conditioned by the philanthrocapitalist edict to achieve scale, this obsession with the screen's metrics centered the campaign not on building a critical mass of engaged residents but instead on meeting with large co-op boards whose properties could substantively contribute to the 150-kilowatt goal.

I want to suggest that the unique spatiality of NYC's dense urban landscape skewed SOL toward this quantitative governmentality, belying the organization's activist ambitions and diminishing the communal affect that animates most other Solarize campaigns. The Solarize model was designed not for vertically developed spaces but instead for clusters of single-family homes with similar rooftops and similar exposure to the sun.[52] Therefore, every home in a typical Solarize campaign can generate about the same amount of solar energy. As such, these campaigns do not privilege some homes over others, focusing instead on recruiting as many homeowners as possible to achieve economies of scale. So a Solarize contractor might charge $4.00 a watt if a community group recruits twenty homeowners for their purchasing pool, $3.75 a watt if forty homeowners are recruited, and $3.50 a watt if sixty homeowners are recruited. Reaching these numerical tiers requires an intimate, interpersonal approach to community engagement in which campaign organizers leverage their familial and neighborly ties.

But SOL lacked these affective and enumerative dynamics because it didn't target single-family homes. Instead, EQUAL targeted buildings with dozens, sometimes hundreds, of tenants, and these buildings were not structurally similar: some had more roof space for solar panels than others, some necessitated more complex installation systems than others (e.g., raised canopy systems instead of ballasted systems), and some had more sun exposure than others (different-sized buildings create an uneven landscape of shadows). As a result, SOL's tiered pricing incentives were not pegged to the number of recruited households, unlike in traditional Solarize campaigns. Instead, SOL's incentives were pegged to the kilowatts of solar power that could be generated on local roofs, with 150 kilowatts representing the highest threshold and thereby the lowest prices. To achieve this goal, SOL didn't need to recruit as many people as possible; it instead needed to target select people who either sat on or could influence the board of their multitenant building. Therefore, EQUAL focused more on navigating the internal building management politics of NYC's residential co-ops—determining how to present financial information to a board treasurer, for instance—than on leveraging neighborly connections across contiguous blocks. Often, EQUAL presented SOL as a technical service for building co-op boards concerned with high electricity bills—not as a feel-good community sustainability campaign centered on empowering friends and family to go green. The quantitative threshold of SOL therefore delineated the campaign's affective contours, setting the terms through which EQUAL organized for solar locally.

The smart board stylized this threshold, magnifying the affective power of 150. A massive screen in the center of EQUAL's main meeting room, it transformed abstract kilowatt metrics into a visual presence, orienting EQUAL's activists to the narrow goal at hand. Indeed, the smart board demanded attention, supplanting eye-to-eye contact and centering the campaign on not the feeling of solidarity but rather the imperative of "objectivity."

As a figure on a flat surface, the 150-kilowatt threshold also shifted EQUAL's activism away from the community's bodily concerns. Instead of discussing heat exhaustion, flooding in public housing, or the other physical precarities that SOL was supposed to address, Janice fixated on the disembodied phantasmagrams of scale and price: "The idea is, we can first get to this amount [150 kilowatts] of solar, and then we can expand [SOL] and get more kilowatts, and really build the [SOL] model so that then other EJ communities can get to even more solar. And the more solar we do, the more we bring down prices . . . put[ting] us on a more sustainable path." Samir synthesized this teleological sentiment in declaring, "The 150-kilowatt goal is just a stepping stone to something bigger." As one community member explained to me, "We're talking 150 kilowatts today; maybe we're talking 500 kilowatts tomorrow; and, who knows, maybe it's one megawatt down the road." In this way, SOL's modest kilowatt metric situated the campaign in a teleology, rendering EQUAL's small-scale solar plans as a part of "something bigger," to quote Samir.

Yet that *something bigger* seemed like just a *bigger something*—a generic aspiration for *more solar* untethered to the specific, on-the-ground benefits that *more solar* is supposed to yield. In other words, the "more sustainable path" that Janice mentioned is paved with immaterial metrics. While 150 kilowatts certainly represented a significant improvement in a community that generated very little solar energy prior to SOL, it by no means signified the widespread adoption of solar locally; most community members still weren't able to solarize their buildings in spite of SOL's best intentions. Thus, for most community members, the 150-kilowatt goal did not point to any direct improvement in their day-to-day lives. Instead, the phantasmagram enabled people who otherwise had no infrastructural relationship with solar to at least perform a connection to it, casting themselves as players in the just transition. In stylizing 150, the smart board gave a particular shape to their general aspirations—something specific to think about in the absence of proximate referents—bringing some clarity to the elusive idea of a renewable energy future. The screen, then, choreographed kilowatts to temporalize solar, gesturing to a greater horizon that one could see in a rising red bar, endowing the spreadsheet's prices with possibility for those who still couldn't afford solar. As an *achievable* goal, 150 kilowatts indexed room for growth, an uncomplicated direction: "one megawatt down the road." The smart board's aestheticization of this teleology had a totemic quality, cultivating a tribal affect detectable in the community's fealty to the almighty kilowatt figure.

Like any totem, the 150-kilowatt threshold was a way for community members to relate to not only each other but also the material world. Specifically, the phantasmagrams of the smart board and spreadsheet connected them with an infrastructural realm they had no corporeal relationship with—technologies they had never before touched. Consider this statement from one of the community members reflecting on the spreadsheet: "We can get 46 kilowatts on our roof, and that's gonna cost us forty or fifty grand after incentives, so I think getting eighteen cents a kilowatt[-hour] for the PPA is gonna be needed for my building, so I think it makes sense to go with the $2.80 rate if that gets us the PPA." In this statement, the kilowatt projection (46) and the price metrics give the community member's words an air of legitimacy—a sense of expertise. But while she sounded like an expert, by her own estimation she knew very little about solar. She knew nothing about the craft of installing panels, about the technology that helps the panels convert sunlight to electricity, about the ballasts that hold the panels down on a flat rooftop. In other words, she had none of the know-how that comes from *hands-on* work. For her, solar was less a tangible substrate and more a set of metrics that she knew through the screen.

Among energy democracy activists, equicratic expertise emerges not from kinesthetic skill but from interfaces of phantasmagrams. Their equicratic affect, then, is decisively removed from the bodily matters that initially impelled them to pursue energy justice work. Yet this affect is also empowering, giving them access to the registers of energy governance that credentialed white men have traditionally dominated—detectable in Janice's animated response to a prosaic spreadsheet. That said, the equicrat's inability to apprehend solar kinesthetically has troubling implications for a climate justice politics concerned with the corporeality of marginalized people. Climate justice is elusive when it overlooks the disconnect between intersectional visions for solar and infrastructural bodies of solar—a disconnect I now explore further.

ON THE MEANS OF REDUCTION

While screenwork has long been a hallmark of the environmental profession, SOL's embeddedness in the smart board and spreadsheet points to a key affective difference between renewable energy initiatives and other forms of modern environmentalism.[53] Unlike most environmental work, including forestry, marine conservation, endangered species recovery, and even many carbon-offsetting efforts, renewable energy initiatives often do not entail corporeal interactions between humans and other forms of biospheric life with the expressed purpose of improving such life. Like all rooftop solar projects, SOL lacked any predetermined effort to *directly* optimize vegetation, animals, microorganisms, soil, or other nonanthropogenic phenomena. That said, forestry, marine conservation, and every other sphere of modern environmentalism resemble Solarize campaigns in the sense that they depend on screenwork—from GIS (geographic information systems) to spreadsheets to social media. But even the most technocratic wildlife restoration project, for example, entails physical labor with the forms of life that said project is supposed to preserve. Although some solar projects engage with biotic forms in some capacity—for example, solar installers transform the grassy fields on which solar farms are built—these projects do not directly support the ecosystems they inhabit. While even many cap-and-trade programs fund initiatives in which humans physically steward other forms of biospheric life (e.g., reforestation efforts), renewable energy programs do not entail, for instance, planting trees, seeding new crops, or protecting endangered species.[54]

Without such corporeal relationships, renewable energy initiatives are deemed environmental due to their capacity to reduce carbon emissions. Put differently, these initiatives focus environmentalist praxis on emissions reductions, not bodily engagement with nonhuman life. Thus, in the context of SOL, community members engage with the environment by reducing prices vis-à-vis spreadsheets so as to reduce emissions and not, for instance, by physically managing forest ecosystems. In this vein, many EJ activists and renewable energy professionals understand the climate primarily through screen-based media that aestheticize emissions metrics. Their work does not draw from their sense of smell, touch, sound, or taste in apprehending the climate or what they perceive to be the environment. Their environmentalist praxis is thus limited to disembodied screenwork (which is nonetheless connected to other people's physical labor).

This process of "informating the environment" paradoxically undermines the energy democracy activism it enables.[55] While energy democracy aims to deploy renewable energy infrastructure in ways that counter exploitative production processes, this infrastructural focus invites a technocratic worldview centered less on the means of production and more on what I conceptualize as the means of *reduction*: the numerical practices through which we apprehend goods and services as capable of reducing social and environmental harms—like a spreadsheet with solar contractors' prices. And it is easy to overlook the extractive ways renewable energy infrastructure is *produced* in marginalized communities, when these quantitative practices render it as nothing more than a way to *reduce* emissions in marginalized communities.

Shifting focus from the means of *reduction* to the means of *production* reveals that solar is not simply *presently* enmeshed in racial capitalism, as I discussed in chapter 1. It is, more radically, *inextricably* enmeshed in racial capitalism—necessarily embedded in an extractive paradigm irrespective of how many poor communities of color collectivize their electricity with it. Specifically, capital is inextricable from the transformation of numerous raw materials throughout solar supply chains.[56] Consider the silicon extracted from the earth, transported to industrial processing centers, heated and purified, and doped with phosphorous and boron, or the extraction of petroleum that is refined to create the plastic in solar cell backsheets, or the deleterious practice of mining lithium for solar batteries.[57] Connecting these processes and many others to manufacture solar infrastructure requires the coordination of human bodies, natural resources, regimes of labor, and sophisticated technologies in ways that have historically occurred through capital and private industry, notwithstanding some Soviet production. Even if worker cooperatives or a communist nationalization scheme improbably took over one facet of solar production—the fabrication of solar cells, for instance—many of the raw materials and industrial processes required to produce solar are so embedded in for-profit industry that it is inconceivable that solar could ever be produced outside of racial capitalism. Yet energy democracy activists often contend that collective action can transform solar into the foundation of a future in which marginalized communities control electricity generation without private capital. But a close look at the means of production reveals that we cannot fully extricate solar from extractive private markets. Even if we radically transformed the energy and technology sectors, solar would still operate through racial capitalism.

This doesn't mean that energy democracy activists should abandon an anticapitalist agenda. While solar is inextricable from racial capitalism, campaigns like SOL can still attenuate corporate power by enabling everyday people to produce electricity, worker cooperatives can loosen the grip of private capital over solar installations, and grassroots organizers can fight for a robust public sector that fully funds equitable energy transitions. Put simply, energy democracy activism can challenge racial capitalism in numerous ways.

Yet this activism can fall prey to utopic simplifications when it occurs through spreadsheets and smart boards estranged from the physical reality of solar's capitalized ecosystem. As screenwork fails to expose the excavators hoisting mounds of silica sand and the furnaces producing molten silicon and the factories haphazardly disposing of silicon tetrachloride in the name of renewable energy, solar appears in a conceptual space, as *alienated*, regardless of one's aspirations to seize the means of production.

This alienation gives form to the differential value of solar production described in chapter 1: a racialized apparatus of labor exploitation in which subaltern groups work under violent, impoverished, and toxic conditions to produce glimmering "clean" energy technologies. While solar is a way for poor communities of color like EQUAL's to fight EJ problems locally, solar, itself, is an EJ problem in many other poor communities of color besieged by industrial production. Most solar panels are made with numerous toxic substances (e.g., cadmium, silicon tetrachloride, arsenic, lead) that decimate the health of factory workers and, in some cases, are recklessly dumped into their workers' communities by solar manufacturing companies.[58] As EQUAL and many other EJ organizations protest toxic facilities in their communities, Chinese communities have similarly mobilized to shut down the corrosive solar production plants that have poisoned their health and environment.[59] At the same time, US federal prisoners manufacture solar for extraordinarily low wages and for corporations backed by investment banks like Goldman Sachs and venture capital.[60] While EQUAL touts solar as a way of creating economic opportunity in communities ravaged by the prison industrial complex, that same violent system exploits Black and brown lives to produce those energy technologies that foreground their utopic vision. The imbrications of the prison industrial complex, the financial sector, and the technologies that community-based campaigns like SOL use elucidate the contradictions of energy democracy campaigns that are inattentive to the means of production.

Toxic waste compounds these problems. Solar arrays last for approximately three decades, at which point they are usually disposed in landfills due to the dearth of solar recycling infrastructure, exacerbating the e-waste crisis poisoning low-income communities of color throughout the world.[61]

Furthermore, while EQUAL praises solar as a clean technology that produces zero greenhouse gas emissions, solar does in fact generate emissions when it is manufactured, installed, and decommissioned. Although these emissions pale significantly in comparison to those of fossil fuels, there are no widely agreed-on data on solar's emissions (i.e., there is considerable variability among emissions estimates).[62] As such, we actually have a limited understanding of solar's contributions to anthropogenic climate change (even if we know, unequivocally, that solar contributes way less than fossil fuels).

But when energy democracy activists apprehend solar through the means of reduction, they overlook the environmental impacts of solar's means of *production*. This alienation is an outgrowth of an equicratic environmentalism in which its adherents have no corporeal relationship with the quantified environment it imagines. Screenwork interacts with solar's quantifiability to animate energy democracy activism that obfuscates the extractivism endemic to *all* industrialized energy. Indeed, an ignorance of "sources, dependences, [and] relationships" facilitates a commercial paradigm of energy production that includes solar—a moral conundrum for activists focused on repairing relations between the earth and those bodies deemed disposable by that paradigm.[63]

At a SOL recruitment meeting in the community room of an affordable housing nonprofit, EQUAL's technical assistance provider presents a PowerPoint centered on the means of reduction—slide after slide on price and scale. But the quantitative visuals fail to convince the community members in attendance to go solar. While the community members understand the data, *they want to actually encounter solar in-person*—not as a set of metrics on a screen—before going through the laborious and costly process of solarizing their buildings. They explain they need to "actually experience it" to "make it real." PowerPoint is no substitute for corporeal proximity; they ask EQUAL staff to organize a community trip to a solarized rooftop.

But the technical assistance provider suggests, politely, that an in-person visit to a roof isn't necessary. The numbers, he insists, are overwhelmingly clear: solar saves money, pays for itself, and reduces your carbon footprint. He's happy to provide additional info that can demystify solar more than the physical presence of installed panels ever could. The community members, though, reiterate their interest in encountering solar in-person. Their physical distance renders EQUAL's hard metrics and glossy visuals inadequate. So, after the meeting, the technical assistance provider organizes a few visits to solarized rooftops.

But these visits do not prove to be enlightening. When we walk out onto the rooftops, there is not much to do besides view the solar panels against the backdrop of the city's arresting skyline. The community members stare at the panels with a forced look of erudition, unsure what to do besides appear engaged. The technical assistance provider talks about arranging the panels in ways that optimize sun exposure, the inverters connected to the panels, and other technical information that most community members appear disinterested in. After staring absentmindedly at the panels for a moment, some of them crouch down and touch the panels, inspecting them from multiple angles, taking photos—but it is evident that this in-person experience with solar isn't as illuminating as some wished it would be. The panels simply sit there; they make no sound, produce no by-product, don't change form, or move. If you reposition your body and look at the panels from different angles, the glimmer on their blue surfaces appears to shift—but other than this, the panels seem static.

Solar does not engage our bodily senses in the same way that fossil fuel infrastructures do. Petroleum refineries, for instance, cannot just mundanely sit in the middle of someone's living space without making their presence felt—their sounds, smells, visible by-products, and mechanized movements capture the senses when we're in close physical proximity to them. These sensory reactions, in turn, reveal a lot of information—the stifling toxicity of fossil fuels, for instance—data points that do not require metrics to adjudicate. Conversely, the stationary blue surface of solar panels does not reveal the material transformations through which they produce electricity, let alone the loud, tactile, olfactory practices of extraction, fabrication, distribution, and installation through which those panels are produced.

Thus, even when EQUAL's community looked beyond the PowerPoint to engage with solar, their sensoria could barely access an industrial commodity that is imperceptible in ways that other energy infrastructures never will be. This imperceptibility informs a mode of activism centered more on the means of reduction than the means of production. There is, then, a graphocentrism at the root of energy democracy—a premium on textual surfaces that pushes from view the bodies in Chinese factories, African mines, and American prisons whose labor makes solar possible.

BODIES THAT (SORT OF) MATTER

Even as EQUAL embraced a technocratic gaze, bodies in the streets remained its brand. The organization's identity and funding hinged on its *reputation* for having a ground game—the popular perception that EQUAL reached community members where they were at. While SOL was focused on recruiting large co-op boards to achieve the 150-kilowatt goal, EQUAL maintained its *image* of grassroots organizing—as though the campaign empowered everyday people to go solar. This image posited Black and brown bodies as a symbol of communal power, privileging the curated appearance of local people over their bodily experiences.

Figure 3.1—the digital reproduction of a photo of me and a Latina student canvassing for SOL—epitomizes this curated appearance. As mentioned, this photo was paired with text on EQUAL's website to simulate grassroots power, creating a look of *local* people of color organizing in the streets for a just transition. But neither the Columbia student nor I was a local. We were, instead, students funded by Ivy League institutions to support EQUAL's work; I was a doctoral student conducting research, and she was in an undergraduate community service program that connects privileged undergraduates with "service" opportunities at community-based nonprofits, including EQUAL. As part of this program, I had been tasked by EQUAL staff to canvass the community with a group of Columbia students to promote a SOL recruitment meeting. The students and I were responsible for hanging flyers in the lobbies of low-income co-ops. But no one at EQUAL actually had much faith in this flyering exercise. The staff at EQUAL were far less interested in getting bodies in the room at SOL meetings than in connecting with individual co-op board members who could get the campaign to the 150-kilowatt goal. Maybe, *maybe*, one person would see the flyer and come to the meeting, but this hypothetical person was unlikely to be a co-op board member. As such, our flyering exercise was really just busywork—something the students could do for their community service.

But at the behest of EQUAL staff, we scrupulously photographed our day of SOL canvassing, taking dozens of pictures of us taping flyers to buildings and chatting with locals who wanted to know what we were promoting. This documentation was par for the course at EQUAL. The organization zealously photographed anything that looked like grassroots action, always seeking new visual content to simulate *bodies in the streets* in reports to their NPIC benefactors. While all of the students, with the exception of the Latina, were white, EQUAL posted the *only* photo of the Latina and me. The image's implication was as clear as it was untrue: we, a young Latina and a Black guy, were local EQUAL activists organizing on the ground for change. Thus, EQUAL's website used our raced phenotypical traits to visualize SOL as a grassroots campaign led by people of color—obfuscating the campaign's calculative determinism.

The image also belied SOL's shifting raced and classed dynamics. The older residents of color who attended SOL's initial meetings slowly stopped participating in the campaign as it progressed. Increasingly, campaign meetings were mostly attended by white, upper-middle-class gentrifiers with an affinity for EQUAL's grassroots image, including Janice. The photos of the white students would have therefore been a more felicitous reflection of the campaign. Furthermore, the old smiling Black man that the Latina and I were speaking to in the photo was in fact uninterested in solar, and he quickly scooted us away after the photo was taken. EQUAL's website, then, privileged our surface appearance over our actual actions, curating us as Black and brown bodies.

Yet I want to suggest that this grassroots simulacrum is not evidence that EQUAL was dishonest. Instead, this misleading visual is counterintuitively felicitous due to the aforementioned ideological muddledness of solar work. As I have discussed, solar gives form to different political sensibilities that don't often comprise a coherent ideological position. While solar's quantifiability demands technocratic screenwork that most laypersons are not proficient in, solar's diffuse spatiality and modularity animate the feeling that everyday people can lead us toward a more sustainable future. These contradictory sensibilities inspired EQUAL's equicratic turn, aligning the organization with the philanthrocapitalist imperative to, on the one hand, quantify and scale social justice through market-friendly mechanisms that require credentialed expertise and, on the other hand, enact this technocratic work through bodies-in-the-streets organizing. I want to suggest that these contradictory sensibilities constitute a NPIC workplace in which people simultaneously believe that organizing is essential to a just transition *and* that organizing *isn't* essential to a just transition, never acknowledging the tensions between these sentiments. Solar is at once aligned with energy democracy and rule by (corporate) expertise, corroborating neither idea in isolation. It is in this contradictory space that the image in figure 3.1 takes form. For this image simulates grassroots organizing that is not actually necessary in a Solarize campaign focused on a kilowatt threshold in a metropolis in which most people live in buildings owned by someone else. In this way, the image exemplifies the ideological tensions of an infrastructure that simultaneously democratizes energy and reproduces differential value. It is a simulacrum of community-based activism that bespeaks not a disingenuous interest in local power but rather a failure to recognize the contradictions of energy democracy.

Despite this failure, the image evinces an effective form of late liberal screenwork that EJ organizations depend on: digital activism. Digital activism negotiates "the twinned formations of neoliberalism and liberal cultural recognition" by amplifying marginalized groups as they confront the deafening fallout from racial capitalism.[64] It is also essential for nonprofits' fiscal survival under philanthrocapitalism, as organizations must abide by the late liberal imperative to paradoxically address differential value through media rooted in extractive structures that produce differential value.[65] As such, nonprofit organizing depends on "digital storytelling"—communication that is disinterested in nuance—content with a simple narrative arc.[66] This is why EQUAL staff beseeched the Columbia students and me to document everything in photos; they needed visual content to tell the story of local, marginalized subjects organizing for energy democracy—a late liberal project that requires centering Black and brown bodies. Insofar as this project focused on rooftop solar—a technology that requires credentialed expertise—the story that emerged disingenuously renders "the community" as a force for change divorced from the technocratic complexity of infrastructural transition.

That said, the SOL canvassing image is also a part of a broader lineage of visual representation and thus not particular to the affective intersections of solar and late liberalism. Specifically, across contexts, the visualized Black body is metonymic of racialized matters that are not immediately legible in the visual field.[67] The organization's aestheticization of me and the Latina student taps into this mode of representation, gesturing to resilient people of color addressing environmental injustices that are outside of the photo's frame. But while the metonymic Black body is not particular to late liberalism, I want to suggest that this image dangerously elides the corporeal experiences of people who actually live in EQUAL's community—a late liberal move in which representations of minorities counterintuitively obfuscate differential value. For as I now show, this simulacrum of community peripheralized the embodied realities of EQUAL's community members, contradicting EQUAL's energy democracy rhetoric.

GINA

Gina, a fifty-year-old Black single mother, is a low-income EQUAL community member whose experience with SOL typifies the problems with the organization's late liberal screenwork. She lives in public housing projects owned by the New York City Housing Authority (NYCHA). As a tenant in a NYCHA high-rise with over two hundred low-income apartments, she has no power to solarize her building. Furthermore, she wouldn't directly benefit from SOL's reduced prices because NYCHA pays for her electricity; the means of reduction meant little to Gina.

As such, she attended SOL meetings for reasons related not to energy data but rather to everyday danger: her experience in an environment that decimated her bodily well-being. Many years before SOL, she became a dues-paying EQUAL member to address "lack of green space, lack of trees, lack of clean air, lack of clean water, garbage and waste in my community, and . . . dirt and soil issues related to the environment." The decrepit conditions of NYCHA buildings inspired these environmental concerns. Mold, asbestos, flickering lights, broken elevators, vermin, leaks, large piles of garbage stacked on adjacent streets, an egregiously poor recycling program, lack of heat or far too much heat depending on the apartment, building code violations, structural disrepair, a dearth of outdoor space, and inaccessible supermarkets are among the many environmental problems that NYCHA residents like Gina encounter every day. The causes for these problems are manifold, but a central cause is disinvestment. Severely underresourced, NYCHA perpetually faces budget cuts.[68] To address these challenges, NYCHA considered leasing out parts of their housing campuses to private luxury developers. This was a nightmare scenario for Gina and many other NYCHA residents, because, if it came to fruition, they'd lose their limited outdoor space—the dilapidated benches, rusted barbecue pits, and trash-infested walkways that, despite their condition, were used for communal bonding, music, dance, shared food and drink, tai chi, and other bodily practices that contribute to their communities' well-being. In Gina's building and in many other NYCHA residences—spaces scourged by asthma, heightened lead exposure, hypertension, and the garbage's daily assault on the senses—these bodily practices constitute some semblance of environmental quality, a degree of comfort in one's everyday environment. For Gina, then, environmental justice is the right of marginalized people to be well within their own communities; her EJ ethos privileges embodied experience over visual data.

In this vein, she initially participated in SOL to preserve her community's spaces and bodily well-being. She believed that if NYCHA could lower their operating costs by adopting rooftop solar, they wouldn't have to lease the community's outdoor public space to private developers. While this calculus hinged on the means of reduction, Gina's aim was not reducing costs or carbon but rather protecting communal space. So, like most environmentalists, Gina recognized that solar could preserve the environment, but this conviction elevated bodily well-being over disembodied metrics—communal rejoicing over calculative rigor. We can situate her solar politics in a long trajectory of Black women's environmental practice focused on collective place-making—"Black geographic togetherness," in Katherine McKittrick's words—amid the violent spatiality of both the plantation and its postemancipation offspring, such as public housing.[69]

In Gina's words, she initially attended SOL meetings because:

I was hoping [to] get involved with an energy cooperative where even if I don't live in a building that's going solar, I'm still investing in and benefiting from more solar coming into the neighborhood. I was hoping that with that model at some point we could reach the capacity of a microgrid and have a local power company that's cooperatively owned by local people. . . . I knew that I might not be able to get solar in my building, but I knew there was a possibility I could bring it to the community, and the investment in [solar] could . . . [make] a larger, stronger community network.

Gina here expands on Samir's community microgrid vision by articulating a commitment to a *body politic*: a cooperative solar enterprise built on strong neighborly bonds. Gina didn't care that solar wouldn't reduce her energy bills; she viewed it in terms of collective power—not consumer power.

Yet after spending hours at multiple SOL meetings, she became disillusioned. The campaign was unconcerned with both her bodily well-being and the body politic she envisioned. As she explained:

I live in public housing; I don't own a home . . . and it seems that [SOL's] emphasis is on homeownership. But many of the [local] people who own homes are hedge fund people, people who are trying to buy a place for their kid in college, and they see how cheap it is, and they buy two apartments when people who have lived in the neighborhood for years are being priced out. . . . As a low-income person living in NYCHA, I'm dealing with all these different day-to-day struggles, in my apartment, in my building, and in my [NYCHA] complex, so I get involved with a lot of interesting things [like SOL], but once I get involved in them, if my energies is being directed to an outcome that isn't really going to affect my life in a way that I can look around and say concretely, "I contributed to that, I did that," or if I know that this isn't changing something concrete for me and my neighbors, then there's only so far I can go with it. . . . I'm trying to engage my neighbors around the problems we experience in the building, and sometimes the solar stuff just goes over their head. . . . I feel like [there's a] disconnect between [EQUAL's] goals and the reality on the ground . . . because [EQUAL is] grant funded . . . and the solar work is not improving my everyday environment in this [NYCHA] complex; it's just extra work [for me], and I'm not a "lady who lunches." . . . So it means a real commitment if I'm involved in these things. And you know I don't mind making a real commitment, but I don't wanna spend three, four, five years working on these things that don't translate into some kind of knowledge about how to bring these things to where I am or how to capitalize on the skills I've learned to develop something on my own from where I am.

In making numerous references to "where I am," her "concrete" circumstances, her "everyday environment," and her NYCHA neighbors and building, Gina assesses solar in terms of her bodily realities and the body politic. She disengaged from SOL because EQUAL failed to situate solar in this material and affective terrain. Along these lines, she argues that EQUAL's solar strategy didn't improve her environment because it overlooked the community's socioeconomic shifts, evincing a "disconnect" between EQUAL's quantitative goals and on-the-ground gentrification. She attributes this disconnect to EQUAL's enmeshment in philanthrocapitalism, as she suggests that the organization is more responsive to their grants' terms than their community's lived experiences (an insightful assessment, as I will show). Thus, when the means of reduction guide EJ activism, it overlooks the "slow violence" of garbage in corridors, leaking pipes in lobby vestibules, mold on bathroom ceilings, and other everyday environmental injustices.[70] I want to therefore suggest that EJ campaigns for renewable energy often struggle to *directly* address the bodily realities of the community members who traditionally formed the bedrock of the EJ movement—Black women, in particular.

This disconnect became jarring as SOL moved closer to its numerical goal. The LMI housing co-ops that EQUAL recruited for SOL were largely occupied by white residents who had moved to the community in the preceding few years as part of a broader wave of gentrification throughout NYC. Many of these white newcomers became co-op shareholders in buildings set aside for LMI residents because they just barely qualified for the moderate-income bracket as young professionals, not because they lacked assets. While not all of these buildings' white residents were newcomers, white people comprised the overwhelming majority of regular participants at SOL meetings after the campaign's first year. Gina and other Black and Latino members—especially NYCHA residents—lost interest in the campaign when it shifted from Samir's inclusive microgrid vision to technocratic screenwork focused narrowly on installing solar on private property.

Furthermore, several recently constructed, market-rate luxury co-ops also participated in SOL, and the majority of these buildings' shareholders were white, upper-middle-class professionals who were new to the neighborhood. These white professionals navigated SOL's spreadsheet with ease and were emotionally invested in SOL's 150-kilowatt goal, as their buildings moved the campaign beyond its quantitative threshold. Their participation in SOL was at odds with the simulacrum of grassroots activism on EQUAL's website not only because it belied EQUAL's bodies-in-the-streets brand but also because it elevated the means of reduction over the lived experiences of people like Gina. As such, the disconnect that Gina identified underscores the problems with both EQUAL's online image and inaccessible screenwork—visual media that reproduced the differential value they were supposed to counter because they overlooked what was happening "on the ground," as Gina put it.

THE EQUICRATIC IMAGE BROUGHT TO LIFE

While the photo of me and the Latina student simulated a grassroots spirit that is misaligned with SOL's technocratic screenwork, that photo and this screenwork (in addition to other similar media) worked together to generate a unified (albeit contradictory) equicratic image: the appearance of an organization that was simultaneously in the streets and into spreadsheets. This equicratic image took form not only through digital platforms and virtual visuals but also through the real-life bodies that EQUAL curated at their monthly in-person all-member meetings. These meetings reminded me that *an image* can refer broadly to "a concept or impression, created in the minds of the public"—a concept that need not be digitalized or rendered by hand.[71] On the surface, these meetings were not image oriented but issue oriented—focused on EQUAL's campaigns. Yet at their January 2018 meeting, in particular, issue-oriented discourse took the chimeric form of an equicratic image.

Around ninety community members attended this meeting, a plurality of whom resembled Gina: Black women and/or Latinas over the age of fifty who lived in either LMI co-ops or NYCHA housing. That said, many of the community's white, (upper-)middle-class gentrifiers had become dues-paying EQUAL members in the previous few years, reflecting the SOL campaign's demographic shift. At EQUAL's meetings, these white gentrifiers often appeared unnervingly comfortable with an EJ politics that challenged the very structural conditions that enabled them to occupy the community in the first place. Their eagerness to take part in activism that should, in theory, unsettle them evinced the "progressive dystopia" discussed in chapter 2: a space in which antiracist politics paradoxically operates through contemporary forms of gentrification embedded in long histories of racial capitalism and state violence.[72]

Nonetheless, EQUAL staff were, in principle, committed to foregrounding the voices of people of color in their monthly membership meetings. *We need to follow the community* resembled the chorus of a song stuck on repeat in EQUAL's office prior to these meetings, with this refrain's final two words standing in for *low-income Black and brown people.* Staff often announced their heartfelt commitment to empowering community members of color through their campaigns, undermining any claim that the organization was simply choreographing a bodies-in-the-streets spectacle for social media and nothing more.

But this sort of choreography was disturbingly present at the January 2018 meeting. The choreographing began when EQUAL's policy coordinator pulled aside four attendees of color who were schmoozing during the premeeting breakfast. He asked if they could help facilitate part of the meeting. They agreed, and he explained that shortly after the meeting began, he would call on each of them to read aloud a different "energy justice scenario"—a hypothetical anecdote on "energy justice" that EQUAL staffers had written and printed in a pamphlet handed out to all attendees. These scenarios illustrated the structural violence of energy production that poor people of color face every day. In inviting four attendees to read the scenarios aloud, EQUAL staffers attempted to foreground the voices of local people of color. Oddly, though, EQUAL didn't ask these four people to speak about their *own* experiences with energy. Instead, the energy justice scenarios operated as a literal script; the four were tasked with reading others' words *verbatim*—orchestrating the meeting like a play, not a participatory forum.

Tellingly, I was one of the four people that the policy coordinator enlisted, even though I didn't live in or come from EQUAL's community. To be sure, I volunteered at EQUAL, and I paid membership dues. But if the organization sought to center community voices, I was, on the surface, an odd choice to speak. I knew, though, that EQUAL targeted members who they believed would follow instructions while maintaining the *optics* of a participatory organization led by local people of color. I—a person with Black skin and a respectable career—was, therefore, a sensible choice to speak. This speech curation, then, focused not on amplifying agential voices of color but instead on staging docile bodies of color. In other words, EQUAL embraced an image-oriented respectability politics centered on the visibility of ostensibly upright Black and brown people.

This conclusion might seem presumptuous; one scripted exercise doesn't preclude a more open exchange of ideas. But as the meeting commenced, it became evident that this exercise was part of a greater "illusion of inclusion"—an *appearance* of certain bodies in the room.[73] Specifically, EQUAL's policy coordinator opened by offering some general context on energy democracy before asking Cliff, a Black community member, to read the first energy justice scenario. Cliff was one of the few Black people who regularly participated in SOL meetings—but he usually sat quietly in the back of the room, as EQUAL staff frequently ignored his intermittent comments, which they perceived to be off-topic. Thus, Cliff's voice didn't seem to matter when it came to crafting EQUAL's energy agenda, yet the optics of his Black body in a space that had grown whiter was an asset worth curating. Or at least this was my impression when the policy coordinator called on Cliff to stand in the front of the room for everyone to see. He then proudly but misleadingly characterized Cliff as a "leader" in EQUAL's energy democracy work, maintaining the organization's image of a Black grassroots base. With this image staged, the policy coordinator passed the mic to Cliff so he could read the energy justice scenario.

Cliff, though, failed to follow the script. Instead of reading the pamphlet verbatim, he paraphrased the scenarios, commenting extemporaneously on the intersectional injustices of green-collar job industries. But before he could get too far, EQUAL's lead organizer interjected. "Cliff, buddy, I think you've jumped the gun; scenario 2 is about jobs; you're supposed to read scenario 1." The spectacle of community participation grew painfully apparent at that moment; the organizer's interjection brought attention to the fact that the whole exercise was curated—*that community members were not even free to comment independently on the topic at hand.* Cliff's voice didn't matter; what mattered was the positioning of his Black body at the front of the room—an appearance of empowerment that the scenarios were supposed to soundtrack.

This simulacrum grew more troubling when Cliff, two other "respectable" Black/brown members, and I shared the energy justice scenarios we were tasked to read. These scenarios addressed many of the bodily challenges that poor people of color face every day: emergency hospitalizations during heat waves, hunger from minimizing grocery purchases to pay energy bills, inhalation of toxic air from power plants. In theory,

EQUAL's community members could attest to this physical precarity by reflecting on their own experiences. If EQUAL were truly about empowering people of color to redress their own bodily challenges, there would be no need to curate a script and a participatory appearance. After we read aloud the four scenarios, Selena led the broader group in a brainstorming exercise to consider how the community should respond to them. This exercise was supposed to determine EQUAL's energy democracy agenda by actively soliciting community members' ideas—or, at least, this is what EQUAL proclaimed. In reality, though, the organization's vice president had determined this agenda *before* the meeting; the community members' ideas would not shape EQUAL's future work. In internal staff meetings after this community meeting, EQUAL staffers did not reflect at all on community input to refine their agenda; they continued to pursue the plan they had mapped out beforehand. The true purpose of this exercise, then, was to create the impression that EQUAL's agenda was community driven—to further curate a grassroots appearance.

Importantly, EQUAL linked this grassroots simulation with a show of technical rigor, choreographing an equicratic activism at the NPIC's behest. For EQUAL was starting the final year of their large three-year grant for energy democracy, and they had all but abandoned Samir's microgrid vision, which, as stated, was never more than a pipe dream. As such, they needed to demonstrate to their funder that they were making progress on a similar but more practical energy policy agenda. Toward this end, EQUAL's leadership developed an energy efficiency policy initiative, and given their grant's impending end, they did not have time to ground this initiative in a community participation process; they had to hit the ground running with their agenda irrespective of community input. But their ability to move legislation still hinged on EQUAL's participatory brand—the *image* of community. This organizational imperative inspired EQUAL's effort to simulate a community forum vis-à-vis the energy justice scenarios script—a dialogical appearance that concealed that their agenda was developed behind closed doors.

That said, I don't believe that EQUAL's staff were being duplicitous. They truly enjoyed talking to local people about their needs and aspirations; they simultaneously wanted to hear the community's input and had no intention of seriously using this input. I believe that this "disconnect," as Gina put it, did not result from a conscious decision. Certainly EQUAL's staff didn't hatch a plan to dupe community members into thinking they were contributing to the organization's energy agenda. Instead,

EQUAL's participatory spirit stood side by side with their closed-door strategizing, as staff betrayed no outward consideration of the contradiction between the two.

I want to suggest that they attempted to reconcile this contradiction by manifesting it as an equicratic image, juxtaposing their participatory performance with a comparably performative display of technical rigor. Specifically, when Selena concluded the "energy justice scenario" show, she positioned herself at a podium with a laptop connected to a giant projector screen and pivoted to a PowerPoint on SOL. The PowerPoint infused the space with a managerial affect antithetical to EQUAL's egalitarian rhetoric; as discussed in chapter 1, a PowerPoint amplifies its narrator's voice at the expense of others. Yet Selena framed this presentation as simply a summary of an allegedly grassroots campaign—an extension of people power.

Unsurprisingly, though, the PowerPoint's content belied the reality on the ground in EQUAL's community. For starters, the first slide characterized SOL as a campaign for "front-line communities"—code for Black and brown bodies—ignoring the fact that white gentrifiers comprised a disproportionate share of SOL's beneficiaries. But the alternate reality truly emerged on Selena's second slide, which had numbers and infographics that read "Progress and Next Steps." "Our goal was to get 150 kilowatts of solar PV [photovoltaic] on rooftops," she announced, centering the kilowatt metric as always. "We have far exceeded that goal. We are at over 300 kilowatts." She then gushed over the slide's various metrics: "303 kilowatts installed," "885+ residents benefit from affordable solar power," "6 tons of greenhouse gases offset annually," and "100+ workers trained in Photovoltaic [solar] installation." These numbers would be impressive, *if they were true*. But they were, at best, a manipulation of the truth. In fact, *zero* kilowatts of solar had been installed at that point—not 303. The organization had simply garnered commitments from buildings that would bring SOL's total number of kilowatts installed to 303 *if* these commitments were brought to fruition. But it was not clear how many of those commitments would culminate in an installation, and while SOL eventually surpassed 150 kilowatts, it never broke 300. Because zero kilowatts had been installed at that point, "885+ residents" had *not* "benefit[ed] from affordable solar power," and "6 tons of greenhouse gases" had *not* been offset annually. Furthermore, the claim of "100+ workers trained in Photovoltaic installation" was also disingenuous. In fact, EQUAL had trained over a hundred workers for a construction safety certification, but this training had no solar-related content whatsoever. While EQUAL was planning an advanced solar training for fifteen local residents, this training would not cover installation, it would not train over a hundred workers, and it had not yet occurred. In sum, EQUAL's metrics of progress were almost entirely fabrications based on creative rationalizations of the work they had done and planned to do. Yet Selena harped on the slide, enthusiastically reiterating its metrics as though a polished visual could make the aspirational math real. She lauded the metrics for making SOL "more tangible" for people in the room, as though we could touch nonexistent solar panels by gazing at a slide. Paradigmatically performative, Selena's enunciation reified its object of discussion, as she appeared to truly believe what she was saying—as though she were occupying the world that the slide simulated.

A selfless young mother from a working-class Latino community resembling EQUAL's, Selena had committed her life to addressing the everyday challenges she experienced growing up—pollution, policing, food insecurity—eventually earning a professional master's degree from an elite university before landing the policy coordinator position at EQUAL. She took her job very seriously, often articulating her sense of indebtedness to her elders. Meticulous, straightforward, and unsentimental, Selena was a hard worker who appeared disinterested in shortcuts. I therefore hypothesized that the ease with which she enunciated the PowerPoint's half-truths bespoke not conscious dishonesty but instead her habituation to a mode of philanthrocapitalist labor in which one effortlessly renders their work in terms of trumped-up PowerPoint metrics—the ingrained practice of hyperbolizing for a philanthropic benefactor. This hyperbole felt especially pertinent to a renewable energy campaign already plagued by a lack of moral precision—a muddled equicratic ideology affectively animated by solar's modularity, quantifiability, and diffuseness.

When she moved to the next slide, she confirmed my hypothesis, revealing the extent to which she was earnestly habituated to the NPIC's misleading accounting practices. This slide revealed the terms of EQUAL's grant agreement for their energy democracy work—terms that aligned with the previous slide's inflated metrics. Taken together, these two slides suggested that EQUAL was abiding by their funder compact. Even so, broadcasting this grant info was bizarre because it revealed the philanthrocapitalist impetus for EQUAL's distortional optics. While nonprofits often acknowledge their funders in PowerPoints, these acknowledgments almost never include the specific terms of a grant agreement. After all, in a grassroots campaign, the community's desires—not the funder's—are supposed to be front and center. In publicizing the grant terms, Selena transparently—and unusually—suggested that EQUAL was primarily accountable to their funder, not their constituents. The organization had nothing to gain from this display of fealty to a benefactor who wasn't in the room; the slide was neither sycophancy nor simply a symptom of financial dependence. Instead, the slide suggested that Selena had taken on the funder's gaze, as though she apprehended the world through the NPIC's numericized eyes. In this world, scale and quantity were ultimately what mattered—"more tangible" than the bodily realities that one could adjudicate without metrics.

While I don't know what Selena was thinking when she presented this simulacrum of solar impact, it was quite evident that she was not willfully distorting the truth. Instead, Selena occupied an alternate reality—a philanthrocapitalist dreamscape that solar's quantifiability made "tangible." As she stood at a podium before dozens of community members and rendered misleading metrics as the factual offspring of an imagined bodies-in-the-streets activism, her own body and the screenwork she gestured to fused into an image in its own right: the grassroots equicrat. In other words, this image encompassed not just Selena's slides but also her in-person presentation of them to people who appeared in its frame as Black and brown bodies—human avatars of the grassroots—passively viewing their "empowerment" via solar on a screen.

THE NPIC'S ENUMERATED HABITUS

The equicratic image that Selena publicly displayed at a podium was congruent with the person she performed when the spotlight wasn't on her in the office. She enjoyed screenwork—its feeling of productivity and "concreteness"—as she compiled spreadsheets that tracked campaign metrics, organized folders with EERE policy research, and mapped local clusters of affordable housing while ensconced at a desk adorned with photos of her and Gina in their yellow EQUAL shirts in the streets during a policy rally outside the state capitol. That desk, her screen, and those metrics were, in her words, a "tangible" orientation to the world—a way of concretely progressing in the face of persistent environmental injustices and the inertia of public meetings where dialogue often meandered aimlessly. Yet for all her screenwork-philia, she always affirmed the necessity of bodies-in-the-streets activism, returning to the monotonous chorus of "community" as a retort to corporate oligarchy, while taking the time to truly listen to community members when they'd shoot the shit at EQUAL's office. Every day she carved out moments for this reparative maternalism, pausing her laser-like focus on her screen—a balance between streets and spreadsheets that she cultivated in grad school long before working at EQUAL.

Yet when she took the reins of EQUAL's energy democracy work, this equicratic comportment morphed into the image I saw at the all-member meeting—a public spectacle of grassroots expertise in which her heartfelt commitment to EJ seemed like philanthrocapitalist theater. A PowerPoint with funder metrics became the centerpiece of her activism not due to a conscious ideological transformation—this was not a matter of *belief.* Instead, Selena's enactment of the equicratic image is evidence that power is not beholden to ideology—that sometimes we apprehend the world through a lens that doesn't perfectly align with our fervent political convictions.[74] This disconnect is often the result of neither overt political force nor purposefully veiled coercion, as power is affective, moving us in ways that exceed the logics of domination.[75] When the quantifiability of solar works in tandem with late liberal screenwork, it generates such affective power, operating through philanthrocapitalism to orient Selena toward an activist politics not fully aligned with her anticapitalist consciousness. In other words, the limited apprehension of solar vis-à-vis screenwork in the context of the NPIC creates a philanthrocapitalist "infrastructure of feeling" that inspires the equicratic image she enacts.[76]

I experienced this infrastructure firsthand when I accompanied Selena to the belly of the NPIC beast: the headquarters of a private foundation that was hosting a convening of their energy democracy grant recipients, including EQUAL. As we rode a gold-plated elevator to a giant corporate workspace on the twenty-fourth floor, I felt a shallow yet substantial sense of importance there before the doors even opened. This space was not simply imbued with wealth. It also had an air of sober professionalism—extravagant but with a purpose. The oak receptionist desk, smooth navy carpeting, floor-to-ceiling windows, and sleek ergonomic chairs suggested that this was a place where you stiffly shake your colleagues' hands, furrow your brow in contemplation, and scribble important thoughts in small notepads. We were only a quick subway ride from EQUAL's community, but the space—which overlooked a wealthy, white section of Manhattan—couldn't feel any further from the bodily precarity of EQUAL's community members. Although the funder had invited their grant recipients to discuss energy poverty, fossil fuel pollution, crushing electricity bills, unemployment, and investor-owned utilities, these things felt like abstractions in that space—affectively inaccessible, something to simulate for our host.

But this doesn't mean that the grant recipients were simply putting on a show for their funder. To the contrary, Selena and her counterparts from EJ organizations across the country were clearly at home in such a professional space. For example, as part of an icebreaker exercise, they were instructed to talk with someone in the room whom they didn't know for ninety seconds about their work, prompting the professional rite of passage known as the *elevator pitch*. As these dozens of nonprofit professionals collapsed the complexity of EJ organizing into poetic strings of social justice sound bites, the adeptness of their pitches was consonant with our corporate surroundings. Turning their affective settings to schmooze mode, they aptly inhabited the room with not only their words—*resiliency, empower, ownership, equity*—but also their measured cadence, contained gesticulations, ready-to-hand business cards. Their air of professionalism had become second nature; they were not simply performing for their common benefactor. For the NPIC's corporate confines invite a particular comportment that is not just a means of securing funding—a performance of professionalism that is an end in itself, untethered from budgetary objectives.

Not coincidentally, they had cultivated this professional affect in the course of becoming energy democracy nonprofiteers; when their organizations pivoted to EERE, they learned to engage closely with the technical particularities of commodified electricity (such as locational marginal pricing). While their previous work also addressed legal-technical matters, energy justice activism demands a more dogmatic focus on screen-work and thus greater distance from on-the-ground organizing—a graphocentrism attributable to electricity's enmeshment in phantasmagrams. This affective quantifiability inspired activists like Selena to adopt the philanthrocapitalist approach to measurable impact. Their newfound technical competencies, then, shifted their professional identities, enfolding their activist bearings into an equicratic appearance. Put differently, solar further professionalized their activism, as evinced by the ease with which they navigated their common benefactor's space.

This solar-powered habituation was most palpable when the foundation summoned everyone to present their work visually. In advance of the convening, the foundation had asked attendees to prepare poster presentations on their respective energy democracy campaigns. As a prominent academic medium and a precursor to the PowerPoint, poster presentations use visuals to facilitate a one-way flow of information that the presenter controls, rendering their data in a cogent narrative and positioning them as a polished expert. In the philanthrocapitalist context, this technocratic format can work toward late liberal ends, framing differential value as a problem that financial investments in social justice can solve. Put differently, the poster endows nonprofit interventions with a managerial affect that suggests that something as multifaceted as structural racism can be administratively ameliorated—its visual conventions eschewing complexity, establishing a trajectory with clear next steps. So when the foundation led us to a wall covered with poster presentations, whatever hints of activist wokeness infused the space during the elevator pitch exercise succumbed to a late liberal managerialism.

Aestheticizing grassroots organizing for the technocratic gaze, the posters used statistics, infographics, and photos of Black and brown bodies in bright organizational T-shirts (resembling the image in figure 3.1) to craft a neat visual narrative connecting the problem of environmental injustices with the "solution" of solar and other EERE infrastructures. The polish of these visuals foreclosed an earnest dialogue on the challenges of energy democracy work, as every organization looked successful and competent. Selena's poster, for instance, offered no access to the jarring noises, putrid smells, and aching joints that Gina and other NYCHA residents experience every day, reducing their bodily precarity to asthma incidence and energy costs. Selena then presented the same disingenuous metrics that she had presented at EQUAL's all-member meeting, talking confidently about the hundreds of solar kilowatts that had allegedly been installed. But these metrics didn't feel discordant with the space (as they did in the NYCHA community room), radiating the sense of purpose and sobriety that philanthrocapitalism feasts on. While Selena's display of phantasmagrams was an appeal to EQUAL's benefactor, she appeared no less moved by the metrics than her hosts, emotionally embedded in the screenwork of late liberalism.

I better understood her affective alignment with this visual mode when we finished our quasi-science fair and the funder summoned one of their consultants to the front of the room. The funder had hubristically tasked this consultant with developing a "model" for an "equitable energy transition"—a meta-calculus for social change that the grant recipients could theoretically apply to their various communities despite their many differences. Standing confidently, the consultant predictably presented their model through a poster presentation comprising visual icons. Each icon corresponded to a different component of an energy democracy campaign: "participation in decision making," "program and policy design," and, of course, "metrics," among many other things. These icons visualized energy democracy as a philanthrocapitalist formula, a set of variables to input irrespective of local context, paradoxically using metonymic abstraction to concretize infrastructural transition. The icons also bore a frighteningly close resemblance to those which Selena used on her poster—their similarities born from a datafied vision of equity in which a technocratic register can make infrastructure accessible to the grassroots. This mode of spectacle was therefore as easy on the eyes as it was inattentive to the bodily precarities of people like Gina—which energy democracy is supposed to address.

In this way, the consultant's iconographic model could be seen as evidence of a philanthrocapitalist apparatus that coercively reterritorializes communities as local test sites of a broad neoliberal project to organize the world in the image of a rational marketplace.[77] But such a macroanalysis overlooks the ways in which equicratic visuals affect not only their viewers but also their progenitors. For as we discussed the energy transition as if it was a conceptual schematic untethered to spatial particulars, the model's icons anchored a shared rationality among the funders and fundees, dissipating the philanthrocapitalist hierarchy separating the two. The visual that these icons composed held the space like an oracle, radiating an equitable future in tension with the gilded space we occupied and thus demanding the deference of everyone in the room. I'm sure some were skeptical or even dismissive of this model, but I'm sure, too, that no one there would deny the feeling of all-knowingness that it exuded. Of course, this mode of spectacle is not particular to energy; one could develop comparable visuals for health care or racism, for instance. But both the quantifiability of renewable energy and its affective distance from people's everyday experiences make this spectacle all the more central to energy transition work, demonstrating the particularly potent intra-actions of renewable energy infrastructures and late liberal screenwork in the philanthrocapitalist sphere.[78] Out of these intra-actions, anticapitalists become habituated to the economizing dispositions of neoliberal racial capitalism, as metric-driven iconography affectively animates a seemingly populist activism.

Even when EQUAL wasn't curating spreadsheets, PowerPoints, and posters filled with phantasmagrams, its staff seemed delusionally preoccupied with SOL's campaign metrics. As SOL progressed, the statistics Selena
projected were not simply optics for their benefactors or a quantitative
baseline for concrete climate goals but also, increasingly, ends in themselves, irrespective of their real-world referents. After years of staying
confined to screenwork, the metrics jumped off the screen, miraculously
becoming a part of the bodies-in-the-streets practice that it had traditionally complemented.

This spreadsheets-in-the-streets approach was particularly evident on
an afternoon in which I left the EQUAL office with Dominic, a SOL campaign leader, and some community members to recruit low-income co-op
tenant-owners for the SOL purchasing pool. Of course, as stated, recruitment didn't really entail engaging with people in the flesh; we simply
taped SOL flyers on the doors of co-op lobbies. While this lack of communal engagement was out of step with EQUAL's public image, I noted other
troubling discordances as well. We drove around the community in an
extravagant SUV littered with cheap social justice–themed T-shirts and
tote bags made in sweatshops abroad, we ate food from Styrofoam con-
tainers, we blasted the air-conditioning on a rather temperate day, and
we cavalierly printed hundreds of flyers, most of which we wouldn't use.
In these and other fossil-fueled actions, I detected a practiced indifference to the EJ principles that SOL was supposed to uphold. This indifference testified to the power of those metrics that EQUAL's screenwork had
ingrained in the campaign, as I would soon see. For as we drove around
in the SUV, Dominic focused not on mobilizing the community to address
climate injustices but instead on the enumerated kilowatt potential of the
co-op buildings we were targeting, verbally reiterating the 150-kilowatt
goal. He'd gesture to a building and say, "That one could get us 50 kilowatts" or "That one's only like 20 kilowatts," as though calculating electrical power trumped canvassing for intersectional power.

The significance of this calculus became evident when a co-op board
member stopped us in one of the lobbies to inquire about the flyers we
were posting by her mailbox. She was pleased to learn that we were
promoting solar. Dominic asked about the building's electricity metering system to determine if the building would be a good fit for SOL. She
explained that her co-op had an unusual electricity billing classification

under which they paid a very cheap rate for common-area electricity (as opposed to individual apartment electricity). This rate was so low that the building could possibly end up paying a little more for electricity if they solarized (through a PPA); it didn't actually make financial sense for the building to go solar (which is exceedingly unusual). On learning this, Dominic stated the price per kilowatt for solar and the building's electricity rate several times, harping on the *minor* gap between these two numbers. But the co-op board member was still interested despite this gap. She thought the building should seriously consider going solar because the additional electricity costs would be very minimal—a small sacrifice for climate resiliency, breathable air, green jobs, and a sustainable community. Solar's potential impact on her community's well-being and livelihood mattered more than its minimal projected impact on her pocketbook.

Yet, to my horror, Dominic insisted that she shouldn't consider going solar because the whole purpose of doing so was financial savings. "If you're not making dollars, you're not making sense," he said. He reiterated the price per kilowatt of solar and her building's electricity rate several times to reinforce this claim, myopically centering his argument on a contrast in numbers. This boggled my mind. We were *actively discouraging* a Black board member of an LMI co-op from going solar! It's one thing to forthrightly disclose solar's downsides; it's another to actively discourage adopting solar. We were there to catalyze a just transition—not provide low-level property management consulting!

Furthermore, Dominic seemed less concerned with the co-op's financial well-being—he swatted aside the board member's assertion that they could afford a *tiny* hike in their electricity rate—and more entranced by the accumulative rationale of racial capitalism. In his words, solar was about "dollars and sense"—and this premium on sense vis-à-vis market value suggests not basic financial prudence but instead *homo economicus* logic. Because he primarily apprehended solar through the market's metrics, the co-op's bottom line eclipsed the bodily value of clean air and climate resilience.

This counterintuitive inversion of value was directly attributable to EQUAL's screenwork, as the organization's dogmatic focus on their smart board and spreadsheet had instilled in Dominic a sense of the almighty metric. Indeed, before Dominic was tasked with leading SOL's latent ground game, he had not worked on energy issues and had instead focused on facilitating community meetings—never fetishizing commercial metrics. But EQUAL's shift from the streets to spreadsheets vis-à-vis solar changed that. I wondered, then: When an invisible yet ubiquitous force like electricity requires datafied screenwork to exist in our consciousness, is its solar form necessarily incompatible with a consciousness of life beyond the screen? I considered Selena, an Ivy-educated policy wonk who aspired to essentially do screenwork for justice. As Dominic heeded her call to focus doggedly on smart board metrics, her equicratic aspirations felt futile—a perhaps inevitable realization given solar's muddled ideological affordances.

SOCIALIST NEOLIBERALISM

The image in figure 3.5 was a welcome rejoinder to my cynicism after flyering with Dominic. Paradigmatically equicratic, this visual reminded me that solar's datafied screenwork is not necessarily irreconcilable with an equity agenda. Specifically, this image is a PowerPoint slide proposing an approach to incorporating the "value" of environmental justice impacts into VDER, the state's algorithm for calculating the price of distributed electricity, described in chapter 2. Developed by a liberal economist in partnership with a coalition of energy democracy activists, the proposed

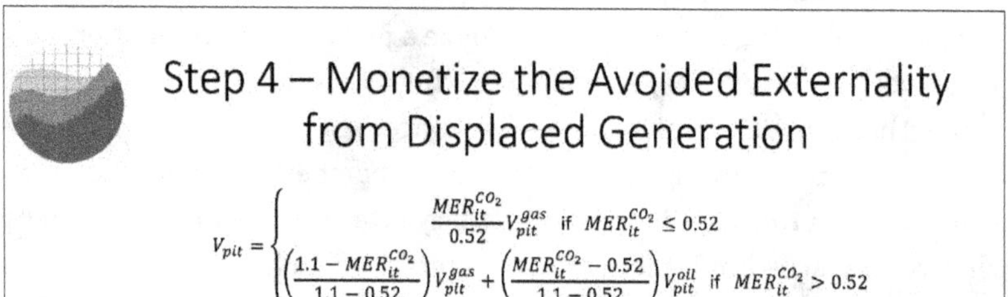

Step 4 – Monetize the Avoided Externality from Displaced Generation

$$V_{pit} = \begin{cases} \dfrac{MER_{it}^{CO_2}}{0.52} V_{pit}^{gas} & \text{if } MER_{it}^{CO_2} \leq 0.52 \\[2ex] \left(\dfrac{1.1 - MER_{it}^{CO_2}}{1.1 - 0.52}\right) V_{pit}^{gas} + \left(\dfrac{MER_{it}^{CO_2} - 0.52}{1.1 - 0.52}\right) V_{pit}^{oil} & \text{if } MER_{it}^{CO_2} > 0.52 \end{cases}$$

- If $MER_{it}^{CO_2}$=0, then V_{pit}=0

- If $MER_{it}^{CO_2}$=0.52, then V_{pit}=V_{pit}^{gas}

- If $MER_{it}^{CO_2}$=0.75, then V_{pit}= 0.6 V_{pit}^{gas} + 0.4 V_{pit}^{oil}

- If $MER_{it}^{CO_2}$=1.1, then V_{pit}=V_{pit}^{oil}

FIGURE 3.5. A PowerPoint slide proposing an algorithm to monetize the environmental justice impacts of solar. Institute for Policy Integrity, "Environmental Value of Distributed Energy Resources," 12.

"EJ value" aimed to address the coalition's concern that VDER was prohibiting low-income communities of color from going solar by precipitating extreme price volatility in the community solar market. Community solar "is a distributed solar energy deployment model" that allows laypeople, small businesses, and community groups to buy or lease part of a large, shared solar electricity system that has not been installed on their individual buildings.[79] As such, community solar can make solar accessible to people who are not homeowners, lack disposable income, or occupy buildings with damaged rooftops—overwhelmingly residents in poorer communities. When these people subscribe to a community solar project, they receive a monthly credit for electricity generated by their project's share, offsetting part of their electricity costs. But VDER caused the price of these credits to fluctuate in ways that were impossible to predict, meaning that community solar developers, investors, and subscribers could not estimate the value of community solar projects with any level of certainty. Thus, VDER injected financial risk into the community solar market, causing developers and investors to shelve planned projects, including many in LMI communities of color. The image in figure 3.5 is part of a presentation that proposed to improve VDER by monetizing the noncommercial environmental value of community solar in areas burdened by environmental injustices. This would raise the price of the credit that LMI communities of color could receive for injecting solar electricity into the grid, aligning the state's energy valuation policy with equity goals.

It countered my sense that solar's enmeshment in quantitative screenwork undermined the ideological commitments of energy democracy organizing, for the activists who demanded that the state develop this equicratic valuation scheme shared Janice's and Samir's trenchant critiques of neoliberalism. But they knew, too, that the state-run community solar marketplace was a promising avenue for empowering marginalized communities to adopt solar under the oppressive conditions of racial capitalism, inuring them to a neoliberal process that they believed could paradoxically distribute financial resources equitably: the monetization of both "the environment" and "justice." While they championed an economist's market-based screenwork, they had not grown so hypnotized by the solar market's phantasmagrams that they lost track of their justice-oriented objectives or forgot that racial capitalism was foundational to environmental injustices. As such, they advocated for a sort of neoliberal socialism: a market-based mechanism for governing energy through consumer prices that nonetheless redistributes wealth through a hands-on public sector.

Even so, I found that when solar's quantifiability works in tandem with screenwork, the equicratic interventions it produces—for example, the "EJ value" algorithm—are so alienated from people's lived experiences with environmental injustices that they reproduce the structures of racial capitalism they are trying to supplant. For the slide in figure 3.5 does not even approximate the EJ challenges that EQUAL community members encounter regularly: trash-filled vacant lots; buildings perpetually encaged in piss-scented scaffolding; treeless streets that are at times thirty degrees hotter than streets only a couple of miles away in wealthier, white neighborhoods in the summer months; a lack of fresh food; and of course the myriad aforementioned problems with NYCHA housing.[80] Instead of addressing these more tangible injustices—the slow violence that Gina endures every day—the slide indirectly renders EJ through a complex formula that even elite policymakers can't understand. Embedded in that formula is an opaque representation of the environmental problems that are caused by fossil fuels and that plague poor communities of color—an abstraction inaccessible to the very people it is supposed to support.

This abstraction, of course, was a proposed amendment to an even more sophisticated formula; VDER was so complex that many of the energy experts who worked with the state to develop it confessed to me that they didn't fully understand it. In contradistinction to SOL's fairly straightforward economies-of-scale neoliberalism, VDER is the outgrowth of a far opaquer form of neoliberal governance: deregulated electricity markets. Neoliberal policies created these markets in the 1990s to enable energy service startups to compete with utilities, turning an already sophisticated pricing regime (for aligning electricity supply and demand) into a labyrinthine web of competing actors who nonetheless had to be delicately coordinated to ensure grid equilibrium.[81] Distributed renewable energy only added more strands into this web, necessitating more market coordination, which the VDER pricing algorithm was supposed to address. But coordinating such a multilayered web is not simple, to put it mildly. Most laypersons cannot seriously comprehend VDER, such that the equicratic effort to improve it by altering the algorithm was not affective in ways that more tangible EJ interventions—such as the decommissioning of a power plant—are. So while VDER prevented communities plagued by environmental injustices from adopting solar, it was immune to the populist outrage of effective EJ campaigns. Although the energy democracy coalition organized protests at the state capitol, they could never get enough bodies in the street to pressure the state to amend VDER because the matter just wasn't something that most laypersons would ever care enough about to protest. And without bodies in the streets, a thoughtful equicratic PowerPoint would never have the political leverage necessary to actually change policy.

Unsurprisingly, then, the state dismissed the EJ value proposal almost immediately after it was presented in thoughtful slides to the New York Public Service Commission; the equicratic aesthetic could only go so far as a political tool without mass mobilization. Energy politics vis-à-vis late liberal visuals that simulate governance by algorithm (or other technocratic forms) too often fail to counter the status quo. For this is fundamentally a politics of elite statecraft that is uninterested in social change. While brilliant minds can curate a PowerPoint with truly innovative equicratic interventions, tweaking the neoliberal state's screenwork cannot, by itself, achieve equitable ends.

To be sure, the energy democracy coalition did not put all their faith in an economist's PowerPoint. They also worked closely with solar companies to push state legislators to propose a bill that both rolled back VDER and prioritized EJ communities in the valuation of distributed electricity. But the bill didn't pass. Much like the EJ value proposal, the bill was centered on a technical matter that was too alienated from people's everyday realities to muster anything approximating grassroots action or widespread outrage. The bill, then, was a thoughtful equicratic intervention that failed because it was constrained by the hegemonic paradigm it was operating in: a world of quantitative abstraction untethered to people's bodily encounters with environmental injustice.

This distance presents an interesting twist on the facile image of the equicrat that I discussed previously. As I have shown, this image is part of a broader regime of "community" representation more concerned with the appearance of Black and brown bodies than the lived experiences of people who are interpellated as those bodies. As such, this image demonstrates the "cunning of recognition" at the core of late liberalism: the prioritization of a "recognition justice" framework that aestheticizes differential value vis-à-vis identity politics over a "distribution justice" framework that meaningfully redistributes the wealth and environmental burdens of racial capitalism.[82] Yet as the glossy "EJ value" slide suggests, equicratic approaches to redistribution can fail precisely out of an *absence* of recognition: a failure to acknowledge the everyday forms of marginalization experienced by the people whom the EJ value is supposed to help. Put differently, an equicratic intervention that attempted to redistribute wealth by valuing solar electricity generated by LMI communities was largely ineffective because it trafficked in inaccessible technical terms in which marginalized people couldn't begin to see themselves. It failed not because energy democracy activists don't know how to speak to the LMI communities they organize but rather because the quantifiability of electricity (and solar electricity in particular) generates complex screenwork that appears divorced from those communities' immediate needs.

While solar's modularity and diffuse spatiality inspire aspirations for an energy democracy, this quantifiability requires an enterprising technocracy, underscoring the muddled ideological affordances of renewable energy technology. Much like gas and oil infrastructure, such technology produces not only material power but also political horizons, animating just-transition activism that will never neatly hew to an anticapitalist politics.[83]

This does not mean, though, that solar is inevitably predisposed to the facile forms of digital representation that EQUAL practiced. An energy democracy campaign need not prioritize the means of reduction over recognition justice. When a spreadsheet, smart board, social media post, PowerPoint, and poster presentation aren't the primary ways we engage with solar, the just transition becomes less about optimizing metrics and more about recognizing matter, redressing the corporeal terrain that comprises environmental injustice. In chapter 4 I lay out such an approach, calling on those of us opposed to the extractive energy paradigm of racial capitalism to expand our apprehension of solar beyond late liberal screenwork.

4

While solar requires abstract metrics, it also requires something more tangible: physical labor. Installing solar panels on rooftops, connecting solar to the grid, deploying battery storage technology, and manufacturing solar infrastructure necessitate numerous hands-on skills and human bodies capable of lifting heavy objects, scaling heights, and operating tools and machinery. This corporeality—solar's dependence on our bodily capacities—inspires ecosocialist imaginaries of "green-collar" worker power: political visions that counter both the economistic bluster of extractive industries and the elitist technofetishism of "cleantech" industries. As solar demands able-bodied blue-collar workers—a group disproportionately comprising men of color—their physical labor has taken on new political significance, reconfiguring the relationship between racial capitalism and environmentalism.

Figures 4.1–4.5 visualize this shift. Extracted from various online/social media platforms, these images render blue-collar workers as avatars of climate action and renewable energy, suggesting that solar infrastructure can amplify worker power. Several of these images, in particular, are focused on the proposed Green New Deal—the expansive policy framework for large-scale, equitable climate action in the United States—positioning public investments in solar as the centerpiece of a pro-worker politics. In presenting blue-collar bodies as icons of the Green New Deal and similar policies, these images aestheticize decarbonization as a means of redressing differential value.

In other words, physical labor appears as evidence that climate action can ameliorate the economic exploitation of racial capitalism. For these images and their affiliated text suggest not simply that blue-collar solar

'As World Teeters on Brink of Climate Catastrophe,' 600+ Groups Demand Congress Back Visionary Green New Deal

4.1

DSA for a Green New Deal 🌱 🔌 ⚡
@DSAecosocialism

A jobs guarantee could be our best bet for a "Medicare-for-All" climate policy—but it's crucial to fit proposals with a climate justice lens, creating a labor force for rapid decarbonization to elevate communities & workers that capitalism has left behind. inthesetimes.com/working/entry /...

4.2

FIGURES 4.1–4.5. Ecosocialists and other leftist policy advocates center images of blue-collar solar workers and their corporeal labor in appeals for the Green New Deal and other climate policies. Figure 4.1 from Organic Consumers Association, "'As World Teeters on Brink of Climate Catastrophe.'" Figure 4.2 from DSA for a Green New Deal, "A Jobs guarantee could be our best bet," Twitter, May 24, 2018, 5:26 p.m., https://twitter.com/DSA_Enviro /status/999763601538904064. Figure 4.3 from PolicyLink Facebook page, February 14, 2019, 10:37 a.m., https://www.facebook.com/share/p/14CBi- iGuYe. Figure 4.4 video still from BlueGreen Alliance, "Manufacturing to Build Back Better." Figure 4.5 from Beck, "Tribal Solar Projects Provide More Than Climate Solutions."

#GreenNewDeal
HTTPS://PLCYLK.ORG/TAKE-ACTION-GND

4.3

TELL CONGRESS TO BUILD BACK BETTER
bluegreenalliance.org
BLUEGREEN ALLIANCE

4.4

Tribal Solar Projects Provide More Than Climate Solutions

4.5

installers have good employment prospects but, more important, that they control their labor power, take pride in their work, and "elevate [the] communities . . . that capitalism has left behind," as "rooftop solar [allegedly] shifts power" to marginalized groups, including Black, Indigenous, and women workers.[1]

While these images are deployed in the United States by people who broadly subscribe to some version of ecosocialism, including left-leaning Democratic politicians, the youth-led Sunrise Movement, environmental justice (EJ) organizations, and the Democratic Socialists of America, we can situate these images in a long line of pro-worker propaganda that visualizes the strength and masculine virility of working-class bodies to support a leftist political agenda. Consider Bolshevik posters depicting revolutionary blacksmiths or Works Progress Administration–commissioned paintings of "laborers as heroic figures."[2] These images represented the male worker's body as a paragon of fortitude and strength, simulating proletariat empowerment to counter narratives of oligarchical omnipotence. Along these lines, the figure of the solar installer offers a visual rejoinder to extractive industries' propaganda, which similarly aestheticizes masculine blue-collar labor but in a way that pits "jobs" against "the environment."[3]

But in positioning decarbonization as an antidote to differential value, the ecosocialist iconography overlooks how solar can generate not only new jobs related to the environment but also new ways of relating to the environment through one's job. For solar projects require labor that affectively animates relations with the built environment that are novel in urban spaces where infrastructures are often construed more as a backdrop to life than beings in their own right. When energy engineers design rooftop solar systems in NYC, for instance, they might assess the shade cast by elevator bulkheads or learn the subtle bends of an otherwise flat surface or crawl on their knees to inspect electrical wiring under drizzly clouds, apprehending the skyline's contours in ways that their deskbound counterparts (and most New Yorkers) don't. Their jobs therefore entail a close attunement to the topography, environmental conditions, and structural composition of their surroundings, (re)orienting them from abstract screenwork to the physical particularities of the city's rooftops.

While the labor of developing most infrastructures requires such an attunement, I want to suggest that energy efficiency and renewable energy (EERE) work is unique because it invites workers to see, touch, sense, consider, and remake the built environment with a certain care that runs contrary to the extractive relations of racial capitalism. As discussed in the introduction, racial capitalism is a structure for configuring relations in ways that organize social difference around the pursuit of surplus value. Fossil energy infrastructure has long been central to such relational configurations, designating both Black and brown bodies and more-than-human ecologies as terrain to extract from and pollute for profit. Thus, energy infrastructures that are not primarily premised on drilling, pummeling, and combusting the earth have the potential to transform not simply humans' impacts on the more-than-human world but also a relational architecture that upholds the white supremacist anthropocentrism of racial capitalism—"the genre of the human."[4]

This chapter suggests that sustainable energy infrastructure can realize this radical potential by connecting an ideological concern for the environment with a *corporeal feel* for the spaces we inhabit. I locate this corporeality in the everyday labor of energy transitions, arguing that it aligns an intersectional, pro-worker politics with ecosystemic care. By theorizing a just transition that centers humans' physical relations with the more-than-human world, I offer a corrective to the late liberal screenwork explored in the previous chapters. This just transition counters the differential value of fossil energy by deemphasizing a narrowly visual apprehension of solar power in favor of the kinesthetic senses and sensibilities of EERE labor.

In mapping out this just transition, I suggest that the neoliberal technofetishism of ecomodernists and the pro-worker politics of ecosocialists are often two sides of the same coin, as they similarly overlook the *relationality* of physical work—the vital interdependencies that racial capitalism obfuscates. A just transition that foregrounds this relationality, then, can begin to extricate solar from differential value, imbuing it with a lively presence that is absent in the digital media, photos, spreadsheets, smart boards, and PowerPoints discussed throughout this book. In recognizing the corporeality of EERE labor, energy politics can become less about economic *returns* and more about ecological *return*: an ecocentric ethos absent in both ecomodernism and ecosocialism, as I show in this chapter.

Yet an overemphasis on *return* can neglect the fact that many marginalized workers have no place to return to—that dispossession is the precondition of racial capitalism.[5] As such, a pro-worker approach to energy transitions must tend to the complex ways in which marginalized workers navigate dispossession through their physical labor, leaving room for the contradictions of environmental conservation in a settler metropolis. The low-income workers of color who are powering NYC's solar revolution simultaneously challenge *and emulate* the capitalist class, demanding a nuanced classed politics of ecological return. To illuminate such a politics, I close the book by discussing two Black and brown worker-owned businesses that specialize in solar installation and solar-powered conservation, exploring how they balance the ecological and economic dimensions of energy transition. I situate this balancing act in W. E. B. Du Bois's conception of "double consciousness," which offers a framework for orienting oneself to structural power through multiple competing understandings of the world simultaneously. Specifically, double consciousness, as I conceptualize it, can enable solar advocates to return to what Larry Lohmann calls "little-e" energies—the forces of life and vital interdependencies that can't be stockpiled or produced—while focusing on the returns of what he calls "Big-E" Energy—the quantified forces that have powered racial capitalism since the transatlantic slave trade.[6] I contend that this little-e/Big-E double consciousness is essential to a just transition that confronts the paradox of rejecting an extractive energy paradigm that it is nonetheless inextricably enmeshed in. Through this closing discussion, I return to the screenwork of late liberalism, underscoring the imperative to broaden our apprehension of the streams of charged electrons that at once elude our sight and illuminate our visions.

Reader, be forewarned: the early part of this chapter harps on the banal activities and run-of-the-mill ruminations of EERE professionals with an ethnographic romanticism that might seem to elide the exploitative political economy of racial capitalism. I beg for your patience, though, as the chapter closes with a more holistic analysis.

FROM RETURN TO RETURNS

SunLight's website desituates industrial energy from the physical world, reproducing differential value through a smart aesthetic that erases blue-collar labor. The website boasts that the company uses "data-driven energy intelligence" to "attack the root causes of wasted energy" to help their corporate clients "improve returns" on building portfolio investments. This language is run-of-the-mill corporate-speak, but it also evokes something not so banal: modern warfare. Specifically, the idea that "intelligence" can launch an "attack" in an effort to "improve returns" bears the rhetorical imprint of the military-industrial complex: an alignment of the so-called intelligence community with corporate interests. While this might sound like a simplistic reading of marketing colloquialisms, the cleantech industry actually embraces militaristic metaphors, valorizing smart warfare in both their public-facing and internal rhetoric. Specifically, smart grids, smart meters, "smart energy technologies," and, more generally, data and intelligence are the centerpiece of the so-called war on carbon, as firms like SunLight often characterize environmental problems as a battle to be won through cognitive ingenuity, demanding, for instance, a "Carbon War Room."[7]

This insidious militarism is rooted in a Cartesian schema that positions intelligence—the domain of the mind—as all-powerful, overlooking the corporeal labor through which energy systems are made. When cognition untethered to physical work can seemingly attack energy inefficiencies, the blue-collar bodies on the front lines of the battle are irrelevant. Energy appears here as a disembodied, nonhuman force that can be controlled with white-collar analytics—not corporeal labor. It's therefore not surprising that SunLight's website lacked images of its blue-collar solar installers. The website didn't even mention that SunLight *had* in-house solar installation crews; solar installation was couched as a "service" that the company could provide, as opposed to the skilled labor of blue-collar workers. The website did, though, highlight both the fruits of those workers' labor and the people who controlled those workers' livelihood, presenting polished executive headshots and images of solar panels on commercial buildings shining under the sun. These images, texts, and representational elisions, then, reproduce differential value, subordinating the working-class body to the Cartesian battle between executives' intelligence and "wasted energy"—mind's triumph over matter.

This focus on intelligence distances the war on carbon from humans' corporeal relations with the ecosystems it is supposed to protect. Specifically, many prominent advocates of "intelligent decarbonization" "reject" the "long-standing environmental ideal . . . that human societies must harmonize with nature to avoid economic and ecological collapse," advocating, instead, for a Cartesian "decoupling" of humans from nature grounded in technoscientific advancements.[8] These ecomodernists regard such a human/nature split as essential to both environmental conservation and human prosperity. They insist that by deploying renewable energy and other "innovations" that reduce humans' physical imprint on the nonhuman world, capitalist society and nature can flourish side by side as distinct entities.[9] While not all cleantech corporations like SunLight explicitly endorse the term *ecomodernism*, we can detect its ethos in their appeals to reduce our "environmental footprint" through new "smart" technologies. In this way, disembodied intelligence is foundational to a neoliberal imaginary in which tech titans save the world through a market-based struggle against carbon. Again, this valorization of intelligence reproduces a mode of differential value in which the *physical* labor of technological transitions is debased through erasure, as executive headshots on SunLight's website position white-collar innovators as the vanguard of decarbonization. This stands in contrast to ecosocialism, which envisions empowered workers sustainably remaking the physical world by controlling their own labor (figures 4.1–4.5).

After reviewing SunLight's website and attending my first meetings at the company, I expected it to be painfully ecomodernist. At these meetings, leadership emphasized "intelligence," "data," and their unending pursuit of doing things in a "smart" way, centering their mastery of information in their "fight" against climate change. In between meetings, I walked through their Data and Software Division's office to observe this information mastery in-person. The deskbound professionals who worked there appeared transfixed by their screenwork, ostensibly divorced from the *physical* contexts that their data referred to. Intelligence seemed to sit in the cognitive space between their eyes and their monitor—an illusion of disembodiment.

When I talked to two of these white-collar workers about SunLight's profitable solar installation work, they explained that they had never been to SunLight's solar project sites and had never met the company's solar installers—alienated from the physical spaces and corporeal labor of the energy transition that their employer championed. They further explained that SunLight's solar installers didn't have their own company email addresses. It seemed telling that such a basic marker of institutional belonging wasn't bestowed on the workers who installed the company's most profitable product: solar. Furthermore, these workers—even the most senior among them—got paid an hourly wage, not a salary, unlike their white-collar colleagues. SunLight appeared to value white-collar intelligence more than physical labor.

But as I got to know the company's office-based employees, it became evident that they were not simply technocrats who fetishized data in some phantasmatic war on carbon (in spite of SunLight's rhetoric). While their day-to-day screenwork more directly engaged their sense of sight than their other sensory modalities, many of these employees regarded their labor as full-bodied—linked to their kinesthetic faculties. In their estimation, energy transitions do not happen behind a screen; they occur when people can sense, feel, and encounter their surroundings outside the data-driven bounds of intelligence. To elucidate this corporeal politics, I now return to the insights of Sean—a SunLight vice president (VP) and the unofficial spokesman for the company's ideological contradictions. In the next passage, I center Sean's voice to amplify the sentiments that his white-collar colleagues also articulated: an ethos of environmental attunement at odds with the company's ecomodernist intelligence.

SEAN'S ONTOLOGY OF ENVIRONMENTAL WORK

Sean and I were chatting in SunLight's employee lounge, staring at the skyscrapers that surrounded us, at 8 p.m., after most of his colleagues had left for the day. Sean engaged in lighthearted banter on how he manages too many people, how his whole life seems like an administrative blur, and how he doesn't have time to do anything with his hands anymore. I asked when he last did something he enjoyed with his hands, and without flinching he replied, "I sharpened my knives last night. It was the best." When I asked why, his response offered a rejoinder to SunLight's intelligence:

Well, good, sharp knives are a pleasure to work with. But if you want the more philosophical answer, it's like, it's a pretty primal thing to do: sharpen your tool. I'm gonna take this piece of steel, I'm gonna rub it against a rock, and then it's gonna be sharp, and then I will chop things and put them near fire and cook. Using tools with fire: it's primal to us—an essential part of being a human. Which is really fucked up in a way. Like, look at "80 by 50" [which is NYC's goal of reducing 80 percent of the city's greenhouse gas emissions by 2050]: it requires that we reduce energy usage, and we know that in order to get there, we need to stop digging the ground up and burning it and releasing it into the air. We have to stop burning things. But burning things: that's what we do—that's a central part of the human experience. And we're saying that we need to stop doing this, whether it's for comfort or for cooking or for light. That's weird! So to save ourselves, we need to stop doing basically the most essential thing that we do. I look around all these [office

skyscrapers], and I just see fire. It's amazing. They all have chimneys
. . . this building [that we're in] puts it out the side; you'll see a big
black puff shoot out if you look outside enough. I look at all these
buildings, and I just see fire sources.

Sean's views on fire and knives challenge SunLight's ecomodernism
in several ways. First, he locates energy in humans' bodily drives and
embodied practices, centering "primal" human behavior—the *manual*
harnessing of fire—in ruminating on the city's energy transition. In the
process, he renders energy as not a disembodied metric that screenwork
controls but, instead, a relational force that emerges from our haptic
engagement with matter. In doing so, he problematizes the city's energy
transition plan, which SunLight supports, on the grounds that it severs
humans from their long-standing, allegedly instinctual relations with
energy. He is *not* suggesting that these relations make fossil fuel combustion unproblematic. Instead, he challenges the idea that a modern
energy infrastructure that powers corporate skyscrapers is somehow
more advanced—smarter—than the energy that we've historically cultivated manually. In pointing to shiny office towers emitting plumes of
smoke, Sean suggests that we are not as evolved as sustainable energy
discourse suggests—that we are still the same species that historically
generated energy through physical work. In this schema, the sustainability phantasmagrams of our most brilliant minds (i.e., "80 by 50") cannot
escape the physical realities of being a human.

As our conversation progressed, Sean continued to ruminate on the
ways industrialized modernity undermines the instinctual skills embedded in our corporeal relations with our environment:

When I have a car, I have maps in it. And that's how I get places.
Real maps, atlases that show me all the backroads of a place . . .
and I resent it when people insist on using GPS because . . . it takes
skills that I have, and it makes them not fun. . . . There's lots of
times when I'm driving . . . and it's very clear what's happening if
you look around, but when you're glued to your telephone, things
get very confusing. I mean, first, you don't have any context because
[the phone is] the size of your palm, so how could you have any context of where you are or where you're going? Second, [the GPS] will
actually be wrong, it'll say you're on a street that you're not on. . . .
There has to be a balance with the technological thing; I don't want
my car to drive by itself.

Here Sean contends that information technology disrupts our embodied relations with space and the skill that this relationality cultivates, offering a further contrast between his habituation and SunLight's technological intelligence. During another conversation Sean similarly critiqued SunLight's practice of valuing screenwork over physical labor, problematizing not only the white-collar/blue-collar divide in compensation but also the existential question of whose work matters (see chapter 2). For him, intelligence is nothing without physical labor, as hands-on skills make the world work in ways that the smartest screen-based abstractions never could.

But for all these differences, Sean identified with one facet of Sun-Light's brand: eliminating waste. "The elimination of waste" was his go-to refrain when he explained why he had committed himself to a company that privileged intelligence over the corporeal relationality he valued so dearly. Yet even his waste reduction ethos deviated from SunLight's. As discussed, the company's website stated that eliminating waste is about "improv[ing] returns"; an accumulative logic drives their efficiency pursuits. In contrast, Sean's stated emotional investment in eliminating waste centered on not *returns* but, rather, *return* in the singular. Specifically, Sean explained that his philosophy of waste is derived from Wendell Berry—a cultural critic and farmer who renounced high-tech modernity—and Berry's philosophy of waste is rooted in a distinction between what he calls "machine-derived energy" and "biological energy"—a distinction centered on the concept of *return*. As Berry explains:

> The moral order by which we use machine-derived energy is . . . simple. Whatever uses this sort of energy works simply as a conduit that carries it beyond use: the energy goes in as "fuel" and comes out as "waste." This principle sustains a highly simplified economy having . . . two functions: production and consumption. The moral order appropriate to the use of biological energy, on the other hand, requires . . . a third term: production, consumption, and *return*. It is the principle of return that complicates matters, for it requires responsibility, care, of a different and higher order than that required by production and consumption alone, and it calls for methods and economies of a different kind. In an energy economy appropriate to . . . biological energy, all bodies, plant and animal and human, are joined in a kind of energy community. They are

not divided from each other by greedy, "individualistic" efforts to produce and consume large quantities of energy, much less to store large quantities of it.[10]

For Berry—and, by extension, Sean—*return* is antithetical to the *returns* that SunLight celebrates; the latter refers to surplus that an individual hoards for their own profit through the differential value of technological innovations, while the former refers to the regeneration that occurs when bodies (human and more-than-human alike) work in tandem with one another in an "energy community" without modern technology's distancing effects. Sean, then, aims to minimize waste not through profit-making intelligence but through embodied *skill*—a concept at the heart of Berry's formulation of *return*. As Berry puts it, "Skill . . . is the enactment or the acknowledgment or the signature of responsibility to other lives; it is the practical understanding of value. Its opposite is not merely unskillfulness, but ignorance of sources, dependences, relationships."[11] The skill of orienting one's body to one's environs (sans GPS) or of cultivating fire manually presents a contrast to white-collar intelligence because it emphasizes corporeal interdependence. While Sean's executive tasks are centered on the extractive pursuit of returns, he rationalizes this "uninspiring" screenwork by situating it in an imaginary of return: the *skill* of SunLight engineers who feel the errant hum of an inefficient heating system before improving it, who learn a rooftop's idiosyncrasies with their limbs before designing a solar system for it, who kinesthetically relate to the parts of the landscape that New Yorkers seldom consider, enacting a "conservation ethic" that Sean claims to have felt "in his hands" when he fixed HVAC systems earlier in his career—a full-bodied care for the dependencies that modern life obfuscates. When he reflects on the work of the people who report to him, he nostalgically romanticizes their skilled attunement to the skyline—a labor of love that his upward career trajectory took from him but that nonetheless reminds him why he committed to sustainable energy to begin with.

Sean's focus on skill is incongruent with his professional habitus; a corporate VP who works blocks away from the epicenter of global capital is obviously not a yeoman like Berry. But this incongruence is especially notable because the elite stratum of labor he occupies is *constituted in opposition* to the sphere of return he theorizes. Social theorists have long called attention to the ways in which white-collar work is defined by an affective distance from corporeal labor (including that of agriculture, industrial production, care work, construction trades, and arti-sanry). This distance—the sense of removal from physical toil—shapes the habitus, experiences, and ideology of white-collar work, creating the epistemological preconditions for an intelligence divorced from corpo-reality.[12] In this way, smart salaried labor can define the white-collar worker not only according to what he is but also according to what he's not. As C. Wright Mills argued, the white-collar slot emerges through a negative contrast to the productive labor of craftsmanship, rendering the office employee in terms of a lack.[13] In obfuscating physical labor, Sun-Light's ecomodernist website evinces this differential value.

But Sean's emotional investment in skill and return suggests that Sun-Light's white-collar ontology does not exhaustively account for the labor politics of the cleantech intelligentsia. Instead, Sean cultivates an ori-entation to physical labor that is constitutively missing from his profes-sional habitus. This is not evidence of Sean's quirkiness. As this chapter will show, energy transitions affectively animate white-collar work that privileges return over returns, embracing that which such work has tra-ditionally been defined against. This tension positions ecosocialist pre-dispositions against ecomodernist intelligence, creating an existential crisis for the professional order through which energy produces differ-ential value.

OFFICE SOIL

I am sitting in the back of SunLight's boardroom, observing a meeting of the company's sales division. The VP of sales exclaims that they just broke their monthly revenue record. The city's solar market is booming, and SunLight is seeing healthy returns on its investment in several blue-collar solar installation crews. Indeed, the company's pivot to blue-collar work has made these returns possible, as low-paid solar installers gen-erate profits that the company has never seen from their office-based intelligence.

The sales team responds to this news with conspicuous satisfaction, maintaining their professional cool but failing to hide that they're ecstatic that their hard work is paying off. *Their* hard work: I wonder if they con-sider that SunLight's underpaid Black and brown solar installers made their returns possible. To be sure, the sales team certainly had a substan-tial role to play in their own success, working tirelessly at their desks deep into the evening, cajoling clients and hammering out project scopes, occasionally glancing out the window or running to the Sweetgreen next door for a salad but mostly focused on life *within* the office—on revenue goals and solar demand. Ten-hour days, a crumpled bag of chips or a half-empty coffee mug by their keyboard, the contours of their ass ingrained in their seat, an open Twitter tab offering momentary respite from the sea of emails—the office constituting not just their livelihood but also their life. How odd it is to hinge your survival on places you've never been, on people you've never met, on labor you've never done, on technologies you've never touched—and yet this is the life of a SunLight sales team member (a "member," as though we're talking about a civic association and not a for-profit, multinational corporation). Indeed, the sales team has no physical contact with the physical labor they sell for a living—labor opaquely coded in their spreadsheets and scoping documents. This is the world of returns, where profit motives circumscribe the spatiality of livelihood.

After the meeting I take the elevator up two floors to the part of Sun-Light's office where I'm temporarily stationed at a makeshift desk among the company's software and data analysts. While every human on this floor scans their screens and punches at their keyboards amid the eerie silence of office work, I spot some out-of-place coworkers: a small swarm of fruit flies hovering over green compost bins only feet from a cluster of desks. The SunLight analysts appear unbothered by the flies' harried trajectories and the rotting refuse they feast on in a space that other-wise has all the trappings of a sterile white-collar office. Organic matter decomposes freely without a profit motive, suggesting that nature, in its putatively pure form, is welcome in this space proudly organized around the free movement of capital. Compost, of course, replenishes soil, epitomizing the "metabolic interaction between man and the earth" that Karl Marx valorized—the *return* that Berry calls for.[14] But if, as Marx argued, "capitalist production . . . simultaneously undermin[es] the origi-nal sources of all wealth—*the soil and the worker*" (emphasis added), then how can a soily return occur in the same office where the sales team cel-ebrates their returns?[15] Can fruit flies and humans really be reciprocally entangled in white-collar spaces if capitalism erodes workers' interdepen-dencies with soil, rendering both as commodities to extract surplus value from? Does office composting entail return if office workers are alienated from the forms of life that they transform into capital—from blue-collar solar installers and the elements extracted for solar panels? Do returns overwhelm and undermine return?

I learn that one of the data analysts—a self-described "numbers geek" who spends her days at her screen—started SunLight's composting operations and that every Monday during lunch a group of employees—several wonkish data experts and two office administrators—voluntarily haul the compost to a compost collection service at a farmers' market a few blocks from SunLight's office. Environmentally minded, these employees happily sacrifice most of their lunch break to dump the compost bins on each floor into a few larger bins, using their hands to scrape off residues from apple cores and rotted banana peels, unbothered by the smell and the slime that drips off their fingers. They then haul the bins to the farmers' market in wagons, dumping the waste in a giant container, further scraping off excess scraps with their hands. Commercial waste management work is not the domain of white-collar labor—it is unevenly relegated to poor workers of color—yet SunLight's composters do not seem out of place hauling waste.[16] They bond over their soily fingers, at ease with organic matter, as they paradoxically perform a ritual of return in the corporate habitus of NYC's Financial District. Clearly, this is not the stereotypical behavior of alienated white-collar workers, not the intelligence of energy technocrats, not the workspace where Marx would expect to find "metabolic" relationality. While critics of white-collar do-gooderism castigate the profit motives and branding imperatives of "corporate social responsibility," those perks are inapplicable to these workers' compost.[17] The compost yields no financial returns, and its slimy, smelly, rotting matter is entirely off-brand—unaligned with the datafied B2B (business-to-business) aesthetic that SunLight strives for. The company makes no effort to extract PR value from it, and the fruit flies are concerning to employees who bring corporate clients to the office. More generally, compost is the stuff of soily hands—not enlightened minds. It emerges out of not skyscrapered intelligence but earthly intuition—the desire to return, not desire for returns.

I study the fruit flies' harried movements beside sterile desks with workers who appear indifferent to the contrast between the two, and while I know that corporations are not monolithic entities committed only to increasing shareholder value, I am still fixated on the incongruence between the sales team's returns and the data wonks' return, confounded by a white-collar space that somehow makes room for both.[18]

In reflecting on SunLight's fruit flies, I consider the insights of Celeste—the Marxist, ecofeminist Google Earth aficionado whom we met in chapter 2. She spends her days priming solar projects on her screen for corporations she doesn't even work for, improving their profit margins in spite of her anticorporate politics. When I asked how she reconciled this, to my surprise she suggested that the screenwork of returns can support the land work of return:

> When I'm in a pessimistic mood about if anything I do at a computer changes anything, at the very least I fall back on: the more we mitigate [carbon emissions], probably the less intense the [climate] impacts. . . . At the most . . . [I think solar can make us more] aware of energy use, our energy production needs, [and] where our energy comes from . . . and that [solar] could therefore help us transition to an even deeper *ecology of living*. I'm not saying that's automatic. I just think that that's a way we could—and *should*—be doing this [screen]work. . . . Like, first you note the solar panels on your roof, then you understand more about where your energy comes from, then you push for 100 percent renewable energy. And this direction is a good way to orient people so . . . [they become] more intimately connected with the resources in their region and their land, and how to live sustainably off the land. I would love if we used our labor more to actually procure food and build things as opposed to all staring at computers [in the] tech economy [she pointed at Google Earth on her screen]. . . . I think that would be good for people's well-being and . . . that's how we actually start to make real transformative change. I want more kids to have hands in the dirt . . . and I think solar will be an important part of that shift, but only if we do it right.

In suggesting that screenwork can help "shift" society toward haptic living with the soil, she proposes a teleology in which the logic of returns culminates in the land work of return: first, Google Earth economizing leads to rooftop solarizing, which catalyzes a greater awareness of one's environment, enabling one to become "more intimately connected with . . . [the] land." Perhaps this teleology—the signifying chain connecting returns and return—affectively animates SunLight's workplace, where soil and sales coexist, shaping the experience of corporate employees who long "to have hands in the dirt," like SunLight's weekly composters. Celeste suggests, though, that this hypothetical teleology does not nullify the existential discordance between "staring at computers" and land-based living; screenwork is perhaps most alienating when it gestures to the ecological. For it often felt as though the fruit flies hovering by my interlocutors' screens could only remind them of the sustainable world they hoped to return to, highlighting their distance from the land as they spent their days working two blocks from the epicenter of global capital.

RECONCILING RETURNS WITH RETURN

I further contemplated the coexistence of return and returns when I glanced at SunLight's bookshelves in my spare time. As expected, the majority of books were technical in nature—textbooks pertaining to EERE and Big-E Energy systems, for instance. The office also had ecomodernist paperbacks, such as Herman Scheer's utopic *The Solar Economy*. But there were also some notably out-of-place titles. The anthropologist in me was struck by the sight of James Ferguson's *The Anti-Politics Machine*, a book that is largely an indictment of technocratic intelligence and not something you'd expect to find at a US corporation. On the shelf beneath this ethnographic classic I found *One-Dimensional Man*, by Herbert Marcuse, an antimodernity polemic against all that is capitalist, technofetishist, and technocratic—a few of SunLight's defining characteristics. On one VP's personal bookshelf I found (1) *The Elusive Quest for Growth*, by William Easterly, another anticapitalist polemic entirely out of place in an office where many employees—and that VP in particular—were preoccupied with company growth, and (2) *Song for the Blue Ocean*, by Carl Safina, an appeal to alienated humans to familiarize themselves with the natural beauty and industrial desecration of earth's waterways, closer to Berry's relationality than ecomodernist rationality. No one commented on these and a few other anti-industry titles; they stayed on bookshelves, nestled in between the expected hodgepodge of neoliberal sustainability paperbacks. This inattention pointed not to the explicit equicratic ethos that employees often wore on their sleeves (see chapter 2)—pragmatic calculations that precluded Marcuse's polemics—but rather to the ways in which a radical ecosocialism could quietly exist in the company's broader ideological patchwork.

Yet while this mélange of employee worldviews helped contextualize the coexistence of Marcuse and Scheer, it did not substantively illuminate how return and returns could effortlessly sit side by side on SunLight's bookshelf. This coexistence began to make sense for me when I focused less on SunLight employees' conscious ideologies and more on their everyday practices of work: the corporeal and emotional labor through which they enacted these ideologies.

Consider the corporeal and emotional labor of Martha, a nerdy data analyst I introduced in chapter 2. As discussed, she is an energy expert, a democratic socialist, and a paradigmatic equicrat who was partly drawn to SunLight because she believes that EERE can empower marginalized communities by attenuating their dependence on investor-owned utilities. Even so, her job requires that she helps several financial and real estate corporations reduce their energy costs to enhance their returns. Yet she also believes that revolutionary change requires labor in opposition to the corporation—work that specifically empowers the blue-collar workforce that renewable energy demands. As such, in her spare time, she organizes and participates in direct actions for economic and climate justice.

While she has a nuanced way of explaining the tensions between her politics and her career, I began to understand Martha's affinity for work that is freighted with contradiction when I shifted my attention from her verbal explanations and toward her lived experience with both her day-to-day job and her out-of-office activism.

As discussed, she spends her days meticulously scrutinizing energy data at her screen. She enjoys this work not simply due to its "tangible" goals but also for its *intellectual challenges*; her work is an edifying *analytical* practice. Thus, she does not engage in prototypically physical labor, but I want to suggest that the pleasure she takes in analyzing data *is* decisively corporeal. For when she plants her elbows on her desk, contemplatively swivels in her chair, stands up intermittently to rub her hands through her hair, does a little joyous shimmy when she cracks an analytical code, munches on clementine pieces while expertly navigating her cursor, nibbles on her hair as she's lost in thought, slaps her desk triumphantly on identifying the cause of a data outlier, she embodies the practice of problem-solving with an effortlessness that suggests it's second nature to her. And in a sense, it is. As an educated millennial in a modern megacity powered by screenwork, she has long been habituated to her desk; it is where she connects with people, performs her identity, and situates herself in the world. Her posture, mannerisms, and habits suggest she is at home in the process of analysis, such that her corporeality corroborates her claim that she enjoys her work's analytical milieu. At the same time, Martha effortlessly adopts the bodies-in-the-streets posture discussed in chapter 3. She enjoys the catharsis of grassroots mobilization, of occupying public space, of making and flailing colorful protest signs, of organizing die-ins, of amplifying voices in the streets that champion energy democracy. For the imperative to transition away from fossil fuels enlivens political work that is as embodied as it is ideological, tapping into an activist ontology that Martha seamlessly enacts through her habitual presence in street protests.

This body-in-the-streets work and her formal occupation are similar in that both demand certain habituated orientations to the world that can't be reduced to her worldview or politics. Her ease in the office and in the streets is shaped by both her conscious worldviews and predilections that can't be understood in terms of her antipathy to fossil fuels—a ready-to-handness detectable in her posture when she occupies both wonkish and protest spaces. Martha's activist and office corporealities suggest that sustainable energy transitions reinforce different ways of occupying space that are not particular to EERE infrastructure. On the one hand, sustainable energy has bolstered a regime of capitalist labor centered on growth, returns, and, by extension, Martha's analytical habitus, and, on the other hand, it has given new life to long-standing struggles against racial capitalism—modes of monopolizing streets that activists like Martha have long been habituated to.

In this context, returns can cohabitate with return: Marcuse can sit side by side with Scheer on SunLight's bookshelf; a for-profit corporation will allow employees to connect with the soil during the workday. The energy transition aligns with not just different ideologies but also different habituations such that this cohabitation emerges less from conscious decisions and more from a codependency between the different forms of labor and living that the energy transition demands. *Codependency*, as there is no technocratic clean energy corporation without activists in the streets demanding clean energy, no mass-produced solar panels without care for the biospheric life that those panels aim to safeguard, no profitable energy transition without blue-collar workers. These codependencies preclude a dogmatic ecomodernism, an unbending ecosocialism, or a stringent neoliberalism, as contradictory ontologies comprise the world of energy equicrats like Martha.

Of course, white-collar workers in *all* industries engage in practices unrelated to their company's bottom line. But my point is that EERE gives shape to corporate spaces in which workers' habits and attunements often stand *in contradistinction to* their employer's commercial ethos, creating odd juxtapositions between returns and return, while reconfiguring the lived experience of white-collar labor. I now further explore these habits and attunements, showing how the corporeality of technocratic work as it pertains to solar infuses a praxis of transformative care into the world of energy professionals.

KINESTHETIC SKILL IN A MOMENT OF MACHINES

Today's corporate boardroom is the perfect token of white-collar intelligence. With its mounted flat screens, built-in microphones, surround-sound speakers, and videoconferencing capabilities, the boardroom, in its twenty-first-century form, is designed to optimize knowledge transfer, perform expertise, and practice data-driven decision-making. At the same time, its lengthy table, leather swivel chairs, and soundproofed walls retain the gravitas of yesteryear's boardrooms, demanding a self-importance and professional machismo constitutive of the C-suite. While corporate titans fetishize "disruptive innovation," the corporate boardroom invites none of that; its staidness inspires status quo comportments.[19] In this space, intelligence does not look the same as it does on open platforms like Slack or in quirky startup offices with Ping-Pong tables and meditation rooms. Intelligence instead appears as the domain of pressed pants and cuff links, masculinist hierarchy, and a lucid goal-orientedness (even when the occasional dad joke surfaces). Finally, the boardroom is physically and affectively sealed off from the outside world, enabling focus on ledgers and laws but not on the myriad lives and landscapes that loom outside the boardroom's windows. The boardroom is not about dwelling in the world; it is purposefully cordoned off from those not privileged to occupy it.[20]

But this affective texture was absent at a particular after-hours meeting. Every Wednesday evening, SunLight's engineers gathered in the boardroom for an educational, team-building event in which one of them would give a presentation on anything broadly related to energy engineering. This weekly event lacked the gravitas of a normal boardroom meeting during work hours in several ways: attendance was voluntary, the company provided a keg of beer, the engineers were often dressed in semicasual clothes, several slouched in their chairs during presentations, their discussion didn't directly address company strategy, and the clock was irrelevant—no one else booked the room for the night. But this absence was especially pronounced when Matt, a well-liked senior engineer, gave a presentation entitled "Visual Thinking and Engineering." While this title suggests ocularcentricity, Matt's presentation was actually focused on the full-bodied dimensions of energy expertise—his lively encounters on rooftops and in boiler rooms that couldn't be understood in terms of boardroom intelligence.

Matt began not with a PowerPoint but with an exercise in which he first gave written instructions to Andy, a solar engineer. He then asked for four "volunteers" to follow these instructions, which Andy would read aloud in a moment. To my chagrin, I was chosen as a "volunteer" for reasons that would soon make sense. Matt then passed out a blank piece of paper and a pencil to us volunteers. Andy proceeded to read the instructions, telling us to draw an image based on a description that he was about to read aloud; he would identify the "thing" being described for us only *after* we finished our drawings. He then described the unidentified thing, talking a lot about lines, elevation, and intersections, and the four of us drew the thing as best we could without context. When we finished drawing, Andy revealed the thing: a solar panel pergola—the infrastructure for elevating solar arrays over obstructed rooftops. He then drew the pergola on a dry-erase board, but it took him fewer than twenty seconds to finish his drawing, whereas the four of us volunteers had labored ineptly for almost five minutes. The four of us then revealed our drawings, comparing them to Andy's. One drawing very loosely resembled Andy's, but the rest looked nothing like his. Matt then explained that he had chosen volunteers who weren't in the solar subdivision and therefore were not familiar with pergola technology. This exercise, he explained, aimed to demonstrate that you can't actually apprehend a physical structure by relying on words; you need to visualize it to understand how it works. But visualization is never just a matter of seeing. After all, the three other volunteers and I had seen solar pergolas before. In Matt's estimation, visualizing an infrastructure is instead a matter of getting to know it in-person, touching its surfaces, familiarizing yourself with its dimensions, scrutinizing how its parts come together, and working with it as if it's your partner. Andy had effortlessly visualized the pergola because he had encountered pergolas in-person many times before, unlike the four of us, who were merely drawing from words.

Matt then began his presentation: an hour-long PowerPoint on "visual thinking" in engineering. While the pergola-visualizing exercise centered on the limitations of words, Matt's presentation focused instead on the limitations of numbers. Specifically, Matt vehemently critiqued the hegemony of numerical data at SunLight and in the EERE industry, contending that engineers fail to actually diagnose, work with, and improve energy infrastructures when they only apprehend them through formulas on screens or an aggregate of measurements. He explained how he never looks at the quantitative dimensions of an infrastructure before he engages with it in-person—how he, instead, draws it while he's introducing himself to it in real time and how, through this "visual thinking," he determines how to improve energy systems in ways that data-driven expertise could never allow him to do. He talked about getting on his knees; contorting his body in hot, cramped spaces; and manually fiddling with boiler room technologies as part of an energy efficiency assessment, elevating this corporeal knowledge production over blueprints produced at a workstation. As such, what he calls "visual thinking" can be characterized as full-bodied apprehension, linking multiple sensory modalities with the manual practice of visual representation. Put simply, Matt was calling for *skill*, counterposing it to the boardroom's disembodied intelligence. So it was odd seeing Matt at the front of the boardroom—a space curated to perform data-driven expertise—railing against the epistemological power of metrics at SunLight. The room's sleek corporate layout and computing capabilities contradicted his contention that his colleagues should minimize their office's visual abstractions and design solar electricity systems based on their experience crawling on their knees on rooftops.

To support his claims, Matt offered a brief engineering history lesson. He explained that engineering consisted primarily of manually drawing systems—using one's *hands*—until the mid-twentieth century, when engineering shifted to quantification practices and, eventually, screenwork like spreadsheets. In many instances, screenwork did not actually deepen the knowledge that engineers developed through kinesthetic encounters. As he put it, "You can make all of the complicated computer simulations and calculations that you want, but it's not really gonna give you more information." He explained that all of the body's sensorial capacities—not just the visual—are integral to developing EERE infrastructure:

> We need to use tactile knowledge; muscle memory is a way of thinking too. . . . For example, we just installed some VFDs at a building—

VFD is variable frequency drive, it allows a motor to run at different speeds—and we were testing them out. . . . But sometimes when you change the speed of the motor, it changes the frequency that it vibrates at, the system can resonate, and you can have problems. So as we were testing it at different speeds, I put my hand on the motor to feel it, and I was comparing how it felt—that vibration of my hands—to how it felt all the other times when I'd put my hands on the pump. . . . And imagine trying to describe those sensations with a drawing or words—you can't. So I really think that's another kind of knowledge too: the way materials feel and the ways they respond to touch. . . . There are lots of other faculties to explore. So draw a lot but also touch stuff and feel stuff.

Tactility here stands in contrast to the boardroom's stilted corporeality; it entails an openness to one's surroundings that is absent when one is ensconced in a swivel chair, eyes glued to a PowerPoint, hands folded deferentially or constrained in a contemplative posture while being serenaded with the "intelligent" words of a VP who has barely left the office in weeks. In Matt's telling, hands can be a medium through which white-collar experts attune themselves to the world beyond their office, learning to care for the built environment in ways that their boardroom presentations preclude. Indeed, EERE expertise requires a corporeal attention to one's surroundings—a kind of touch—that is incompatible with the corporate logic of returns.

I wondered, then, if this touch, this skill, could facilitate return: a praxis of reciprocity with our environments—a haptic ecocentrism—that contests the differential value delineating the genre of the human from his extractable others. The naivete of this romanticism wasn't lost on me; while Matt railed against intelligence, he still stood in a boardroom built on the returns of racial capitalism. But I continued to contemplate the potential for return after his presentation when I joined his engineering colleagues as they went from the confines of the office to the openness of the rooftop—experiences to which I now turn.

THE EYE OF THE SUN

In NYC even the smartest experts cannot design solar infrastructure with only intelligence. Because the city's rooftops have numerous structural idiosyncrasies, one cannot simply calculate empty roof space and map out solar electricity systems accordingly. Instead, engineers and other solar experts physically assess the roof when a project has been green-lit, venturing from their office to the field to take note of curvy dimensions, oddly sized plumbing vents, bulkheads, pocks of bitumen, and other physical particularities that require in-person attention. In doing so, they draw from corporeal capacities—mobility, touch, manual sketching, lower body balancing—that elude numeration and screenwork. Solar is therefore inseparable from both the built environment and these corporeal capacities.

This entanglement is exemplified by SunEye technology: a central actant in the design of solar projects (figure 4.6). According to its manufacturer, SunEye "is the world-leading shade measurement tool for solar site assessment. This hand-held electronic tool measures the available solar energy by day, month, and year with the press of a button by determining the shading patterns of a particular site."[21] As I will show, the "hand-held" adjective in this description is key, but the manufacturer further describes their product in ways that obfuscate SunEye's dependence on human touch: "The SunEye® enables the best solar designs and the highest ROI [return on investment] for solar professionals and their customers. It helps win competitive bids, saves time and money, and gives confidence in performance guarantees. The award-winning SunEye incorporates a calibrated fisheye camera, electronic compass, tilt sensor, and GPS to give immediate measurements. . . . The one-handed operation, rugged enclosure, and outdoor readable display make it a reliable partner in the field."[22]

I highlight this product description for a few reasons. First, most solar professionals in NYC (and, I'd imagine, elsewhere too) would largely agree with it; the SunEye is essential to designing a solar installation project. To my knowledge, no projects in the city move forward without a SunEye assessment because the technology generates a range of helpful data—from projections of annual electricity output to measurements for optimally positioning solar arrays. I also highlight this description because it locates omnipotent agency in the machine, suggesting that a SunEye can design a system "with the press of a button." We see here a technofetishism reinforcing the idea that solar technocrats docilely follow their precise technology's data—as if design is a fully automated process—echoing the industry's discourse of disembodied intelligence. Finally, this description suggests that the SunEye is first and foremost focused on

FIGURE 4.6. The SunEye, a handheld solar assessment technology that requires a thoughtful physical orientation to rooftops. Photo by Myles Lennon.

returns (i.e., return on investment). We see nothing resembling return—nothing pointing to the corporeal expertise that Matt called for.

But the description hints at something otherwise when it characterizes the SunEye as "a reliable partner." This suggests that the technology is not an all-knowing source of intelligence but instead an actor in cahoots with the solar expert, gesturing to a crucial dimension of SunEye assessments that it does not describe: the SunEye requires a human's *bodily* dexterity; the technology does not simply work "with the press of a button." When using it, solar engineers and site assessors demonstrate a corporeal consciousness of rooftops—a physical orientation to space that is absent in their screenwork. They first position themselves at a point on the roof where solar panels could potentially be installed. With the SunEye in their hand, the technology renders a "skyline image," which simulates the "annual sunpath"—the projected sun exposure over the course of a year—for the specific point where they are positioned. The SunEye takes longitudinal and latitudinal measurements of that point, identifying the optimal orientation of the proposed solar panels. But the engineers and assessors do not passively receive this data; the technology requires their active participation. Specifically, they have to very precisely position the SunEye so that it projects that the solar panels will face south with a 180-degree azimuth. Toward this end, they crouch low (where the panels would be on the roof), orienting their body and hands in a very particular way that enables the SunEye to take the right measurements. This bodily coordination entails very slowly turning their wrist to position it just so, bending their knees at different angles, bowing awkwardly as if they're about to pray, taking a step backward or a step forward, following a little ball on the SunEye screen with their eyes—a full-bodied attempt to orient the technology in a way that ensures alignment between the sun and solar panels. This is a delicate little dance, as the handheld technology, the user's body, and the earth's axes must momentarily cooperate to generate the needed data. Once they capture data for one point of the roof, they have to do this little dance again several more times, moving to a point close to the previous point, and then another point close to that point, and then another point, until they capture data for several points on the roof.

How many points? And how far should the points be from one another? These questions can't be answered with definitive metrics or prescribed rules. Using a SunEye requires a sensibility gleaned from doing—a *skill* in Berry's parlance—not a precise science. SunEye users demonstrate a ready-to-handness—a sense of where to position themselves based on their experience with the technology—capturing as much data as they feel is necessary. After one has captured data at a few different points, the process of positioning one's body becomes quicker and less awkward— there's an intuitive corporeality to getting the positioning right—what Matt called "muscle memory." Bodily contortions morph into a smooth sequencing as you learn to interact with the technology and orient yourself with the earth's axes. The technocrat is not merely using a machine to generate and analyze data. A reflexive engagement with physical space—an embodied expertise that can't be learned at the office—makes the SunEye functional.

In addition to using the SunEye, site assessors manually sketch the roof and shadows on an unruled notepad, roam around the roof while taking photos at different points, and crouch to take measurements of parapet walls, chimneys, the top of the stairwell, the roof's slopes, and any obstructions—familiarizing themselves with the sun-exposed surface in different ways. While they spend most of their days in offices behind screens, their expertise is not confined to the white-collar habitus, demanding knowledge that exceeds Cartesian intelligence. Indeed, solar/EERE professionals have a corporeal relationship with the infrastructural terrain of everyday spaces that those spaces' occupants lack—they are attuned to the environment in ways that belie their vocational positionality.

Of course, numerous workers intimately know crawl spaces, rooftops, electrical rooms, and other facets of the built environment. But what distinguishes the labor of a telecommunications worker, for instance, from that of a solar engineer conducting a SunEye assessment is that the latter is reflexively focused on their environmental impact while the former is not. Even so, this labor is intended to yield returns and is often evaluated through the means of reduction—desituated from the environmental interdependencies that constitute return.

Yet return is immanent in this labor, even if it's often overlooked. For the expert's body during a SunEye assessment is a medium of ecological alignment, conjoining the earth's axes and the sun's path across time such that their physical comportment links the machine in their palm to the cosmos. What if we recognized this corporeal expertise as a means of not merely taking solar measurements but also cultivating an ontological attunement to the earth and the sun? A means of moving beyond Cartesian intelligence and toward reciprocity with our environments? What if the SunEye was celebrated for facilitating greater awareness of the interdependencies of the universe? These would be impractical aspirations if solar experts lacked an environmentalist ethos. But both the political affordances of their labor and its corporeality create the preconditions for imagining solar expertise as a form of environmental stewardship. The bodily alignment with the sun and the earth that the SunEye demands suggests that solar can reorient our whole relationship with the nonhuman world—that this relationship is irreducible to the means of reduction. I am suggesting, then, that solar professionals' corporeal attunement to space can focus more on return than returns—a premise I will now flesh out by further exploring solar's physical labor.

THE CORPOREALITY of SunEye assessments, Matt's penchant for haptic skill, Martha's pro-labor protests in the streets, Celeste's manifesto on soil, the data analysts' compost, SunLight's anticapitalist paperbacks, and Sean's attunement to fire all suggest that white-collar EERE expertise is neither defined in contradistinction to physical labor nor productive of an elitist managerial politics. Many of my interlocutors characterized their professional pursuits in terms of kinesthetic skillfulness and more-than-human encounters, on the one hand, and skepticism, antipathy, or ambivalence toward corporate intelligence, on the other. In these respects, their enactment of expertise countered the differential value that positions white-collar screenwork above physical labor under racial capitalism.

But privileged technocrats with soily hands, physical finesse, and pro-labor politics are not evidence that sustainable energy is returning humanity to a putatively purer state that existed before racial capitalism positioned corporeal skill beneath corporate smartness. More specifically, my interlocutors' affirmations of bodily labor do not disentangle hands-on work from a regime of differential value that codes bodily exertion, manual trades, and toiling outdoors as Black/brown, poor, and less-than-human. An educated white person crawling on their knees on a roof or taking out the office compost does not, in isolation, disrupt the structures of power that banish minoritarian subjects to work in the field and that situate the land beneath the domain of intelligence. While EERE can cultivate kinesthetic care for one's surroundings among office professionals who have traditionally been defined by hegemonic screenwork, that care will not unsettle racial capitalism if it maintains the symbolic matrix that relegates physical land stewardship to the savage slot.

I want to suggest, though, that EERE—and solar, in particular—can, in fact, destabilize this symbolic matrix, challenging the raced and classed character of environmental work. I recognized this potential when SunLight's diversity efforts created an unlikely opening for skillful labor that bridged white employees' screenwork and Black employees' physical work—a transformation I explore in the next sections.

BLUE-COLLAR BLACKNESS IN THE WHITE-COLLAR OFFICE

As discussed in chapter 2, the diffuse spatiality of EERE brought both marginalized social groups and a progressive politics into the realm of energy professionals. This, in turn, injected a late liberal praxis into the once-insulated world of electricity. This late liberal praxis was perhaps best evinced by Sean's unnervingly blunt confession to me one day that "everyone wanted to have a party when we hired a Black senior project manager!" He was talking about Ron—the only Black employee in a senior position at SunLight at the time. "We were just excited and relieved [about Ron's hiring]—like, *Finally!*" Sean's comments revealed the cosmetic workings of late liberalism—the corporation's preoccupation with the *appearance* of a single Black employee in the face of structural critiques.

But Ron's appearance cannot be reduced to optics; his hiring disrupted the company's intersectional character, shifting SunLight's attunement to the built environment in the process. This disruption was evident from the moment he stepped off the elevator on the thirty-fourth floor. While he was clearly qualified to work for SunLight's engineering division—he had fourteen years of senior experience in the solar industry and decades of work experience as an electrician—he was the only Black person, the only baby boomer, and the only nonengineer on the engineering team, lacking his colleagues' elite credentials and networks. He was also the only SunLight employee who grew up in public housing—and he did so during the city's fiscal crisis and crack epidemic. This class background was not hard to read from his comportment among his whiter, educated, "bro-y" colleagues; in his speech, dress, and appearance, he did not resemble a nerdy *Family Guy* stan. His folksy affability, his lack of smartness— that is, his failure to performatively render intelligence as cultural capital in an elite space—and his baggy workman clothes comprised a masculine blue-collar habitus discordant with SunLight's brand.

This out-of-place class character was painfully discernible at a meeting in which the company's solar engineers reviewed their roster of current projects. With unapologetic curiosity, Ron often interrupted speakers to ask questions and gather information, disturbing SunLight's staid, hypercollegial workspace. Other than his animated inquiries, the meeting was unremarkable until one of SunLight's solar installation inspectors (who wasn't at the meeting) texted a photo of one of their solar installation projects to Claire, a solar engineer. Spread across eighteen multifamily buildings in a low-income community of color in the Bronx, this project was one of SunLight's largest. Because of its size, SunLight had hired a solar installation subcontractor to complete most of it. The photo that Claire received, though, suggested that the subcontractor wasn't doing their job. Her blueprint for the project called for eleven-inch gaps between the solar arrays, but the photo showed gaps that were around six inches—a considerable discrepancy. "Oh shit," she said, gasping before recomposing herself.

She turned to Ron, who was responsible for overseeing part of the project and who had chosen a subcontractor he was personally familiar with. She explained the installation problem to him, underscoring the severity of his subcontractor's error. Ron's other engineering colleagues also shared their consternation while portraying no surprise; they had grown accustomed to blue-collar installation teams failing to follow their meticulous measurements. Ron, though, seemed unbothered by the error, nodding indifferently in response to Claire's worried words. "I need you to call the subcontractor immediately and have them fix this," she said sharply. Ron again nodded and unenthusiastically said he'd call the contractor. Clearly, Claire wasn't getting through to him, so she ratcheted up her tone—call the contractor ASAP!—while SunLight's director of solar engineering chimed in with his own urgent appeal.

In retrospect, I can say that it was telling that a photo of solar panels installed by people Claire had never met at a site she had never visited had provoked such a strong reaction in her. Only later would I understand that her lack of experience installing the solar systems she designed and her physical distance from a space whose photographic rendering upset her were symptomatic of how SunLight's screenwork reproduced differential value. But on the surface, her response was well warranted. Subcontractor errors usually delayed projects and diminished returns, generating distrust between SunLight's office-based engineers and their less refined blue-collar counterparts. To bridge this classed divide, SunLight had recently held communication trainings for engineers focused on working with field staff. They had learned that proactive communication with workers in the field prevented problems and kept projects running smoothly. They therefore believed that Ron's subcontractor snafu could spiral out of control if he didn't proactively communicate.

Yet Ron continued to appear unfazed. He nonchalantly reiterated that he'd call the contractor in response to his colleagues' dismay. But Ron wasn't going to call; he had another plan in mind. "I'm going to the Bronx tomorrow morning," he told me, referring to the problematic project site. This rogue move was rooted in an idea that Ron knew well and would soon instantiate: infrastructural transformation is as much about intersectional affect as it is physical labor. He clearly wanted to impress this unspoken idea on me, the anthropologist, when he devilishly raised his eyebrows and said, "Wanna come?"

At 7:58 a.m. I arrived at the building complex where SunLight's negligent subcontractor was set to begin work for the day. Ron was already there waiting for me; he had cultivated extreme punctuality over years as a union electrician. As we walked up seven flights to the roof, Ron explained that none of his SunLight colleagues was ever at a jobsite this early—they were accustomed to arriving at the office at 9:10—another sign of their class differences.

When we reached the roof, the subcontractor's foreman and one of his workers, both of whom were Afro–South Asian, were prepping jobsite materials for the day. The roof was covered with the inaccurately installed ballasted solar racking system that Claire's photo exposed. Ron inspected the racking, noting the flawed dimensions immediately. As we made our way to the other end of the roof, where the foreman was working, I felt nervous about the forthcoming altercation, not sure how Ron was going to confront the foreman about his errors. I was certain, though, that Ron would lay down the law with the sort of machismo characteristic of every all-male solar installation site I'd been to. There was clearly a reason he didn't want to confront the subcontractor on the phone; an in-person confrontation, I assumed, would convey the magnitude of the subcontractor's faults, positioning Ron as the boss.

But, once again, Ron defied my expectations. He greeted the foreman with a hearty smile. While Ron is *always* friendly, as he trained his attention on the foreman, his amicability bespoke a subtle code-switch. He appeared lighter than usual—gone was the workplace intensity that he displayed in his arduous line of questioning at the previous day's meeting. After exchanging pleasantries, Ron and the foreman shot the shit as though they were friends with nothing else to do, making no mention of work or their contractual relationship. This lighthearted banter indexed a raced, gendered, and classed solidarity—an unspoken appeal to their shared background as men of color in the blue-collar trades.

Finally, after a few minutes, Ron mentioned the faulty solar installation, starting off complimentary: "Y'all got a lot done! Y'all got a lot done. But the spacing [between the solar arrays] . . . " He then raised the matter of the dimensions—but instead of reprimanding the foreman, he spoke as though he was just a messenger. "Listen, man, you know, the way *they* do things, it's different—they like things done in a particular way." *They* referred to his colleagues and his employer; he elided the *we* that people usually use when representing their company. "I know you're doing a great job here, but they're engineers, they're not like us, so we gotta just make some changes here." *They're not like us.* Once again Ron marked his intersectional alignment with the foreman, characterizing the engineers as a raced and classed other—not as his colleagues.

The foreman then explained that he was simply following the manual and the industry norms in positioning the arrays six inches apart. The racking has two different holes—one that creates a six-inch gap between solar arrays and another that creates an eleven-inch gap—and SunLight is unconventional in designing systems with the eleven-inch gap. Ron and the foreman then reviewed SunLight's project blueprint and realized it was unclear because it included *both* measurements. Ron proceeded to exonerate the foreman, laying the blame on SunLight's classed charac-ter: "Our company is engineers, they overthink. You need one measure-ment. You didn't need five measurements. . . . Our engineers: they give you way too many measurements." Again, the word *engineers* positioned Ron's colleagues as a classed (and, by extension, raced) other, distancing the (Black) blue-collar worker's common sense from the (white) techno-crat's bizarre obsession with metrics.

He later expanded on this insight when the two of us left the roof: "Engineers think if you design it, then it's gonna work. . . . [But] I know I'm dealing with human beings so at no point in time am I just gonna come up with a plan and just go like this, 'Here.'" He mimed handing a blueprint to a subcontractor. "Instead I'mma get your [i.e., the subcon-tractor's] input, 'cause you're the one who has to deal with it, and then I can see where your resistance might be, and then I can push things that I know need to be done. There's always play area. Engineers don't have play area." He explained that his younger colleagues seldom left the office to come to the field like him, failing to engage with both the phys-ical rooftop—its odd curves or small pipes, which Google Earth doesn't reveal—and the people tasked with implementing their design. The engi-neers primarily apprehend rooftops though their blueprints—a quint-essentially technocratic visual genre—privileging images conjured with the aid of satellite photography over the experience of physically inhabit-ing the space at hand. Ron suggested that these spatial dynamics were at the root of SunLight's white-collar/blue-collar disconnect, arguing that the engineers' unwillingness to learn and interact with installations in-person set the stage for the company's project challenges (as evinced by Claire's inclination to reprimand the subcontractor without visiting his jobsite). Ron implicitly rebuked screenwork in this assessment; if the engineers left their desks to experience the spaces and people they worked with, their subcontracting relations wouldn't be rife with acrimonious miscommunication.

Ron and the foreman agreed on an aggressive timeline for fixing the problem, affirming their commitment to supporting one another—a pact informed by their intersectional solidarity. "I know you guys do quality work, man," Ron avowed. "That's why I went with you, quality. I trust you, man." Ron always works with subcontractors from underrepresented racial backgrounds for two reasons. First, he wants to create opportunities for people who face structural barriers like he does. Second, when he works with subcontractors of color, he can cultivate an organic interpersonal connection that strengthens project coordination. This race-conscious approach to subcontracting proved fruitful in the case of the misaligned solar arrays; the subcontractor rearranged the arrays in a few days—a very quick turnaround directly attributable to his personal relationship with Ron. Thus, Ron's raced, classed, and gendered positionality as a Black man from a poor background with blue-collar experience was instrumental in the project's expeditious completion, correcting Sun-Light's distrustful off-site subcontractor management.

More crucially, though, his minor act of face-to-face empathy points to an intersectionally situated skill that unsettles the differential value embedded in infrastructural subcontracting. As I have only begun to show, Ron intentionally cultivates an in-person relationship with the spaces he's tasked with solarizing—a spatial attunement affectively animated by his intersectional identity. Through this attunement he performs his paid position as an energy expert in ways that renounce the logics of spatialized separation and corporate hierarchy that constitute energy expertise. As I show in the next section, solar is at the heart of this rogue positioning; a Black man from poverty with a blue-collar background would likely have greater trouble securing an engineering-equivalent position in the *non*-EERE realm of white-collar energy professionals. As such, Ron centers an intersectionally situated skill in a position historically defined by spatialized labor stratifications, redefining the corporeality of energy expertise.

In the months after our rooftop visit, Ron began to transform SunLight's solar practice, familiarizing himself with the particularities of jobsites, building relationships with SunLight's blue-collar installers, getting a feel for the rhythm of each project—the flows of labor, weather conditions, contractors, construction materials, building management, and government inspectors. He engaged in screenwork, but he didn't center it, focused as he was on in-person observation, personal rapport, tactile knowledge, and a more proximate sense of place. While he never envisioned an ecocentric return, his attunement to space gave a real-life form to Celeste's nebulous aspirations for corporeal care via solar—the haptic energy transition Sean imagined—bringing a reflexive relationality, an "ecologics," to the alienated world of energy expertise.[23]

But his skill could not, in isolation, upend the differential value at the heart of blue-collar environmental work. He made this limitation painfully clear when I bumped into him in the lobby of SunLight's office as I was leaving on a sunny spring afternoon. The sun was clearly not cause for contentment; usually affable, Ron was noticeably stressed. He told me he was returning to the office from a protracted solar site visit in Queens—a surprise to me, as I thought his Queens project should have been long done at that point. When I asked why he was still working on that project, he looked both ways and said, "Come with me," and we walked out of the building. "You wanna know why it's taking so long? It's [SunLight's solar installation] crew, man. I'm frustrated with the quality of their work. They move so slow, it takes them forever to get anything done." As a former union electrician, he was accustomed to a higher quality of work. I asked if the crew lacked either a good work ethic or training. Ron considered this for a second. "Neither." Sounding contemplative, he continued:

> The problem is that they don't have good mechanical aptitude and hands-on dexterity because they're younger, and they've been raised on screens, and they were not instilled with hands-on skills at a young age—they learned primarily through computer screens and simulation. . . . They don't know how to be creative with their hands 'cause they're used to being creative on the screens. When I was growing up, we didn't have screens like they do now. I remember banging on pots and pans with wooden spoons when I was kid

when I was trying to be creative; there was no screen, you had to use your hands and body.

He explained that SunLight had fired many young solar installers after only a few weeks on the job because they lacked the "hands-on dexterity" he had started cultivating as a child. He argued that solar installation jobs require not only an able body, a willingness to work, and technical tutelage but also a *haptic* knowledge of rooftops and solar technologies. In this analysis, infrastructural transformation demands corporeal attunements to the built environment that exceed job training and work ethic—physical labor that is illegible to the labor market. Ron thus underscores that a blue-collar worker is not simply a physical unit of labor power, countering "intelligent" management practices that ignore the social and affective conditions under which the physical work of energy transitions occurs.

While Ron blamed screen culture for his younger colleagues' lack of skill, I suggested that it might make more sense to situate their skill differentials in a historical intersectional framework that aligned with his own autobiographical analysis. For in Ron's telling, his skill reflects his positionality as an able-bodied, straight, Black male baby boomer who was raised in the projects and came of age in the 1970s. While Black women had disproportionately served as the primary breadwinners of their homes for decades after abolition, this changed when the workforce slowly diversified in the aftermath of the civil rights era, offering Black men a blue-collar path out of poverty.[24] Specifically, construction trade unions begrudgingly opened up their ranks to young Black *men* like Ron who had cultivated manual skills in their homes and communities as part of the gendered division of labor in domestic and informal economies. That said, penetrating these unions required not just hands-on skill and a good work ethic but also a social adroitness at finding classed and gendered solidarity with white blue-collar men. Ron began to cultivate this interpersonal common ground as an apprentice in the International Brotherhood of Electrical Workers, practicing a congenial fortitude in the face of structural racism partly by leaning into their shared class character and connections as the beneficiaries of institutionalized patriarchy. This brotherly solidarity afforded him access to hands-on training and labor protections as a union member, which enabled him not only to rise out of poverty but also to adopt a professional identity rooted in craftsmanship—a sense of pride and a respectable career built on his physical labor. This professional identity, then, conjoined his "hands-on dexterity" with his raced, gendered, and classed positionality, as the intuitive feel of working a construction site is, for him, inextricable from his lifelong struggle navigating structures of power and the construction trades' heteromasculinist brotherhood.

This orientation to physical labor continued to anchor his career as he transitioned to solar installation work. Drawing from his decades of experience as a union electrician, he developed a sophisticated understanding of the ins and outs of solar, which enabled him to rise in the solar industry's ranks and make a name for himself as an expert in a still-nascent field. His intersectionally situated corporeal labor, then, positioned him to become a white-collar professional at SunLight, as his expertise is evidence of not his company's intelligence but instead his dexterity in both working with his hands and navigating raced, classed, and gendered structures of power. Ron's professional trajectory in the solar industry is arguably as much about his experiences as a Black man in the post–civil rights era as it is about his kinesthetic capabilities. When he practices his craft on rooftops, forms relations with installers, and attunes himself to a project's rhythm, he displays affects that he cultivated over years struggling for something other than a paycheck: the feeling of respect and pride that comprised his coming of age. Therefore, his work is less about scraping up the pittances of his employer's returns and more about asserting values that racial capitalism denies: one's inherent worth as a Black man.

While some might dismiss this as a problematic respectability politics in which employment is the basis of one's value, Ron locates a transgressive pride in making something of himself with his hands—a refusal to economize his care for both his craft and colleagues. When he transforms energy infrastructures, he rejects the reductive ways of relating to one another and our surroundings choreographed by racial capitalism. He therefore practices a form of skill that has resonances with a broader regime of return, dislodging the strenuous physical work of energy transitions from the extractive relational architecture of differential value.

Yet this means little if, as he suggests, his younger colleagues lack such intersectionally situated skill—if they are positioned as disposable bodies to generate corporate returns. For the dearth of "hands-on dexterity" that Ron attributes to screen culture cannot be extricated from broader technological innovations that displaced skilled labor and padded profit margins, reversing the small twentieth-century gains in Black men's labor power in the generations after Ron's. Perhaps, then, his skill can only counter the extractive status quo and facilitate return if it is paired with pro-worker organizing—an ecosocialist approach that I now consider.

RETURNING TO THE ROOFTOP

When ecomodernists valorize intelligence, they endorse differential value, privileging corporate smartness over blue-collar labor by suggesting that the cleantech intelligentsia drives energy transitions. Ecosocialism offers a corrective to this working-class erasure. Ecosocialists generally call for rapid, large-scale decarbonization through the democratization of energy generation systems; a robust public sector that protects the environment; and the redistribution of petro industries' wealth to everyday people. Ecosocialists recognize that physical labor is essential to energy transitions and that efforts to mitigate climate change must therefore strengthen labor rights, empowering blue-collar workers.[25] In their analysis, energy transitions require an organized workforce with the hands-on skills, compensation, and power to collectively transform how we, as a society, live and prosper. As such, ecosocialism explicitly targets the extractive relations of fossil capital.

Yet as an ideological project, ecosocialism resembles ecomodernism in its limited attunement to workers' on-the-ground experiences with energy transitions. Just as these workers undermine the ecomodernist myth of intelligence, they complicate the ecosocialist ideal of an empowered green-collar workforce. Specifically, I want to suggest that the corporeality of solar installation labor in the context of racial capitalism affectively animates a working-class praxis that is at odds with ecosocialist principles; many solar installers do not subscribe to ecosocialists' political agenda. This physical labor, then, destabilizes utopic visions of a green-collar workforce equipped with the relational skillfulness that I have equated with an abstract ecological return.

Like all physical work, installing rooftop solar isn't easy. At its most grueling, it can entail scaling a steep pitched roof for hours in ninety-five-degree weather with direct exposure to the sun, precariously harnessed, sweating uncontrollably in thick work boots, thick overalls, a construction vest, and a hard hat, skin chafing in the heat, while delicately keeping your balance as you carefully drill and install racking. It is no more pleasant on flat roofs, though; you might have to carry an eighty-pound ballast block up seven flights of stairs before running down to the ground floor to get another block, and then back up again before grabbing another block, the scalding sun greeting you with every return to the roof. And while there are some temperate days, there's also snow, sleet, hail, and freezing winds at the top of twenty-story buildings in the dead of winter, to say nothing of intermittently rainy days and air pollution advisory days. Even when it's not so physically taxing, this work can be incredibly tedious. Imagine climbing a stepladder to install a series of wires on a solar pergola, climbing down the ladder, moving it three feet, and doing the same thing again, and again, and again, and again for hours. While using power tools or installing your first array or learning an inverter's electrical workings can be edifying at first, after a while these practices become, at best, routinized—just another thing you do to make a living—and, at worst, mind-numbing, boring, strenuous. None of the installers I spoke to—all of whom had been on the job for at least a few months, and in some cases for almost a decade—enjoys the sheer physicality of solar installation. They don't wake up every morning excited about lugging solar panels up a hundred feet of stairs for the millionth time; they're not particularly thrilled when they realize that the racking, arrays, and wiring they've spent the past four days installing need to be moved slightly due to an engineer's miscalculations or their foreman's inattention—these "slight" modifications meaning, in effect, that everything they've done so far is for naught, that they're effectively starting from scratch. No, none of this is particularly fun. "The hard part is drilling and keeping your balance on a roof like this," explained Joe, an installer, as he prepared to scale a forty-five-degree-tilted roof. "And it's scary, honestly. On a roof like this, I've fallen off once, broke some bones, and it's so fucking hot out here, and with all this work gear on, it's like being in a fucking furnace."

Yet most of these installers genuinely enjoy their work—many claim to love it. Perhaps they wouldn't have such a rosy assessment if their physical labor was cordoned off from social context. But the grueling corporeality of solar installation has a particular affective texture in an emergent industry that can feel alive with possibility in ways that comparable but more established industries don't. In other words, the social context of the renewable energy field endows this physical labor with affect that differentiates it from similar work in different fields, notably the construction trades. I want to specifically suggest that three facets of this field—its technocratic solutionism, entrepreneurialism, and environmentalism—converge with a working-class individualism to generate a praxis of physical work that is less aligned with collectivism and organizing than that of comparable NYC industries. As such, solar installers' orientation to their labor does not align with ecosocialist analyses. While installers and workers in comparable industries encounter very similar labor challenges, the three facets of the renewable energy field that I just identified inflect their on-the-job experiences in ways that complicate ecosocialists' understanding of the blue-collar job market.

Before exploring how these three facets—that is, solutionism, entrepreneurialism, and environmentalism—affectively animate the bodily work of solar installation, I want to first discuss the ideological glue that holds these facets together: individualism. Most of the thirty-four installers I interviewed and observed during my fieldwork regularly espoused narratives of upward mobility focused on the power of the individual to shape their own destiny, linking their solar careers to their personal guile and hard work. "When we was coming up in the projects, everything around you tells you that you're a nobody, that you can't make it, like there's nothing out here for young Black males," explained Shane, a forty-five-year-old installer who previously worked as a general construction laborer and repairman for decades. "My mission every day, when I . . . come to the jobsite, is to prove them wrong, to show them what I'm made of, what I can do . . . [to show] that I can use *my own two hands* to make something of myself, to provide for me and my own. No matter what you ask me to do—like, you want me to install these raceways [i.e., the wire management system] in a few hours? Done; I got you. Like, I got a fire in me for excellence; everything I touch: excellence." In Shane's telling, his physical labor indexes his personal ingenuity in the face of structural barriers—a testament to his ability to persevere and prosper despite the odds. Shane cites his manual exertion as evidence that he is not beholden to societal constraints—his hands like a cipher of self-sufficient individualism.

Crucially, Shane contends that when he worked in construction and as a repairman, the haptic nature of his labor was just as symbolic of his personal ingenuity as the hands-on work of solar installation. Most of the installers I spoke to echoed his ethos of work; their job's physical challenges indexed their individual perseverance as working-class men (mostly of color). This blue-collar individualism complicates the labor politics of ecosocialists. It is challenging to mobilize workers en masse against poor occupational safety conditions, low compensation, lack of benefits, and the absence of collective bargaining when workers see themselves as rugged individualists capable of providing for themselves no matter the structural odds—when they take pride in difficult work. Of course, effective labor organizing campaigns make the case that corporations should value and respect the labor that workers take pride in. Even so, workers' rugged individualism is an affective and ideological hurdle to collective action.

Solar installers' compensation is at the heart of this individualism. *Relative to other jobs they've done*, solar installation work pays well. "It's good money, and they let us work overtime," explained Bobby, an installer, when I asked him why he subjects himself to the physical challenges and discomfort that he said his job brings him every day. "I don't see myself doing this forever," he added, "but for now it's a decent living compared to what's out there, better than my other options." According to Glassdoor, NYC-based solar installers make $57,700 annually on average, and while this number does not come from formal industry data, it is consistent with my interlocutors' hourly wages.[26] Although installation work is not financially lavish, for many of the working-class men employed by the solar industry, it represents a source of financial stability that they couldn't easily find elsewhere. This is especially the case on larger solar projects on government property, which, in NYC, pay prevailing wages—three times the workers' normal wages, considerably more than what they would earn doing anything else in the formal job market with their skill sets and education. These wages buttress many workers' sense that they are self-sufficient and that they have used their corporeal skills to escape the precarity of their previous lives. While most of these workers strive to earn more, their wages inspire upwardly mobile teleologies, gesturing to greater prosperity down the line. Thus, solar installers' relatively decent pay contributes to the blue-collar individualism of hard physical labor, informing individual progress narratives centered on their skilled work as opposed to collective organizing.

While this blue-collar individualism is not particular to solar, three facets of NYC's solar industry configure this individualism in particular ways that problematize ecosocialist imaginaries: the aforementioned solutionism, entrepreneurialism, and environmentalism, which I now discuss before returning to the matter of return.

Several NYC solar companies like SunLight cultivate an organizational culture in which social capital and office life revolve around *solving problems* through technical expertise. These corporations doggedly use the language of *solutions* not only because it commonsensically characterizes their pragmatic responses to both the macroproblem of climate change and the local challenge of solarizing a dense vertical skyline but also because this language is affective. It is *affective* in the sense that it animates the lived experience of office work and gives form to the cultural identity of companies in which employees often conceptualize themselves as problem-solvers. At solutionist corporations, office workers express an affinity for not simply the end goal of a solution but also the *process* of finding solutions—the feeling of using one's intelligence and expertise to navigate and overcome challenges.[27] As such, this solutionism has imprints of the Weberian spirit of capitalism in which work is personally fulfilling.[28] At SunLight and elsewhere, many office-based employees seek to achieve personal growth by using their smarts to find solutions, framing problem-solving as a matter of intellectual fulfillment irrespective of the specific ends they're working toward.[29]

This solutionism has classed dynamics that shape the blue-collar individualism of SunLight's solar installers. Specifically, problem-solving is often understood as a "cognitive process," and it is therefore evocative of professional spaces rooted in smartness and intelligence.[30] There are of course phenomena that are labeled as problems and solutions in *any* classed context, but problem-solving, as an *emic* characterization of one's work, is often affiliated with technocratic workplaces—not physical labor.[31] With their analytical minds, white-collar workers are paradigmatic problem-solvers; blue-collar workers, in contrast, fix, make, or build things, but their minds are not mobilized to innovate—or so the popular classed distinction of work goes.[32]

Yet SunLight's solutionism infuses all levels of the corporate hierarchy, including its lowest rung: the solar installation crew. Specifically, SunLight's solar installers use problem-solving rhetoric to characterize their manual labor—something I first noted not in their use of the word *solutions* but rather in their generic ruminations on the personal improvement they experience at work. For instance, Karim, a former general construction worker, explained what sets apart his often tedious and physically challenging labor in the solar industry from his previous blue-collar jobs: "I like the fact that solar combines a lot of trades; a lot of the skills required . . . is the same as framing, carpentry, [etc.], so just a lot for me to learn and keep learning. There's always new skills to learn and bring together, develop a craft." In this statement, solar installation is not simply a source of decent income, like comparable construction work, but also a medium for personal improvement. Many SunLight installers echoed this sentiment, differentiating their present work from their previous similar physical labor on the grounds that the former offers new opportunities for learning and growth. While this differentiation partly reflects the industry's nascence, it is also derivative of the solutionist discourses that rationalize labor in the company's technocratic headquarters. For many installers characterize this drive to "keep learning" in terms of problem-solving and innovation—not just doing a good job, in the generic sense, but, more specifically, improving outcomes through technical expertise. As Raymond, another installer, explained:

> I'm innovatively inclined. I like learning new things, things that challenge me or that are new. I like little things to keep my gears spinning. Like the hardware, and how to serve a little buttering [i.e., apply waterproofing sealants to a roof]—I wanna add a little moderation to that. . . . Maybe if the raceways were oriented a different way or clicked on a different way, it would make things more efficient, speed up the process of building [a solar electricity system]. . . . Innovation and solving problems is ultimately why . . . I work at [SunLight].

When Raymond characterizes his physical labor in terms of "innovation and solving problems," he echoes the solutionism of SunLight's office, as the language of problem-solving expanded from headquarters to rooftops through the close coordination of white-collar engineers and senior installers working on the same projects. Notably, SunLight's subcontracted installers *never* characterized their labor in terms of solutions during my extended conversations with them; only installers whom SunLight employed spoke of problem-solving, pointing to their inculcation into the company's technocratic habitus. Yet despite its office origins, this rhetoric aligned well with the installers' blue-collar individualism— their sense of pride in overcoming challenges. For example, at the chaotic worksite of a mismanaged project, Juan, the project foreman, gestured panoramically to the various technologies, wires, plastics, and detritus that cluttered the rooftop, earnestly declaring, "I believe there's no prob-lem, just a lack of a solution." The next week on the same site, the crew struggled with screws due to a corporate office mistake, but Rick, a crew member, remarked on their conundrum optimistically: "I don't focus on the problem, I just focus on the solution." In Juan's and Rick's statements, *solution* is an articulation of an entrepreneurial, optimistic spirit in the face of a professional challenge—a teleological will to improve that affec-tively animates both the technical work of SunLight's office *and* the cor-poreal labor of working-class installers. From the rooftop to the C-suite, employees identify as problem-solvers, performing a professional per-sona fixated on fixing things—what Ilana Gershon calls "the self as a bundle of previously successful business solutions."[33]

Along these lines, SunLight's installers often characterize their cor-poreal labor as not so much hard physical work but instead technical problem-solving—referring specifically to the office's analytical imper-atives. Consider this reflection from Darwin, another SunLight foreman:

> I've been trying to standardize everything, coming up with uniform ways of building [solar projects] so that it's super-efficient. . . . When we first started building, we would only run raceways along the east and west sides and then trace the wires through them under the rows however we could, but you can see there's lots of obstruc-tions that prevent us from running wires directly under the mod-ules, so we've just started putting these raceways between those spaces to help make the circuitry really simple. . . . I came up with this to make it better. You always have to innovate. They put a lot of the design onto the foremen in terms of how we arrange our sys-tems. I've been trying to push for more coordination between the forepersons and the PMs [project managers]. . . . I wanna make a more streamlined approach; sometimes they [the PMs] don't design things well because they partially don't know how the systems actu-ally work and feel out here in the field. . . . *I like feeling my mind is useful, and people want to know what my design ideas are, and I want to be a part of the process.* (Emphasis added)

Darwin here states that his job is most fulfilling when he's valued not merely for his physical prowess but also for his mind—when he engages in work that often occurs in the office, not the field—situating himself in the putatively cognitive domain of white-collar labor. For many install-ers like Darwin, the analytical facets of developing solar electricity systems inflect the otherwise monotonous or physically draining experience of this work, solidifying their sense of personal ingenuity—their blue-collar individualism. Indeed, many installers enjoy developing processes and improving strategy more than they enjoy the corporeal work aestheticized in ecosocialist visuals of blue-collar bodies (see figures 4.1–4.5). These technocratic predilections do not preclude worker organizing, but when field workers identify with their corporate higher-ups, they often eschew a working-class solidarity centered on redressing physical exploitation.

2. ENTREPRENEURIALISM

When installers characterize their physical work as a catalyst for white-collar careers, their upwardly mobile imaginaries bear the imprint of not only the solar industry's technocratic solutionism but also its rudimentary organization. For the solar industry is particularly embryonic in NYC, as the city's diversity of buildings, rooftop particularities, range of ownership/tenancy arrangements, and bureaucratic codes create challenging business conditions (even as government investment in renewables has driven remarkable growth in recent years). As such, many companies fold within a few years, no company has established a stronghold on the market, and solar entrepreneurs dream of beating the odds. In this context, many installers with a general knowledge of the industry aspire to establish their own shop, subscribing to a gold rush mentality that regards empty rooftops as the next frontier. While many physical laborers in the construction trades are similarly entrepreneurial, the solar industry's incipience generates a particularly acute entrepreneurialism among even the lowest-level workers. They regard the solar market's rapid growth as a means of not greater labor power or worker solidarity but instead professional advancement and business ownership.

"I like being in a new industry, a growing industry, 'cause there's lots of opportunity to move up in the working world," explained Igor, a solar installer-turned-foreman. "People [working] in disruptive businesses and disruptive technologies [i.e., white-collar professionals] are there for the same reason: opportunity to move up the corporate ladder. If you're in an established industry [unlike solar], the steps to career progression are slow, so career progression is a key reason why I do solar." Igor here links his professional aspirations in the solar industry with those of white-collar professionals, eschewing a labor politics predicated on working-class solidarity. Gabriel, an especially skilled installer, expanded on this sentiment by espousing a go-it-alone entrepreneurialism that many other installers echoed: "I want to be working by myself. I want a business like this [i.e., his employer's company], but myself—own my own small business. And this [solar] work is the only way I see it for me to start my own shop. . . . One of my dreams is to be working for myself and be caring for myself."

Many installers envision attaining self-sufficiency and personal autonomy through ownership, as the solar industry's small-business landscape configures their blue-collar individualism. Straying from anything resembling collectivism, most installers I spoke to posited accumulation and enclosure as correctives to their precarity. Kevin, a carpenter-turned-installer, aptly synthesized this sentiment: "I've been coming up [from poverty] for a while in the construction and real estate game, and now [that I'm doing] solar, one of my goals is to create my own [incorporated municipal] *town*—start a solar real estate empire and eventually buy a town that I own powered by solar. . . . The information I'm learning from doing solar is gonna help me build my town."

Kevin's stated desire to start a corporate "empire" that enables him to not merely own property but to more specifically "buy a town"—to become a market-based sovereign—bespeaks something more insidious than generic aspirations for upward mobility: the colonial logics of racial capitalism. For many installers, this imperial entrepreneurialism is aligned with the corporeal work of solar installation, as the sense of personal resilience that animates physical labor among working-class men of color like Kevin coheres with capitalist dreams of overcoming their classed position with the master's tools. This neoliberal ethos often makes their labor bearable; it is something to hold on to during the slog of the workweek.

3. ENVIRONMENTALISM

Solar's environmental impacts also shape how installers relate to their corporeal work. Specifically, installers often situate their labor in climate action narratives, performing a pride in their profession's emissions profile. As Michael, a Bronx-based installer put it, "[I'm] proud to be out here sweating, lifting, fine-tuning—just busting my ass—for something that matters: sustainability"—an assessment he offered on a muggy summer day in which he struggled to hoist several hundred-pound ballast blocks up five flights of stairs. While I cannot speak to Michael's psychic life, his narrative suggests that his labor's environmental impacts endow the *experience* of physical work with moral gravitas, as many installers like him profess to enjoy drudgery with a higher purpose.

Yet for all its eco-affordances, their work is decisively *not* about anything resembling an ecocentric return. For Michael and many of his peers, those affordances are instead enfolded into a broader, intersectionally situated pride in their craft: the sense of accomplishment against the odds, of being a responsible man, of "living a positive life" despite proximate poverty, crime, and drugs. The environmental implications of their labor burnish these bona fides, germinating the sense that they're using their physical prowess to do something good for the world. So Michael's environmentalist ethic is embedded in the classed and gendered feeling of sacrifice that accompanies the physical act of installation: a high-minded machismo—the pride in "busting my ass for something that matters: sustainability."

Consider the zealous testimony of Kendall, a fairly young installer:

> I'm out here doing this job because I can provide for my family and because we need it for the planet. This solar stuff: this is how we fight global warming, and we need that if our kids are gonna grow up and live on this planet. So I'm proud of my work every day because when I . . . go to work, I know I'm helping the planet, helping my kids survive, and my grandkids too, doing something positive for the community. And if that wasn't enough, I'm lowering the community's energy costs. . . . Saving [on] your energy bills and saving the environment at the same time.

Echoing Kendall, many installers specifically espouse this sort of environmentalist ethic when they share their pride in beating the odds and making something of themselves by doing valuable labor.

I want to suggest, then, that solar's environmental impacts endow their work with a positive affect that is discordant with their underacknowledged labor grievances, generating a sunny facade that silences talk of occupational safety hazards, questionable compensation, or even physical displeasure. Although I did speak to one installer who tied his labor struggles—his relatively unsafe labor conditions and low pay—with the anticapitalist platform of climate justice activists, most installers performed an environmentalist ethic that uncritically heralded their physical labor. My point is not that these installers are indifferent to their terms of employment but rather that the pride of helping the environment with their hands can feel somewhat incongruent with an unapologetically pro-worker politics—to say nothing of the ecocentric praxis that I have called *return*.

For while solar is a means for blue-collar workers to corporeally engage with the terrain of life that we often call *the environment*, this corporeal engagement is primarily defined by the lived experience of being an able-bodied, working-class man of color—not by a desire to return to the more-than-human world. The intersectional character of this labor, then, complicates my romantic suggestion that installing solar on city rooftops can entail a reciprocal form of care between humans and our ecosystems. While the constitutive exclusion of Black and brown people from the "genre of the human" has inspired many forms of reciprocity between demattered Black/brown lives and nonhuman life, racial capitalism has, at the same time, conditioned forms of labor among working-class, able-bodied men that do not engender such more-than-human relationality even when this labor is mobilized toward renewable energy futures.[34]

In sum, technocratic solutionism, entrepreneurialism, and environmentalism shape solar installers' working-class individualism, presenting ideological and affective hurdles to realizing an ecosocialist energy transition that centers blue-collar labor power. This does not mean that installers are disinterested in labor organizing or socialist reforms; as I will discuss, many have in fact agitated for greater worker protections. My point here is that the corporeality of solar installation is colored by feelings of technological progress, enterprising initiative, and environmental improvement that affectively counter traditional pro-worker politics. Put differently, the solar industry's particular material, ideological, and economic conditions orient installers to their physical work in ways that problematize efforts to implement an ecosocialist agenda.

Labor organizing in the solar industry must therefore transform not only how installers are paid and valued but also how they conceptualize themselves as classed, gendered, and raced workers—how they *experience* their position in a corporate habitus conditioned by the pursuit of returns. This observation is applicable to *all* blue-collar fields, but it is particularly pertinent to a sustainability industry that operates through technocratic solutionism, entrepreneurialism, and means-of-reduction environmentalism—an industry in which manual laborers echo their employers' white-collar values, corporate aspirations, and normative politics.

A pro-labor energy transition must therefore challenge the affective dimensions of returns, targeting technocratic dispositions that aggravate a working-class individualism productive of differential value. While I have suggested that working toward this end could entail reorienting the physical labor of solar installation toward reciprocal relations with more-than-human lives, this idea is obviously extremely abstract—especially given solar installers' intersectional contexts. It's no coincidence that well-compensated, predominantly white clean energy technocrats dream of something akin to return, while poorer installers of color concern themselves with thriving against the odds. When you have some personal distance from intergenerational poverty, structural racism, and a housing crisis, you can focus your sustainability work on the forms of life that the city's harsh rhythms drown out. This is not to say that privileged white people care more about environmental issues—the EJ movement and public research have shown that this couldn't be further from the truth—but rather that they have greater latitude to define their source of livelihood in relation to matters that are unrelated to their most immediate lively needs.[35] This sort of ecocentrism often overlooks Black and brown lives, forgetting not simply that environmental problems disproportionately harm marginalized people but, more crucially, that the separation of human society from the nonhuman world occurred through the transatlantic slave trade and European imperialism.[36] How, then, can a regime of corporeal work rooted in a vestige of those violent legacies—that is, differentially valued Black and brown labor—be reimagined as a means of return? Following the EJ movement's contention that environmental injustice occurs not only where people of color work but also where we live and play, this question impels us to examine not merely labor but also the realm of life that racial capitalism dialectically counterposes labor to: leisure—the consummate mode of return, as I now discuss.[37]

FROM THE FOREST TO THE DUMPSTER

SunLight occupies a skyscraper two blocks from the New York Stock Exchange—the epicenter of maximizing returns. But in a sign of the company's penchant for return, they held one of their annual three-day employee retreats at a wooded sleepaway camp in the Catskills without Wi-Fi or cell service—effectively preventing screenwork—a back-to-the-land moment if ever there was one. While this location aligned with Sun-

Light's environmental brand, it also revealed the company's affinity for a nature built on the differential value of settler colonialism. For the Great Outdoors, an imagined site of leisure, was engineered by genocidal white male settlers intent on creating a space of natural respite from polluted industrial life in the nineteenth century.[38] Since then, it has remained a staple of white male recreation, shaping normative American masculinities.[39] Of course, people of all genders and races look to the wilderness for leisure—but parks, conservation areas, and camping grounds have long been prevalent sites of anti-Black racism, Indigenous dispossession, and patriarchal exclusion.[40] Such intersectional violence coded these spaces as sites where "man" can escape the pressures of modern work life, allowing him to return to his natural self. As such, the outdoors is a site of leisure in contradistinction to labor, erasing (Indigenous) land-based livelihood and the historical inextricability of human work and well-being in spaces coded as nature. Yet many white-collar men perform outdoor leisure in ways that cultivate an alpha affect that they practice in the boardroom, such that the wilderness animates the raced masculinities that dominate office spaces even as it is imagined as discrete from work.[41]

SunLight's decision to hold their retreat at a rural sleepaway camp reflected these intersectional dynamics. This decision, then, signaled not a decolonial conjoining of leisure, labor, and land but rather the corporate strategy of utilizing leisure to enhance labor productivity vis-à-vis the land.[42] Indeed, the company's CEO alluded to "productive" leisure in presenting his venue choice; the campsite would allegedly enable employees to leave the stresses of the workplace and have fun while they plotted strategy to maximize returns. Unsurprisingly, SunLight's white male employees were the retreat's most stalwart champions. For these employees, the outdoors was at once an enlivening pause from work and a site to affirm their workplace values—a space where leisurely bodies can contribute to corporate intelligence.

While some women and/or people-of-color employees were as excited for the retreat as their white male counterparts, most didn't want to camp in the woods for three days. Many of them had spent years navigating the white male hegemony of corporate America, acclimating themselves to white-collar microaggressions, learning to fit in at the water cooler. And now the office-based careers that they had given so much of themselves to demanded that they camp out in rustic cabins? When they were told to reimagine outdoor leisure as an offshoot of indoor labor, many suggested that their superiors could not see the intersectional exclusions linking their workplace with the wilderness. For them the unremarkable act of white men speaking resolutely in a boardroom and the equally commonplace act of white men backpacking through the woods are in many respects two sides of the same coin, evincing the many ways these men have been habituated to perform a hegemonic masculinity through both labor and leisure.

Consider Matt, the "visual learning" engineer—a white man who effortlessly occupied both the boardroom and the backpacker trail. Before the retreat even began, he had scoped out the campsite on Google Earth, where he found a gazebo in the site's surrounding woods. He imagined transforming the gazebo into an outdoor bar powered by a solar panel and battery, showcasing SunLight's technological ingenuity in the middle of the woods. Put differently, he sought to create a conduit between outdoor leisure and SunLight's brand of indoor labor—the "bro-y" intelligence that Sean bemoaned (see chapter 2). Toward this end, he brought solar technology, Christmas lights, a speaker, and a truck full of liquor to the retreat and quietly ventured alone into the woods during lunch on the first day to transform the gazebo. There he created an outdoorsy speakeasy that he named the Forest Resiliency Bar, branding a space of leisure by referencing his technocratic labor. (*Resiliency* referred to *climate* resiliency—a not-so-subtle allusion to the solar panel and battery.) Mind you, SunLight's leadership, the camp owner, and the unknown party who owned the gazebo had not given Matt permission to set up the bar. The ease with which he claimed a piece of the forest for his company vis-à-vis solar pointed to the effortlessness with which privileged white men occupy both the office and the woods—this audacity buttressed by energy expertise. Thus, Matt displayed an understated coloniality that linked his professional comportment with an outdoor affect—a spatialized entitlement that codes nature as possession.

Importantly, though, this nature is twofold: there is the nature of Matt's leisure (the outdoors that he enjoys dwelling in) and the nature of Matt's labor (the climate system that his solar supports). These two natures work together, as the bodily experience of outdoor leisure inspired him to address climate change through cleantech labor. The Forest Resiliency Bar deepened their alignment, forming a "technonature" of sorts, as Matt's gazebo project was a shrine to both the environment of transcendentalists and the environment of techno-scientists: a show of deference to the nonhuman world in the form of humans' "sustainable" innovations.[43] The bar, then, gestured to return, enabling moderns to dwell *in* the environment through technologies of care *for* the environment.

This return, though, was classed and raced in ways that became apparent around 10 p.m. on the first night of the retreat when dozens of SunLight's energy engineers, data analysts, and executives traversed the woods with iPhone flashlights to get to the off-grid bar powered by renewable energy. Absent from this gang were the two solar installers who had voluntarily come to the retreat. In a show of inclusivity, SunLight had invited its blue-collar crews to attend despite the retreat's corporate focus, predictably eliciting little interest from installers who got paid an hourly wage and therefore would not be compensated for their time in the woods. These installers also lacked the professional credentials and social bonds that would make the retreat more worthwhile than time with their families. Of the two installers who *did* attend, though, one was very clear that he hoped to grow professionally as part of his bigger plans for upward mobility in the industry. But the other one, Manny, was reserved and stoic at the campsite, dutifully participating in the various retreat sessions but with no apparent sense of purpose or enthusiasm. I wondered where these two men were as the Forest Resiliency Bar filled with the distinct chitchat of white yuppie millennials. Did they find the raced and classed affect of solar-powered leisure in the woods alienating?

The bar's patrons were having a blast, enamored with Matt's drinks and their collegial time in the forest under the moon. But the solar panel and battery were the stars of the show, centering everyone's attention on Matt's novel twofold nature: the forest ecosystem coalescing with the technological environment. Commentary abounded on the trees' contiguity to off-grid solar, as the patrons suggested that the bar's renewable energy allowed them to leisurely hang in nature without harming nature—to appreciate nature while giving back to nature—theorizing a reciprocity akin to return. Yet return, as Berry envisioned it, entails a skillful attunement to the nonhuman world that in theory exceeds the spectacle of solarized wilderness. Was Matt's invention a means of living reciprocally with nature or a gimmick with good sustainability optics subsidized by the labor of precarious installers?

This tension surfaced when a finance bro didactically explained the bar's novelty: "Usually parties like this have a negative environmental impact, but the solar offsets that." While several other patrons humored this banal commentary—others had already said the same thing—Suzie, an energy engineer, contested it: the solar would do nothing to help the flora and fauna whose homeland we had invaded for a loud party in the woods. Citing her previous work inspecting nascent solar technologies in a national research lab, she described the corrosive impacts of purifying silicon and mining lithium for these technologies, debunking any delusion that the solarized forest spectacle was unequivocally good. Her hands-on experience deflated the staged scene of return. Her colleagues sheepishly agreed with her assessment: the bar was neither resilient nor sustainable; it was instead a showcase of cool human technology framed by the forest's beauty.

Having acknowledged this, they resumed their praise of Matt's tech-nonature. Even Suzie was struck by the bar's symbolism—its Christmas lights adorning moon-tinted trees and powered by the sun vis-à-vis the battery. This space was an invitation to people who stared at screens all day to instead dwell in the forest—to smell its lush leaves, feel the midnight breeze, hear the sound of squirrels rustling in the bushes—without distancing these sensorial encounters from the technological complexity of the modern world. The bar, then, felt like a cipher of ecocentric care despite its limitations. Our boozy appreciation of (un)natural beauty comprised an ambivalent stewardship pertinent to a dangerously altered climate. Immanent in this gratitude was some semblance of return: a forest-human communion but perhaps one that Berry wouldn't recognize.

The bar's affect stayed with me the next morning when I struck up a conversation with Manny, inquiring about his absence in the woods. "I dunno, I mean . . ." he started reluctantly. "The bar thing was kinda late for me. I'm used to being in bed by 10:30; I usually gotta leave my home by 6 to get to the jobsite on time." These words exposed SunLight's ugly classed divide. Whereas most employees arrived at SunLight's centrally located office between 9 and 9:30 a.m., the solar installation crew had to get to their various jobsites all over the city by 8. Whereas most employees lived in gentrifying or upper-middle-class neighborhoods relatively close to the office, SunLight's solar installers lived on the periphery of the city's outer boroughs without personal vehicles. Whereas many SunLight employees ate their breakfast at their desks on arrival, solar installers had to begin working when they reached the jobsite, without recourse to meals until lunch. For Manny, this classed labor demanded discipline that was incompatible with a late night at a bar.

But the timing only partly explained his absence. As we kept talking, he explained that he had never camped out in the woods. A twenty-four-year-old Afro-Dominican who had lived his whole life in a poor community in the Northwest Bronx and in the projects in central Brooklyn, his outdoorsy experience was limited to field trips to Central Park when he went to summer camp at the YMCA, trips to the Rockaways' crowded public beaches, and trips to his cousin's modest suburban home in Long Island, which had a small backyard, "but at least it had some grass and it was quiet." He had come to SunLight's retreat because it offered him an opportunity to experience "the environment," and while he thought it was "really nice," he felt intimidated by everyone else's ease in the woods—including that of the other Black installer. Although he had never flinched at doing dangerous construction work atop skyscrapers, he felt anxious about possibly encountering spiders, poison ivy, and, tellingly, hound dogs at the campsite. *Hound dogs?* He chuckled at my incredulity. "Well, no, I mean, I'm kinda joking, I just mean like, I'm not tryna get lynched out here, you know?" He smiled with a hint of self-deprecation to suggest that his fear was irrational. But in evoking Jim Crow to explain his discomfort in a northeastern forest, Manny spoke to white supremacy's grip on nature in the United States—the specter of slavery ingrained in our surroundings.

Not wanting to seem like a complainer, he continued to force a smile, adding, "But it's really nice out here. I like it. Just not what I'm used to." I asked what he's used to, and after beating around the bush for a while, he eventually told me about growing up in the projects—how everyone kept their apartments super clean, and yet it always seemed like a giant mess: how a giant dumpster sat adjacent to their building and how it was always overflowing with garbage, how he and his friends would sit on a bench next to the dumpster, and how commonplace this perpetual mound of trash seemed yet how unnerving it was. How the dumpster's visibility felt like a violent exposé: everyone's shit literally out in the open for anyone to see. How when he walked from the subway to Central Park, he'd marvel at the multi-million-dollar apartments not only because they were glitzy but also because you couldn't see their shitty guts. "Rich people make trash like everyone else, and I would bet you that my mother kept our apartment as clean as anybody on the Upper East Side, but the difference is you don't see their shit like you see ours."

Several weeks after SunLight's retreat, I contemplated this visibility when I ventured to the projects where Manny had once lived to see if the dumpster was still there. Indeed, it was. And it was as corrosive as he suggested: this giant stain of stinky trash impeding the entranceway to an old brick skyscraper whose lobby did, in fact, seem very clean in spite of its state of disrepair. The building's trash wasn't the only hyper-visible thing there. A protruding NYPD surveillance station hovered over the entranceway, and an NYPD car idled vigilantly across the street, while the sterile pavement on the building's periphery lacked even a single tree, accentuating the residents' visibility as they went about their day. Their environment was one in which they were constantly seen with limited control over what they displayed. While Manny's community worked hard to keep their environment orderly, the waste infrastructure constrained this agency—their space marking their subordination. To occupy this space is to be visible on someone else's terms, and while this visibility has never stopped public housing residents from seeing one another outside of the carceral state's gaze, it is a testament to the ways in which differential value constitutes their environment—setting the terms through which they live, work, and play.

When this visibility mediates your attunement to your surround-ings, it is difficult to reciprocally engage with nonhuman life there—to *return*. Of course, the sensorial texture of Manny's neighborhood can-not be reduced to surveillance. There you can encounter the flow of water from fire hydrants opened by teenagers on scorching summer days or the sound of elders' old-school jams blasting from antiquated boomboxes on benches or the flutter of pigeons' wings as they descend on the crumbs of a discarded bag of chips. But even with this environmental dynamism, life as an anonymous spectacle offers little recourse to a give-and-take with your environment in the way Berry envisioned it.

Standing on Manny's block from his adolescence, I considered a point of contrast: the sensorial texture of the Forest Resiliency Bar. In the bar's juxtaposition of soil and solar panels, leaves and laughter, botanical life and battery life, SunLight's office professionals felt a liveliness that could not account for life in Manny's environment. Matt's technonature would never redress the spatial legacies that instilled Manny with anxiety about the woods, no matter how attuned with life the bar made his white-collar colleagues feel. As Manny worked hard to reduce emissions through phys-ical labor owned by a predominantly white cleantech company, such labor could not eradicate the spatialized differential value that had rendered the forest foreign to him.

Imagine if Manny's labor wasn't controlled by white men who, in spite of their best intentions, could never understand everyday life by a dump-ster or under the carceral state's gaze. Now imagine that this labor sovereignty could unearth the sense of possibility that many of us felt at the Forest Resiliency Bar, enabling Manny to connect with his environment in ways that a back-to-the-land retreat could never do. As I now show, this scenario is not actually farfetched. For other workers have gone further than Matt in practicing ecological care through technological prowess—resituating modern infrastructure in a broad ecology that includes destitute urban spaces. While the energy transition has been driven by Google Earth gurus and spreadsheet specialists whose screenwork curates Manny's labor, others are seeding solar in ways that center the marginalized lives and livelihoods that even the imagery of figures 4.1–4.5 overlooks.

FINDING LIFE IN LANDSCAPES ANNIHILATED BY LEAD

Racial capitalism casts a long shadow over the financial capital of the world, but that shadow is patchy, giving way to pockets of sunshine that presage return.[44] The shadow, though, makes it difficult to see those sunny patches in a skyline that is a shrine to differential value, molded as it was by labor deemed subordinate to intelligence. A year after SunLight's retreat, I tried to find those sunny patches above that skyline, hoping for the best when Manny and his installer comrades reached a breaking point. SunLight had prioritized returns over their rights, forcing them to toil in inclement weather and unsafe conditions while paying them just barely more than NYC's minimum wage. After several preventable on-the-job accidents, including one that was almost fatal, the company's crews began to unionize. But SunLight then uncoincidentally decided to subcontract *all* of their solar installation work, essentially firing all of their unionizing installers. This union busting was an obvious breach of the company's equicratic brand.

While their disregard for their employees was inexcusable, it reflected not only an immoral antipathy to workers' rights but also the company's external economic constraints, including venture capital investors' unwavering demands, a largely unregulated blue-collar solar labor market, competition among solar companies to minimize costs, and the high operational costs of a fragmented private industry lacking the public sector's economies of scale. In these ways, racial capitalism cast its shadow over every facet of SunLight, impelling the company's leadership to adhere to the crude calculus of maximizing returns. I raise this not to minimize SunLight's culpability but rather to suggest that there are structural limitations to how much good a corporation can do beneath that omnipresent shadow, even when they embrace an equicratic ethos. Although the internet is full of feel-good stories about corporations that thrive by embracing workers' rights, the extractive machinations of differential value often preclude such win-win success—especially in NYC's solar industry.[45]

But while racial capitalism cast its shadow over SunLight, EQUAL (Environmental Quality for All) was beginning to expose the sunny patches. When EQUAL launched their Solarize Our Lives (SOL) campaign, they planned to train local underemployed workers in solar installation and then place these workers in jobs with their solar contractors. But they grew frustrated when those local jobs didn't materialize. Instead of appealing to private corporations who didn't care about the community, EQUAL launched their own worker cooperative: a solar installation enterprise completely owned and operated by local workers.

The cooperative model rejects the differential value separating owners from workers, creating new sunny patches in the shadowy realm of monetized labor. Specifically, EQUAL's solar cooperative enables community members to work collaboratively to secure their own livelihood, steering investments in solar energy markets away from a fixation on returns. The cooperative is governed through a consensus-building process in which all participating workers listen to one another, affirming their solidarity. No decisions are made without the input of a majority of the cooperative's workers, while an EQUAL employee—who is also a cooperative member—provides administrative and logistical support. Most of the cooperative's worker-owners have second jobs in construction and the service sector to supplement the cooperative's part-time work, but they hope that their burgeoning project portfolio will create opportunities for full-time employment.

The cooperative's good compensation, safe working conditions, worker solidarity, and democratic control have improved not only their well-being but also their attention to skillfulness. In their telling, labor sovereignty accentuates their care for the craft of solar installation, fostering a sense of purpose that was less present when they labored for someone else's bottom line. "Now that I'm a co-op worker-owner, I don't really do this for the money," explained Jackson, an erstwhile construction worker and real estate broker. He continued:

> I mean, the money is good, I definitely want the money, but to be
> honest . . . I just wanna do a really good job, and I want every instal-

lation to be so tight 'cause this is so much bigger than a job for me.
... It's not about the individual; it's about the *collective* giving back.
... I do [this work] to put a smile on people's faces. I know that when
we finished that four-megawatt system, there were eight hundred
households that were able to get power from the work that we did.
... Look how many households we saved from coal energy and oil
energy. We know that we making a big dent in their carbon footprint, and that's what I focus on now more than when I was just
focused on ... hustling for a paycheck.

In Jackson's telling, the cooperative's collectivist ethos and living wages
instilled his work with greater accountability to communities and the
environment—accountability that even the most equicratic corporate
employment precludes. Put simply, the cooperative model reconfigures labor around care and craft. By shifting work away from extractive
returns, it presents an opening for an inclusive return—a sunny patch
in capital's shadow.

Yet realizing this potential vis-à-vis solar will require linking cooperative work with other forms of environmental care; worker ownership, by
itself, is not enough to position labor as a medium for attunement with
more-than-human life. Cooperatively installed solar can only go so far in
disrupting those corrosive environmental relations epitomized by Manny's surveilled, trash-filled surroundings. In addition to worker ownership, a solar-powered return requires the sort of labor that reconfigures
how we interact with the world around us.

A Brooklyn-based waste composting organization led by young people
of color, BK ROT, offers a blueprint for this work.[46] In spaces like Manny's where surveilled people of color are partly marked by the visibility
of their refuse, this organization transforms waste through labor that
is partly powered by the sun and eludes the state's panoptic gaze. For a
small fee, BK ROT's young workers of color haul food waste with solar-
powered cargo trikes from businesses and homes to a community garden
called Know Waste Lands in a low-income but rapidly gentrifying neighborhood. They then use solar-powered mechanical sifters to convert this
waste into compost, which they use to regenerate the garden and sell to
gardeners throughout the city. Revenues from compost sales and waste
hauling, combined with nonprofit funding, enable BK ROT to pay their
youth workers a living wage for labor that is usually unpaid.

Community-scale composting might seem like small potatoes when viewed through means-of-reduction screenwork on carbon and climate collapse. But when we understand climate change as a matter of not only emissions but also relations, not only fossil extractivism but also racial capitalism, not only differences in temperatures but also differential value, BK ROT can teach us to live and labor amid a climate system that is unsettling for the same reasons that Manny's space is surveilled. Specifically, their composting approach can inform an inclusive return that disentangles energy transitions from extractive returns in four ways. First, it combines physical labor and solar power to generate reciprocity between human bodies and more-than-human life. Second, it disrupts the hypervisibility of communities of color through vegetation that shrouds their street. Third, it employs a paradoxical nonprofit model that privileges corporeal work over the screenwork that sustains nonprofits. Finally, it holistically links energy transitions to our physical terrain, moving solar from the means of reduction to skillful stewardship. I now unpack these four facets of return, considering a solar-powered economy grounded in corporeal relations with our everyday spaces.

I. RECIPROCITY

The means of reduction suggest that solar is important because it reduces our carbon emissions, minimizing humans' environmental impact. This idea is contiguous with the ecomodernist assertion that we must disentangle our lives and labor from the natural world to ensure a habitable planet. In this schema, renewable energy should maintain an organization of life in which the human and nonhuman are separate domains.

BK ROT takes a different approach, using solar as a means of integrating human and more-than-human life in urban landscapes that are fundamentally lively ecosystems of which humans are just one part. By tapping the power of the sun, their cargo trikes enhance the power of their quads, glutes, hamstrings, and hips to haul organic waste throughout the city, combining their corporeal strength and solar electricity to collect and drop off food scraps. They further bring together the "little-e energies" of the human body with the "Big-E Energy" of electrical technology when they lift mounds of food scraps and dump them into solar-powered mechanical sifters that transform waste into reusable compost.[47] The soily product of these solar/human interactions is then used to (re)generate vegetation, both in the community garden off an otherwise treeless avenue in a low-income community of color and in gardens throughout the city.

Thus, when the power of our limbs and the power of the sun come together, we can sustain life in ways that posit human impact as regenerative—not as an environmental scourge that must be reduced. This mutually beneficial entanglement of bodies and solar exemplifies return: reciprocity between humans and their environment facilitated through physical skill, flourishing with organic energy. It lacks the white-washed nostalgia for a return to a pure nature predicated on Indigenous genocide. Instead, BK ROT historicizes their work as a return to Black and brown ancestors, citing Ghanaian women's composting practices in the thirteenth century, Cleopatra's edicts designating earthworms as sacred to ensure soil fertility in ancient Egypt, and George Washington Carver's regenerative Black farming in the US South at the turn of the twentieth century, among other ancestral references. However, they enact this return not by visiting diasporic origin points but, instead, by returning to spaces they already inhabit—spaces to which we now turn.

2. SHROUDEDNESS

The former executive director of BK ROT, Ceci Piñeda, notes that spaces suffering from "disinvestment in the soil" are also spaces suffering from "disinvestment in community," as redlining and other racial capitalist projects situated Black and brown people in polluted areas besieged by petro industries, waste transfer stations, and commercial truck routes—infrastructures that contaminated the land, prevented the cultivation of life, and, consequently, deprived those people of trees and vegetation.[48] The suffering soil in the city's marginalized communities, then, is kin to humans of color plagued by health defects linked to pollution and waste—part of a toxic entanglement of different forms of life: organic matter, microorganisms fossilized as fuels, Black and brown lives, and the wasteful disavowal of the life cycle. This entanglement gives form to communities like Manny's: landscapes in which the absence of trees contributes to a hypervisible space where the state's panoptic gaze easily detects racially marked bodies.

BK ROT disrupts this hypervisibility through their 2,500-square-foot composting oasis at an intersection of treeless streets: a "wildlife garden" enclosed by tall vegetation doubling as a green sanctuary for local people whose movements are often surveilled on the impermeable pavement adjacent to the garden. Know Waste Lands is therefore a space to freely move one's body without surveillance, to dwell in community without being watched by the state—evocative of the maroon settlements and Black geographies in which runaway slaves worked with more-than-human life to settle outside the gaze of white supremacists.[49] The sun makes possible this "interruptio[n] in the space-time rhythms of mastery and dominion," as J. T. Roane might put it—through both the timeless power of photosynthesis and the modern power of solar technology.[50] Put differently, these solar powers collectively enable the forms of life that keep the garden dwellers imperceptible to the surveillance apparatus. Yet modern solar technology here necessarily facilitates a return that does not preserve a romantic *before* settler colonialism but rather invigorates relations within the hidden floral pockets of concrete spaces where Black and brown people *presently* dwell.

3. REGENERATION

By regenerating soil, BK ROT rejects the extractivist ethos at the heart of differential value, eschewing surplus to return life to the earth. Yet the organization is not immune to the logic of yielding returns; as a nonprofit, BK ROT is dependent on both capitalized labor and the non-profit industrial complex (NPIC)—and they therefore need monetary revenue. Thus, BK ROT must strategically engage with racial capitalism while simultaneously embracing other ways of valuing the Black, brown, and more-than-human lives that racial capitalism falsely designates as waste.

Solar technology is instrumental toward this end, aligning with both capitalist and regenerative imperatives. Specifically, BK ROT's solar-powered cargo trucks and mechanical sifters vastly increase their workers' output, allowing them to collect more food scraps and produce more compost to sell, generating more revenues. But because they use solar in the interest of return—not returns—BK ROT leverages this increased output to care for their workers by investing more in their wages and postponing compost pickups on days with physically taxing weather—actions that prioritize bodily health over their bottom line. In this way, the productive efficiency of Big-E Energy regenerates the organization's little-e energies—their workers' vitality.

The means of reduction are correct when they show that replacing fossil fuels with renewables can greatly reduce pollution and emissions. But by viewing fossil fuels as something that can simply be replaced with new technology, the means of reduction fail to illuminate how fossil fuels are embedded in structures of differential value that require more than technological innovations to supplant. For example, fossil fuel infrastructure has not only produced the highest rates of respiratory diseases and the greatest vulnerability to climate change in NYC's communities of color but also harmed these communities by depriving them of vegetation, robust soil ecosystems, and permeable surfaces. As such, these communities' fossil-fueled precarity requires not simply new technology but also a new approach to stewarding landscapes deprived of life, transforming the matter we treat as waste and providing for the people who live in these "waste"-full landscapes.

BK ROT offers one such approach, tapping into the sun in ways that uproot the extractive conditions of their everyday spaces. With solar technology as an aid, *not a panacea*, the organization's young leaders live, breathe, bike, hold, touch, and haul a tangible alternative to fossil-fueled waste, tending to the literal roots of their vulnerability. In the process, they make physical spaces that clean the air and safeguard against climate disasters, protecting their well-being in ways that simply reducing their carbon footprint through new infrastructure could never do.

IN BK ROT's reciprocity, shroudedness, regeneration, and holism, I detect a return that is neither nostalgic for a colonial nature nor reductive in its approach to environmental care. In this return, the power of the sun coalesces with the power of our soil, bodies, imaginations, and communities to move beyond fossil fuels in a way that centers life in all of its intersectional and interspecies multiplicity. This lively assemblage counters a regime of differential value that deems energy as something that is extracted from the earth by reconceptualizing it as something that is *enacted with* the earth. This challenges the anthropocentric hierarchies through which racial capitalism denies dark soil and dark skin access to the domain of the human. This work, then, rejects the late liberal spectacle of equity in favor of a lithe, sentient energy—not simulating equality on a screen but choreographing alliances in the soil. Yet this earthly labor is not outside the day-to-day life on concrete streets that Google Earth and surveillance cameras reduce to data points. As such, BK ROT stewards the earth in solidarity with the sun in the shadows of racial capitalism, tending to both botanical lives and bottom lines, locating roots in landscapes annihilated by lead—the productive Big-E Energy of solar technology flowing through affective little-e energies of soily entanglements.

CONCLUSION: DOUBLE CONSCIOUSNESS

On the surface, return is at odds with transition. The latter, as it relates to energy, refers to progress, futurity, teleology—a rupture in fossil fuel hegemony moving us toward a more sustainable epoch. The former, inversely, refers to a cyclicality without disruption, recuperating pastimes, rejecting progress as it's normatively understood. While the energy transition embraces technological innovation and novelty, ecological return focuses on long-standing skills and rhythms. Yet, as I have begun to argue, any transition from an energy paradigm rooted in the extractive metaphysics of racial capitalism must be grounded in return. To underscore the essential symbiosis of transition and return, I now recapitulate the two central arguments of *Subjects of the Sun* with a concluding reflection on late liberal screenwork.

Those arguments are as follows: when solar advocates enact the just transition through screenwork and cloud-based platforms, they unwittingly reproduce the structures of racial capitalism that they aim to counter. For instance, their shiny images of solar farms push from view the racialized extractive zones embedded in solar supply chains; their Google Earth renderings of solarized rooftops position corporate actors as a force of social equity, obfuscating cleantech corporations' dependence on differential value; and their smart boards aestheticize energy through the means of reduction, disentangling EJ campaigns from the on-the-ground environmental injustices those campaigns are supposed to redress. The just transition, then, must not be limited to late liberal screenwork; it must transform how we relate to energy, reorienting us to the transformations in matter that sustain modern life. More specifically, I have argued that a just transition must move from a graphocentric apprehension of the world toward a skillful engagement with the forms of life that racial capitalism renders extractable, tapping into the sun's power with our senses: haptic dexterity, interpersonal care, pride, feelings of sovereignty.

To deepen these points, I conclude by considering the two images

FIGURE 4.7. A poster for a renewable energy documentary on a NYC subway platform, 2016. Photo by Myles Lennon.

in figures 4.7 and 4.8, which offer competing visions for a clean energy future. The first is a photo of an ad I took on a NYC subway platform. The ad is promoting a renewable energy documentary sponsored by General Electric and filmed extensively with drones. The second is a computer rendering of a rooftop of Via Verde, a sustainably designed, solar-powered affordable housing complex in a poor community of color in the South Bronx that was built on a remediated brownfield. The rendering features Via Verde's residents stewarding rooftop gardens with elevated solar panels in the backdrop. This rendering appears on the website of the company that designed and installed Via Verde's solar infrastructure. While there are several notable contrasts between these two images, I want to specifically focus on the dissimilar ways they relate energy to differential value.

The first image is a futuristic spectacle whose affective power is rooted not only in the energy infrastructure it displays but also in the contrast between this display and its grimy subway platform surroundings. The soiled off-white tiles and the defiled poster directly to the image's right point to the NYC subway system's unkempt conditions—a filthy splice of underground urbanity that couldn't be further from the image's aerial

FIGURES 4.8 AND 4.9. A rendering of Via Verde's solarized rooftop in the Bronx, and an aerial view of Via Verde's solar panels, circa 2017. From Bright Power, "Via Verde."

view of a mountainous landscape hosting a massive wind turbine surveyed by futuristic drones. The contrast here between a pristine "future of energy" and the run-down subway platform positions the drone image as a spectacle—a larger-than-life sight that shines bright above the banality of everyday life. Put differently, the image aestheticizes high-tech futurity by highlighting the difference between the bleak urban present and a promising ecomodernist tomorrow. The image, then, maps out a horizon that *starkly* contrasts with the present, leveraging this difference to extract value from the ecomodernist dream. This futurity is not interested in late liberal inclusivity; it elevates the technological innovations of the white male visionaries that the documentary centers (e.g., Elon Musk) over the everyday world that the losers of racial capitalism live, work, and play in.

In contrast, the Via Verde rendering situates clean energy in the everyday space of poor people of color who live in an affordable homage to sustainability built on a once highly polluted industrial site. Crucially, it centers these people in the field of view. This sight is juxtaposed with a second image on the website in which solar panels overlook these people's newly greened landscape (figure 4.9). Taken together, these images support the claim that energy transitions can actively redress the differential value that positioned Black and brown communities as extractive zones. Thus, while the drone spectacle suggests that the energy transition reproduces the differential value that separates the spoils of urban modernity from a pristine high-tech future, the Via Verde images suggest that the energy transition can collapse these differences, rendering solar as an antidote to structural inequality.

But despite this contrast, figures 4.7 and 4.8 are similar in that they both reveal the symbiotic relationship between visual forms of representation and racial capitalism. As Susan Sontag argued, "A capitalist society requires a culture based on images," and numerous scholars have theorized the ways in which modern visual media spur capitalist consumption, shape consumer subjectivities, and insidiously position corporations as a force for social good.[51] Many scholars also contend that modern images facilitate commodity fetishism, aestheticizing goods in ways that suggest they have a life of their own.[52] Figure 4.7 arguably exemplifies this fetishism, deifying the wind turbine as "the future" by desituating it from the social relations of the present-day subway platform. But the Via Verde images reveal a different linkage between modern media and racial capitalism. While these images stage solar-powered sustainability as a countervailing force to differential value, I want to suggest that they in fact demonstrate the enmeshment of capital and imagery by presenting a corporate simulation of a late liberal fantasy through not just any old visual medium but digital media in particular. Artist and visual theorist Douglas Davis argues that the contemporary "age of digital reproduction" dramatically accentuates the impacts of mechanical reproduction that Walter Benjamin theorized, erasing the boundary between original and copy.[53] This "realer reality"—Davis's name for what he contends is misleadingly called "virtual reality"—is the modus operandi of late liberal images, reifying racial equity by simulating it on a screen.[54] We can detect this form in the Via Verde images, which, as mentioned, are juxtaposed on a solar company's website. It is telling that the website features an aerial photo of solar panels overlooking a green landscape, while also relying on a computer simulation—not a photo—to center the residents. The simulation adds some human dimensions to a compilation of photos and texts that frames Via Verde's solar in terms of the means of reduction—for example, "130,000 lbs CO_2 offset per year."[55] Put differently, the simulation aestheticizes the real-world impact and everyday realities of a "sustainable" rooftop—*that which the photo doesn't capture*. The simulation, then, makes Via Verde's solar infrastructure real in ways that the photo can't, foregrounding humans' relations with their environs and positioning the panels as a component of an ecosystem, not a fetishized commodity. Yet like any late liberal visual, this realer reality simulates the *appearance* of minoritized subjects, not their lived experiences. For Via Verde's rooftop garden is not, in my estimation, as populated as the image suggests it is—which perhaps explains why the company featured a computer simulation on their website instead of a photo. While I observed the rooftop only for two afternoons, I saw no one on it—and this squared with my observation of other "sustainable" rooftops throughout the city; the roof amenities of NYC's new residential buildings are very underutilized. I partly attribute this to the engineered affect of these spaces. When I visited Via Verde's roof, I didn't feel the lushness and warmth of the rendering—and not because the rendering misrepresents the roof's colors, textures, or dimensions (it does not) but rather because the garden feels like it was designed on a screen. It is a sterilized nature marketed as an amenity, not a space of vital respite for people who live, work, and play in the South Bronx. The frenetic rhythm and everyday exigencies of a gentrifying megacity don't lend themselves to leisure in the architectonic confines of a rooftop that feels realer than real life. The rendering, then, privileges a visual simulation of everyday Bronxites over their lived experiences, casting nature as a service that technocrats design for the less privileged in an attempt to humanize the high-tech futurity they specialize in. Therefore, this "postindustrial nature" does not transform the differential value that positions those Bronxites as beneficiaries of a corporation's sustainable innovations; the divisions that constitute racial capitalism stay in place even when racial capitalism works to redress its sins.[56] My point is not that Via Verde is racist or extractive (it most certainly is not). A towering feat of innovative design, it is an unequivocal improvement over the now-remediated brownfield it sits on. But for all of its sustainability bona fides, it still operates through the structures of racial capitalism that position the white-led corporation as an engine of social governance, detached from the everyday relations that comprise local environments. This is the status quo that the glossy rendering reproduces, simulating bodies to infelicitously humanize a technocratic futurity. In this way, the rendering resembles the wind turbine poster, as both elevate the technocrat's dream of tomorrow over the world as it is experienced today.

Along these lines, this rendering also resembles the far more insidious forms of digital reproduction I explored in this book: other media that produce a realer reality, misleadingly dissolving differential value through megapixels. Among white equicrats, this realer reality often creates a sense of solidarity with the communities of color whose energy performance they manage—as though their screenwork enables them to enact equity in real time in the cloud. Celeste's feeling that Google Earth is a medium of "concrete" change (chapter 2)—that it is more effective than on-the-ground organizing—exemplifies this realer reality. This screenwork, then, accommodates a late liberal form of racial capitalism that grapples with climate collapse, as it affectively animates a world in which for-profit corporations can redress the differential value of an extractive fossil fuel order. As I have tried to show in *Subjects of the Sun*, this late liberal sense of redress simultaneously provides cover for a transnational solar industry that is no less contingent on racialized extraction than its fossil fuel predecessors, and foments an equicratic order that is often indifferent to environmental injustices as they are experienced on the ground by the marginalized people it is trying to help.

This does not mean, as some polemics would have it, that solar is the greenwashed equivalent of fossil fuels. While the means of reduction are a reductive way of conceptualizing environmental change, they do make clear the important benefits of a renewable energy transition. My point, then, is not that we should abandon solar but rather that we should actively steward it in ways that acknowledge the entanglements of racial capitalism and digital imagery and thus prioritize offscreen relations with energy—the sort of on-the-ground, solar-powered skill that BK ROT demonstrates. Admittedly, this is much easier said than done. As I have argued, because the charged electrons that power society are invisible, visual renderings—graphic images, digital platforms, aestheticized metrics—are the primary way we engage with electricity in a graphocentric society organized around the screen. It is therefore easier to visually simulate solar-powered life in a rooftop garden (as figure 4.8 does) than to tap the sun's power in a way that animates our relations with the life around us. Furthermore, such a simulation is essential to actually designing solar infrastructure in marginalized communities like the South Bronx; there is no just transition without the visual expertise of architects and engineers.

Connecting this transition with return, then, demands a broader framework for engaging with energy—one that acknowledges the necessity of screenwork while also underscoring the necessity of forming relations with electrical power unbeholden to screenwork. I am therefore calling for a sort of double consciousness of energy. W. E. B. Du Bois coined *double consciousness* to conceptualize Black people's schizophrenic orientation to the world: "The Negro is . . . born with a veil, and gifted with second-sight in this American world. . . . It is a peculiar sensation, this double-consciousness, this sense of always looking at one's self through the eyes of others, of measuring one's soul by the tape of a world that looks on in amused contempt and pity. One ever feels his twoness,—an American, a Negro; two souls, two thoughts, two unreconciled strivings; two warring ideals in one dark body, whose dogged strength alone keeps it from being torn asunder."[57]

Here Du Bois theorizes Black life as the simultaneous occupation of conflicting scales—the embodiment of contradiction. While he metaphorically evokes vision to illustrate this "two-ness," I detect in his words not so much a dual lens as a tension between the hegemonic white gaze that "the Negro" has internalized and their "dark body, whose dogged strength" resists the discord that this gaze inspires. In other words, double consciousness is less a dissonance between two ways of seeing and more a dissonance between the dominant way of seeing and the truth you can feel in your body when you're seen as other—the dialectical recognition that your existence requires its negation. This ability to both see the world through the master's eyes and hold on to the reality that simmers uncomfortably in your bones is a sensory know-how that can recuperate what the hegemonic gaze peripheralizes.

An energy transition rooted in return requires a double consciousness: the ability, for instance, to, on the one hand, see the world through spreadsheets that contain prices for solarizing public housing, and, on the other, care for this housing with the communal joy that Gina suggests is missing in spreadsheets (chapter 3). Although those spreadsheets' grid-like ledger is embedded in a mode of visual accounting that has historically seen Black lives as units of extractable energy, double consciousness reminds us that this dominant datafied gaze need not be the only way we apprehend the world.[58] I detect this double consciousness when Gina imagines solar as a way of reducing management costs, employing means-of-reduction logics for the explicit purpose not of calculating carbon footprints or cost efficiencies but instead of preserving communal space and caring for her neighbors. I detect this double consciousness when BK ROT taps the sun's power to increase their efficiency, growing their composting business with the intention not of enhancing returns but of regenerating roots: cultivating vitality in concrete surfaces whose lifelessness renders Manny's community hypervisible. And I detect this double consciousness when Celeste frames her screenwork as a practical first step toward a future in the soil, sensing in solar technology an untapped potential to foster an awareness of the ecological webs we are embedded in.

When we recognize that Manny's and Gina's experiences have everything to do with the extractive energy paradigm—that living amid waste on treeless blocks that the cops surveil is symptomatic of the violent ways we extract life to power society—these experiences necessitate we find some common ground between climate models that call for industrial-scale renewables and people's everyday wishes to exist without the normal indignities of life in the shadows of racial capitalism. While Du Bois rightfully characterized double consciousness as a burden, I also sense something generative in it: a comfort with the sort of contradiction inherent to a just transition centered on technologies that are inextricable from an extractive paradigm that is anything but just. Out of this comfort, solar, soil, and skin can coalesce in those spaces that are mistaken for being nothing but sacrifice zones in ways that reconstitute our energy landscapes. What I'm modestly proposing, then, is an energy transition that is as much about remaking the spaces we dwell in as it is about decarbonization—focused more on producing life with organic matter than on reducing emissions as if nothing else matters.

I am beyond grateful for all of the insights, suggestions, edits, selfless-ness, and warmth of the brilliant people who took the time to read this book (and its precursors): Dominic Boyer, Michael Dove, Cymene Howe, Luke Forrester Johnson, Michelle Murphy, Kalyanakrishnan "Shivi" Si-varamakrishnan, Dan Smith, Nicole Starosielski, Deb Thomas, and Sarah Vaughn. It is such a privilege to be in community with all of you. I must also note that this book simply would not have been possible without the heartfelt support, patience, and faith of my editor at Duke, Courtney Berger. I am forever indebted to you.

At the same time, these specific callouts are offshoots of a more dis-tributed gratitude that I feel for a much broader network of selfless, car-ing souls who brought this book to fruition in ways that even they might not be fully aware of. Hackneyed as it might sound, *it takes a village* is an especially apt aphorism for the process of writing a scholarly text. This book, then, is a testament to the *totality* of guidance, generosity, care, and critical feedback offered by so many people over the years. Trying to parse out this vast group into categories of support feels misaligned with my profound feeling of appreciation. Which is to say, this book emerged less from what so many of you did to support me, and more from the patch-work of support that you wove yourself into. In this spirit of patchwork, I want to acknowledge several of you who inspired me, taught me, lis-tened to me, worked with me, cared for me, rooted for me, and made *Sub-jects of the Sun* possible, without speaking to your specific contributions: Juan Aguilar, Daniel Aldana-Cohen, Juan Alduey, Niko Alexandre, Jafari Allen, Janet Ayeni, Eddie Bautista, George Bayuga, Lizzy Berk, April and Tony Biggs, Tony Bogues, Samara Brock, Paul Burrow, Joan Byron, Isa-iah Carter, Rebecca Carter, the whole Cassin/Johnson gang, Elena Conte, Cecil Corbin-Mark, Jamie Cross, Bathsheba Demuth, Mara Cerezo, Lavon Chambers, Deepti Chatti, Kim Cobb, Casey Cronin, Jalle Dafa, Jason Devenny, Andrew Dowe, R. Erica Doyle, Alexander Dunlap, Taylor Ellowitz, R. F., Justin Farrell, Janelle Farris, Paja Faudree, Greg Fay, Annelise Grimm, Mac Folkes, Mike Fung, Lina Fruzzetti, Jack Dafoe, Cara Daggett, Mike Degani, Joanne Derwin, Kate Dudley, Aunt Elaine, Maya Funaro, Deseree Fontenot, Emmaia Gelman, Sharon Griffith, Karen Hébert, Diana Hernandez, Keith Heyward, Mette High, Mawuli Hormeku, Brian Horton, Kevin Hubbard, Laura Jaramillo, Ryan Jobson, A. K., Katrina Karkazis, Paul Kockelman, Zoe Kupetz, Michael Andrade Lalan, Thomas Lax, Filbert Lee, Winston Lennon, Jessa Leinaweaver, Julia Liu, Gabe Lopez, Rebecca Lurie, Aunt Lynn, Tom Lyons, Mina Magda, Zavé Martohardjono, Meredith McLaughlin, Pascal Mensah, Nizal Mohammad, Sage Morgan-Hubbard, Alex Moore, Amelia Moore, Dustin Mulvaney, Giselle Muñoz, Rashad Muñoz, Ali Najmi, Chandni Navalkha, Bruni and Daniel Pizarro, Elizabeth Ochs, J. P., Alyssa Paredes, David Pellow, Leah Penniman, Sara Perl, Emily Nguyen-Perperidis, Stuart Perkins, Dana Powell, Ishmael Reese, Christine Riggio, J. Timmons Roberts, Doug Rogers, Ruth and Steven Rosenthal, Laura Rubin, Adrien Salazar, Sarah Sax, Mindi Schneider, Glenda Self, Irene Shen, Peggy Shepard, Byron Silva, Jessica Smith, Benjamin Sovacool, the Steinhauer/Wyatt clan, J. Terrell, Luis Torres, Irene Tung, Ceci Piñeda, Eric Walker, Andrew Wallace, Kay Warren, Keahnan Washington, Vicki Weiner, Damian White, Adam Willems, Elizabeth Yeampierre, Zamboni, Jerry Zee, Amy Zhang, Charles Zhang, Cynthia Zhang, and the many ancestors whose histories I lost in the shadows of racial capitalism.

My gratitude extends to the Lenape ancestral homelands where New York City was violently settled—a place that nurtured me, challenged me, raised me, and cultivated in me an appreciation for the messy and contradictory that is at the core of this book.

PREFACE

1. For an in-depth ethnographic theorization of screenwork, see Boyer, *Life Informatic*.

2. Debord, *Society of the Spectacle*.

3. T. Mitchell, *Carbon Democracy*.

4. Ferguson, *Anti-Politics Machine*.

5. For a related analysis, see Bond, *Negative Ecologies*.

INTRODUCTION. A MICROGRID ON THE MARGINS

1. New York City Department of Health and Mental Hygiene, *Brooklyn Community District 16*.

2. Throughout this book, all organization and individual names are pseudonyms, unless otherwise noted.

3. Wanger, Ferwerda, and Greenberg, "Perceiving Spatial Relationships," 45.

4. GDSNY, "#GDSNY were very proud to be a part of the official reopening this week of 684 affordable homes at Marcus Garvey Village in Brownsville, Brooklyn," Instagram, June 10, 2017, https://www.instagram.com/p/BVKwT1lhLpD.

5. Broadly speaking, Tyler Wall and Travis Linnemann deploy the term *visual economy* to refer to "the production, organization, and circulation of images and the values culturally ascribed to them." Wall and Linnemann, "Staring Down the State," 138. Also see Campbell, "Geopolitics and Visuality."

6. American Petroleum Institute, "Intro—2022 State of American Energy."

7. This ad presents a *quick* shot of an oil rig and images of petro industry workers posing in unspecified facilities, but like most fossil fuel propaganda, the ad largely abstains from showing energy production infrastructure, subordinating it to prominent visuals of what this infrastructure powers. Many other similar ads show no energy production infrastructure whatsoever.

8. Brooklyn SolarWorks, "#SolarPower in the Big 🍎! With our tilt rack mounting solution, flat roof installation is a breeze. Get ready #NYC, we're coming for ya," Instagram, November 18, 2022, https://www.instagram.com/p/ClHkdaq OCHz.

9. Solar Uptown Now Services, Instagram, September 28, 2019, https://www .instagram.com/solar.uptown.now.services.

10. NASEO, *Diversity*.

11. See, for instance, NASEO, *Diversity*.

12. This sentiment is aptly laid out in Luke and Heynen, "Community Solar."

13. De Onís, *Energy Islands*; Luke and Heynen, "Community Solar."

14. Lennon, "Decolonizing Energy."

15. Mulvaney, "Solar's Green Dilemma"; Lennon, "Energy Transitions"; Nugent and Sovacool, "Lifecycle Greenhouse Gas Emissions."

16. Lerner, *Sacrifice Zones*.

17. On the centrality of screens, surveillance, and, by extension, sight to the modern world, see Browne, *Dark Matters*; Scott, *Seeing like a State*; Kavanagh, "Ocularcentrism and Its Others."

18. Robinson, *Black Marxism*; "differential value" is from Melamed, "Represent and Destroy."

19. Wang and Lloyd, *Sins of a Solar Empire*; Lennon, "Postcarbon Amnesia."

20. For arguments that solar can make the capitalist status quo sustainable, see, for instance, Hawkin, Lovins, and Lovins, *Natural Capitalism*; Ikerd, *Sustainable Capitalism*; Matthews, *Greening of Capitalism*. For arguments that solar can disrupt the dominant political economy, see, for instance, Carroll, "Fossil Capitalism"; Schweickart, "Is Sustainable Capitalism Possible?"; Klein, *This Changes Everything*; De Onís, *Energy Islands*.

21. Strathern, *Relations*.

22. Strathern, *Relations*.

23. Yusoff, *Billion Black Anthropocenes*.

24. Robinson, *Black Marxism*.

25. Melamed, "Racial Capitalism."

26. Lennon, "Decolonizing Energy." For more on the ways in which energy infrastructure (and solar infrastructure, in particular) materializes, exacerbates, and gives form to differential value and social hierarchy—especially under the auspices of capitalism—see Cross, "Capturing Crisis"; Dean "Uneasy Entanglements"; Winther, Ulstead, and Saini, "Solar powered."

27. Larkin, "Politics and Poetics of Infrastructure"; Harvey, "Cementing Relations."

28. On energy as a capacity immanent in certain things, see Berry, *Unsettling of America*. On energy's ability to transform matter, see Stone, "Neglected Topic." On the slave trade as the first industrial-scale energy infrastructure, see Lennon, "Decolonizing Energy."

29. Lennon, "Decolonizing Energy." For an in-depth look at the ways in which racial capitalism continues to reproduce long-standing differential value in contemporary times, see Pellow and Park, *The Silicon Valley*, Pulido, "Geographies of Race."

30. Larkin, *Signal and Noise*; Starosielski, *Undersea Network*.

31. On Blackness as the paradigmatic form of racial difference, see Fanon, *Black Skin, White Masks*; Wilderson, *Afropessimism*.

32. Povinelli, "Social Projects."

33. Povinelli, *Cunning of Recognition*.

34. Boyer, *Life Informatic*.

35. Fleetwood, *Troubling Vision*.

36. Turow and Tsui, *Hyperlinked Society*.

37. Hernández and Bird, "Energy Burden."

38. J. Adler, "Origins of Sightseeing"; Tuttle, "Trip to the Zoo"; Pratt, *Imperial Eyes*; Fullagar, *Savage Visit*.

39. On the sightseeing tour as a colonial mode, see Carville, "Photography, Tourism."

40. Nagel, *View from Nowhere*.

41. Ducarme, Luque, and Courchamp, *"What Are 'Charismatic Species.'"*

42. For a classic account of the relationship between positionality, power, and the illusion of contextless knowledge, see Haraway, "Situated Knowledges."

43. Solar Foundation, *2017 U.S. Solar Industry Diversity Study*.

44. Nader, "Politics of Energy."

45. Nader, "Politics of Energy."

46. Massumi, "Autonomy of Affect."

47. Berlant, *Cruel Optimism*, 16.

48. Ngai, *Ugly Feelings*, 27.

49. Boyer, *Energopolitics*; T. Mitchell, *Carbon Democracy*. In the case of solar, this technological agency often operates through late liberal screenwork, affectively animating specious imaginaries of a solar-powered world beyond racial capitalism. Each chapter, then, attends to the interplay of this screenwork and solar's material properties to elucidate the ideological affordances of infrastructures often narrowly conceptualized in terms of physical and economic needs.

CHAPTER 1. SHINE

1. C. Adams, "Appreciative Engagements with Slideware."

2. Gabriel, "Tyranny of PowerPoint."

3. Robles-Anderson and Svensson, *"'One Damn Slide.'"*

4. King, *Black Shoals*.

5. For a cross-cultural history of sun worship see Singh, *Sun*; for an overview of the sublime see Cronon, *Uncommon Ground*.

6. Cronon, *Uncommon Ground*.

7. For more information on how renewable energy reproduces the extractive order, see Howe, *Ecologics*; Rignall, "Solar Power, State Power"; Folch, *Hydropolitics*.

8. Marx, *Economic and Philosophic Manuscripts*.

9. Wang and Lloyd, *Sins of a Solar Empire*.

10. L. Murphy and Elimä, *In Broad Daylight*.

11. L. Murphy and Elimä, *In Broad Daylight*; Wang and Lloyd, *Sins of a Solar Empire*.

12. See, for instance, Wollerton, "US Solar."

13. Wang and Lloyd, *Sins of a Solar Empire*.

14. Wang and Lloyd, *Sins of a Solar Empire*.

15. Gross, "'Modern-Day Slavery.'"

16. Gross, "'Modern-Day Slavery.'"

17. Gross, "'Modern-Day Slavery.'"

18. Faucher, *Social Capital Online*; Leach, *Land of Desire*.

19. See, for instance, Klein, *This Changes Everything*; De Onís, *Energy Islands*.

20. For more on solar's positive social impacts see Klein, *This Changes Everything*; De Onís, *Energy Islands*.

21. Bowen, *After Greenwashing*.

22. Significantly, the few young people I spoke to in the community while canvassing and the newer white community residents whom I met at community meetings very often raised the matter of climate change. However, I very rarely encountered these people when I canvassed the community because I primarily did so during weekdays, when only older, nonworking, and longtime residents were home.

23. Sivaramakrishnan and Vaccaro, "Postindustrial Natures."

24. Cheng, "Shine," 1034.

25. Cronon, *Uncommon Ground*; M. Berger, *Sight Unseen*.

26. M. Berger, *Sight Unseen*; Sato, "Settler Colonial Projections."

27. Deger, "Shimmer," 1.

28. Seltzer, *Bodies and Machines*.

29. Throughout this book, I intentionally use the gender-specific "man," "man-made," and like gendered terms to stand in for "the human" so as to underscore the patriarchal dynamics of modern anthropocentrism. This rhetorical move is indebted to Sylvia Wynter's helpful elucidation of the gendered nature of "the human" in political discourse. Wynter, "Unsettling the Coloniality of Being/Power/Truth/Freedom."

30. Cronon, *Uncommon Ground*.

31. Sluyter, "Colonialism and Landscape."

32. Cheng, "Shine," 1034.

33. Nye, *American Technological Sublime*.

34. Nye, *American Technological Sublime*.

35. Meskell, "Objects in the Mirror."

36. Kanoi et al., "'What Is Infrastructure?'"

37. Nye, *American Technological Sublime*.

38. See, for instance, Thoreau, *Walden*.

39. Meskell, "Objects in the Mirror," 64.

40. Meskell, "Objects in the Mirror," 62.

41. Quoted in Meskell, "Objects in the Mirror," 65.

42. Meskell, "Objects in the Mirror," 64.

43. Meskell, "Objects in the Mirror," 66.

44. For an account of how the pyramids were designed with attention to the sun's orbit, see Dash, "Occam's Egyptian Razor."

45. Stewart et al., "Underground Lives."

46. Boerum Hill Association, "Schools YES, Towers NO."

47. Singh, *Sun*, 12.

48. Singh, *Sun*.

49. Debord, *Society of the Spectacle*.

50. J. Berger, *About Looking*; Deleuze, *Francis Bacon*.

51. Olafsson et al., "Social Media," 271.

52. That said, "agrivoltaics"—the codevelopment of solar farms and agriculture—does demonstrate that land stewardship and solar farms are not mutually exclusive (Dinesh and Pearce 2016). However, solar farms are very often stewarded exclusively as renewable energy projects, transforming grassy fields into a visual spectacle for those who do not steward land.

53. Marx, *Capital*, 1:164.

54. Cross, "Solar Good."

CHAPTER 2. SPACE

1. Borunda, "'Nature Deprived' Neighborhoods."

2. Shakur, *Rose*.

3. Lennon, "Energy Transitions."

4. Gurevitch, "Google Warming," 88.

5. Gurevitch, "Google Warming."

6. For more on the democratic attributes and potentiality of solar, see Fairchild and Weinrub, *Energy Democracy*; De Onís, *Energy Islands*.

7. On the relationship between the material properties of energy infrastructure, political ideology, and social configurations, see Boyer, "Anthropology Electric"; Abram, *Electrifying Anthropology*; Strauss, Rupp, and Love, *Cultures of Energy*; Winther and Wilhite, "Tentacles of Modernity."

8. Gurevitch, "Google Warming," 88.

9. On surveillance in twentieth-century statecraft, see Scott, *Seeing like a State*.

10. See, for instance, Fairchild and Weinrub, *Energy Democracy*; De Onís, *Energy Islands*.

11. For a similar analysis of *nonrenewable* electricity generation markets, see Degani, *City Electric*.

12. Mulvaney, "Solar's Green Dilemma."

13. Welker, *Enacting the Corporation*.

14. Graeber, *Bullshit Jobs*.

15. Green, "(A)Woke Workplaces."

16. Indeed, screenwork, as Dominic Boyer conceptualizes the screen-based habitus of contemporary office work, impedes white collar workers' focus on "anything offscreen." See Boyer, *Life Informatic*.

17. Appel, *Licit Life of Capitalism*; Nader, "Harder Path"; Nader, "Politics of Energy"; Lohmann, "Carbon Trading"; Gusterson, *Nuclear Rites*; Cecelski, *Rethinking Gender and Energy*; Nader and Milleron, "Dimensions"; Thompson, "Among the Energy Tribes"; Mason and Stoilkova, "Corporeality of Consultant

Expertise"; Hudgins and Poole, "Framing Fracking"; Özden-Schilling, "Economy Electric."

18. See, for instance, Anand, *Hydraulic City*; Appel, *Licit Life of Capitalism*; Boyer, *Energopolitics*; Carse, *Beyond the Big Ditch*; T. Mitchell, *Carbon Democracy*; Vaughn, *Engineering Vulnerability*; Whittington, *Anthropogenic Rivers*.

19. Reitman, "Uncovering the White Place"; Fancher, "Composing Artificial Intelligence."

20. Graeber, *Bullshit Jobs*.

21. On white male nepotism, see Bertrand and Mullainathan, "Are Emily and Greg"; Pager, "Mark." On tech startups, see Reitman, "Uncovering the White Place"; Fancher, "Composing Artificial Intelligence."

22. Yang, "Market Governance and Polycentrism."

23. W. Brown, *Undoing the Demos*; Collier, *Post-Soviet Social*.

24. Ganti, "Neoliberalism"; Peck and Theodore, "Still Neoliberalism?"

25. W. Brown, "What Exactly Is Neoliberalism?"

26. See, for instance, M. Murphy, *Economization of Life*; M. Adler and Posner, "Happiness Research"; International Bottled Water Association, "Demand for Bottled Water."

27. See, for instance, Zapata and McCormick, "How Tech Can Boost."

28. M. Murphy, *Economization of Life*. On the robust role of the state under neoliberalism, see Wacquant, "Crafting the Neoliberal State"; Collier, *Post-Soviet Social*.

29. Scott, *Seeing like a State*.

30. For more on the role of nonstate actors in the management of everyday life under neoliberalism, see, for instance, Özden-Schilling, *Current Economy*.

31. Clemmer, "Federal Renewable Energy Tax Credits."

32. John Farrell, *Feed-In Tariffs in America*; Alliance to Save Energy, *History of Energy Efficiency*; Clemmer, "Federal Renewable Energy Tax Credits."

33. Choose Energy, "About Us."

34. Thill, "Chicago Nonprofits"; Wolske, Gillingham, and Shultz, "Peer Influence."

35. On concern for climate change in the 2000s, see McCright and Dunlap, "Politicization of Climate Change." While a minority of Americans linked Hurricane Katrina to climate change in a noted public opinion survey, this was still a *sizable* minority (39 percent) that suggests that Katrina shaped many, although not most, Americans' views on climate change. See Guskin and Dennis, "Majority of Americans."

36. See, for instance, Apollo Alliance and Green for All, *Green-Collar Jobs*.

37. Apollo Alliance and Green for All, *Green-Collar Jobs*.

38. See, for instance, Jones, *Green Collar Economy*.

39. Du Bois, *Philadelphia Negro*; M. Murphy, *Seizing the Means*.

40. Ottinger and Cohen, *Technoscience and Environmental Justice*.

41. On the problematization of a self-evident political left and right, see,

for instance, Lewis, "Problem of Donald Trump." On the win-win calculus, see Giridharadas, *Winners Take All*.

42. See, for instance, Hannah Appel, *Licit Life*, and Laura Nader, "The Harder Path," for a snapshot of the politics of traditional white-collar energy professionals.

43. Wilmott, "Surface."

44. Heidegger, *Being and Time*.

45. On "hard" analytics, see Mayhew, Saleh, and Williams, "Making Data Analytics Work."

46. Taylor, "Progressive Caucus."

47. Romele and Rodighiero, "Digital Habitus."

48. On smartness, see Ho, *Liquidated*; on the will to improve, see Li, *Will to Improve*.

49. Romele and Rodighiero, "Digital Habitus."

50. M. Murphy, *Economization of Life*.

51. M. Murphy, *Economization of Life*, 24.

52. Ho, *Liquidated*.

53. Giridharadas, *Winner Takes All*.

54. Williams, "Care-Full Justice."

55. Hartman, *Scenes of Subjection*; Bloomquist, "Minstrel Legacy."

56. Shange, *Progressive Dystopia*.

57. Sturken and Cartwright, *Practices of Looking*, 453.

58. On different ways of seeing, see Campt, *Listening to Images*; MacDougall, "Visual in Anthropology."

59. There is no shortage of organizational literature suggesting that corporations that actively embrace internal diversity foster more supportive, caring, and welcoming workplaces; see, for instance, Frost and Alidina, *Building an Inclusive Organization*. That said, this is not a theory that I sought to adjudicate or disprove through my research.

60. BlocPower, "About Us."

61. Amazon Web Services, "BlocPower Decarbonizes Buildings."

62. Glassdoor, "All Hot Air," October 19, 2021, https://www.glassdoor.com /Reviews/BlocPower-Reviews-E1453559.htm.

CHAPTER 3. MODULES AND METRICS

1. *Queens Gazette*, "Power of the People."

2. For more on managerial audit culture, see Fortun, "From Bhopal."

3. Pulido, "Rethinking Environmental Racism."

4. Kharrazi, "Resilience," 416.

5. The site is discrete in the sense that the panels and other on-site electrical equipment work together to generate electricity without connecting to off-site infrastructure; however, most solarized buildings still partly rely on the central electrical grid.

6. Wheeler, *Fire, Wind, and Waves*, 7.

7. Bullard, *Dumping in Dixie*.

8. Fairchild and Weinrub, *Energy Democracy*, 6.

9. M. Murphy, *Economization of Life*; Evans, "Talking about Money"; Spencer, "CSR for Sustainable Development."

10. Evans, "Talking about Money"; Spencer, "CSR for Sustainable Development"; also see Ferrer, "Ditch the Stakeholder Discourse."

11. Wynter, "Ethno or Socio Poetics."

12. Strings, *Fearing the Black Body*; J. Brown, "Body."

13. Roberts, *Killing the Black Body*; McKittrick, "Mathematics Black Life"; Sharpe, *In the Wake*; Campt, *Listening to Images*; Thomas, *Political Life*; J. Brown, "Body"; R. Jackson, *Black Masculine Body*.

14. Campt, *Listening to Images*.

15. V. Adams, *Markets of Sorrow*.

16. INCITE!, *Revolution Will Not Be Funded*.

17. Bishop and Green, *Philanthrocapitalism*.

18. INCITE!, *Revolution Will Not Be Funded*.

19. INCITE!, *Revolution Will Not Be Funded*.

20. Robinson, *Black Marxism*; Lipsitz, "What Is This Black."

21. INCITE!, *Revolution Will Not Be Funded*.

22. Ottinger and Cohen, *Technoscience and Environmental Justice*.

23. International Climate Justice Network, "Bali Principles."

24. Hulme, *Why We Disagree*; Murtugudde, "10 Reasons."

25. See, for instance, Shepard and Corbin-Mark, "Climate Justice."

26. Tishman Environment and Design Center, "Environmental Justice and Philanthropy."

27. These circumstances include the following: (1) the electricity utility paid for the microgrid because the main electrical grid was very weak and expensive to maintain in the microgrid's host community; (2) the buildings involved in the microgrid had been built at the same time and had minimal roof obstructions *and* unobstructed access to the sun; and (3) the community had ample space for a solar electricity storage battery that met the very specific and daunting requirements of the Fire Department, among other things.

28. For some context, New York State generated 147,803,038 megawatt-hours of electricity in 2016; see Energy Information Administration, "EIA State Electricity Profile." If we make an *extremely aggressive overestimate* and suggest that EQUAL's projected 150-kilowatt local solar installations will have five hours of strong exposure to the sun *every single day*, it would generate 273 megawatt-hours in a year, or 0.00018 percent of the state's electricity generation—and, again, this is an *extreme* overestimate.

29. Halawa and Olcoń-Kubicka, "Digital Householding."

30. Foreman and Mookerjee, "Computing in the Dark"; Bruns et al., "Spreadsheets, Service Providers"; Byrne, "Open Data Visualized."

31. Tillery, "Plain Style."

32. Agrawal, *Environmentality*.

33. Foucault, *Birth of Biopolitics*.

34. For more on material-discursivity and mind/matter imbrications, see Barad, *Meeting the Universe Halfway*.

35. Özden-Schilling, "Economy Electric."

36. Özden-Schilling, "Economy Electric."

37. Özden-Schilling, "Economy Electric."

38. M. Murphy, *Economization of Life*, 24–25.

39. *Electricity prices* here refers both to prices per kilowatt-hour of electricity consumed and, more generally, to utility bill charges, although the latter meaning is usually not affiliated with "prices" in Big-E Energy discourse. I purposely conflate prices per kilowatt-hour and utility bill charges to call attention to the multiple ways in which humans apprehend electricity in terms of monetary values.

40. Degani, *City Electric*.

41. Özden-Schilling, *Current Economy*; Hargreaves et al., "Making Energy Visible."

42. This is attributable not only to a failure of imagination but also to a failure of literacy; we, as a society, lack anything resembling an energy civics education.

43. M. Murphy, *Economization of Life*, 24–25.

44. Porter, *Trust in Numbers*; Macintosh et al., "Accounting as Simulacrum"; Widick, "Flesh."

45. Porter, *Trust in Numbers*; Macintosh et al., "Accounting as Simulacrum"; Widick, "Flesh."

46. *Adders* are added expenses for addressing idiosyncratic building structures, like hoisters to lift solar panels onto roofs in buildings without elevators.

47. Pulido, "Rethinking Environmental Racism."

48. On the fallacy of a pure free market, see Tsing, *Mushroom*; Elyachar, *Markets of Dispossession*; Çalışkan and Callon, "Economization."

49. Peters, Elliot, and Cullenberg, "Economic Transition."

50. As David Graeber argues, markets are inextricable from state power. See Graeber, *Debt*.

51. Stuelke, *Ruse of Repair*; Povinelli, *Cunning of Recognition*.

52. See, for instance, US Department of Energy, *Solarize Guidebook*.

53. Fortun, "From Bhopal"; Gottlieb, *Forcing the Spring*.

54. A notable exception is agrovoltaic projects, which integrate agriculture into solar farms—but most solar farm projects do not entail such thoughtful land stewardship. Of course, many cap-and-trade programs *are*, effectively, renewable energy programs; such programs, then, are no more focused on interacting with biospheric life than any other EERE initiative.

55. Fortun, "From Bhopal."

56. As Karl Marx argued, capital works to standardize and align various stages and fields of technological production in ways that make possible complex technologies (like solar). See Marx, *Capital*, vol. 3.

57. Lennon, "Postcarbon Amnesia."

58. Lennon, "Decolonizing Energy"; Lennon, "Energy Transitions"; Zehner, *Green Illusions*.

59. BBC News, "China Solar Panel Factory."

60. Groom, "Prison Labor."

61. For instance, when the lead from decommissioned photovoltaic arrays accumulates in landfills, it can contaminate drinking water supplies, exacerbating the water-quality problems already squeezing low-income communities of color, as evinced by the disaster in Flint, Michigan.

62. Hertwich et al., "Integrated Life-Cycle Assessment"; Nugent and Sovacool, "Lifecycle Greenhouse Gas Emissions."

63. Berry, *Unsettling of America*, 91.

64. Povinelli, "Defining Security," 31. On how digital activism is enmeshed in the complications of capitalism, see Ingraham and Reeves, "New Media, New Panics."

65. Amrute, "What the Facebook Files"; Christin and Lu, "Influencer Pay Gap."

66. On digital storytelling, see Canella, "Social Movement Documentary Practices." On the lack of nuance in digital activism, see Ingraham and Reeves, "New Media, New Panics."

67. Fleetwood, *Troubling Vision*.

68. Navarro, "Public Housing."

69. McKittrick, *Demonic Grounds*, 13.

70. Nixon, *Slow Violence*.

71. *Oxford English Dictionary*, s.v. "Image" (n.), https://www.oed.com/dictionary /image_n.

72. Shange, *Progressive Dystopia*.

73. Jackson, quoted in Friedman, "Perversity of Diversity."

74. On the ways in which power is unbeholden to ideology, see Massumi, "Autonomy of Affect."

75. Bourdieu, "Social Space"; Bourdieu, *Theory of Practice*; Deleuze, *Foucault*; Thomas, *Political Life*.

76. Gilmore, "Abolition Geography."

77. K. Mitchell and Sparke, "New Washington Consensus," 727.

78. On intra-actions, see Barad, *Meeting the Universe Halfway*.

79. National Renewable Energy Laboratory, "Community Solar," 2.

80. On urban heat inequality, see Leland, "Why an East Harlem Street."

81. Özden-Schilling, *Current Economy*.

82. Povinelli, *Cunning of Recognition*; Fraser, "From Redistribution to Recognition?"

83. T. Mitchell, *Carbon Democracy*.

CHAPTER 4. BODIES

1. DSA Ecosocialists, "A jobs guarantee could be our best bet for a 'Medicare-for-All' climate policy—but it's crucial to fit proposals with a climate justice

lens," Twitter (now X), May 24, 2018, https://twitter.com/DSA_Enviro/status
/999763601538904064; for the aestheticization of the phrase "rooftop solar shifts
power," see Creative Action Network, "Rooftop Solar Shifts Power!," accessed
October 23, 2024, https://creativeaction.network/products/rooftop-solar-shifts
-power-by-marcacci-communications.

2. On Bolshevik posters, see Bonnell, *Iconography of Power*; on US WPA paintings, see Doss, "Iconography of American Labor," 53.

3. Daggett, "Petro-Masculinity."

4. Wynter, "Unsettling the Coloniality."

5. Robinson, *Black Marxism*.

6. Lohmann, *Energy Alternatives*, 26.

7. See, for instance, SAV, *War on Carbon*; Rocky Mountain Institute, "Carbon
War Room."

8. Asafu-Adjaye et al., *Ecomodernist Manifesto*, 6. The phrase "intelligent decarbonization" is from Silicon Valley Clean Energy, *2020 Integrated Resource Plan*, 43.

9. Asafu-Adjaye et al., *Ecomodernist Manifesto*.

10. Berry, *Unsettling of America*, 85.

11. Berry, *Unsettling of America*, 91.

12. Mills, *White Collar*.

13. Mills, *White Collar*.

14. Marx, *Capital*, 1:638.

15. Marx, *Capital*, 1:638.

16. Feldman, "Why Private Waste Management."

17. Giridharadas, *Winners Take All*.

18. On how corporations are not solely focused on maximizing shareholder
value, see Welker, *Enacting the Corporation*.

19. Christensen, Raynor, and McDonald, "What Is Disruptive Innovation?"

20. On dwelling in the world, see Ingold, *Perception of the Environment*.

21. Solmetric, "SunEye-210 Shade Tool."

22. Solmetric, "SunEye-210 Shade Tool."

23. Howe, *Ecologics*.

24. On Black women as breadwinners, see Hartman, *Wayward Lives*; on changes
to the Black labor market after the civil rights era, see Windham, *Knocking on
Labor's Door*.

25. DSA Ecosocialists, "DSA's Green New Deal Principles."

26. Glassdoor, "How Much."

27. Lennon, "Problem with Solutions." Also see Gershon, *Down and Out*.

28. Weber, *Protestant Work Ethic*.

29. Gershon, *Down and Out*.

30. Mayer, "Problem Solving."

31. Angouri and Bargiela-Chiappini, "'So What Problems.'"

32. Schreurs et al., "Job Demands-Resources."

33. Gershon, *Down and Out*, 11.

34. For the "genre of the human," see Wynter, "Unsettling the Coloniality"; on

reciprocity between Black/brown lives and nonhuman life, see Z. Jackson, *Becoming Human*; Yusoff, *Billion Black Anthropocenes*.

35. Ballew et al., "Which Racial/Ethnic Groups Care."

36. Yusoff, *Billion Black Anthropocenes*.

37. At the first National People of Color Environmental Leadership Summit in 1991, organizer Dana Alston aptly summarized the EJ movement's ethos: "Our vision of the environment is woven into an overall framework of social, racial, and economic justice . . . The environment, for us, is where we live, where we work, and where we play," Gottlieb, *Forcing the Spring*, 3.

38. Finney, *Black Faces, White Spaces*; Cronon, *Uncommon Ground*.

39. Whyte, "Settler Colonialism"; Boggs, "Material-Discursive Spaces."

40. Finney, *Black Faces, White Spaces*; Whyte, "Settler Colonialism"; Boggs, "Material-Discursive Spaces."

41. Shortsleeve, "Adventure Trips"; Justin Farrell, *Billionaire Wilderness*.

42. Arora, "Leisure Factory"; Gahin and Chesteen, "Executives Contemplate"; Shortsleeve, "Adventure Trips."

43. White and Wilbert, *Technonatures*.

44. On "patchy" capitalism, see Tsing, *Mushroom*.

45. Giridharadas, *Winners Take All*.

46. BK ROT is *not* a pseudonym.

47. Lohmann, *Energy Alternatives*.

48. Mulgaonkar, "Testimony."

49. McKittrick, *Demonic Grounds*; Roane, "Towards Usable Histories."

50. Roane, "Towards Usable Histories."

51. Sontag, *On Photography*, 178; Benjamin, "Work of Art"; J. Berger, *Ways of Seeing*; Coleman and James, *Capitalism and the Camera*; Debord, *Society of the Spectacle*.

52. Benjamin, "Work of Art"; J. Berger, *Ways of Seeing*; Coleman and James, *Capitalism and the Camera*; Debord, *Society of the Spectacle*.

53. Davis, "Work of Art."

54. Davis, "Work of Art," 381.

55. Bright Power, "Via Verde," accessed October 23, 2024, https://www.brightpower.com/projects/via-verde.

56. Sivaramakrishnan and Vaccaro, "Postindustrial Natures."

57. Du Bois, *Souls of Black Folk*, 38.

58. On the relationship between Blackness and ledgers/accounting, see McKittrick, "Mathematics Black Life."

Abram, Simone, ed. *Electrifying Anthropology: Exploring Electrical Practices and Infrastructures.* London: Bloomsbury Academic, 2019.

Adams, Catherine. "Toward Appreciative Engagements with Slideware." In *Proceedings of ED-MEDIA 2007—World Conference on Educational Multimedia, Hypermedia and Telecommunications*, edited by Craig Montgomerie and Jand Seale. Vancouver, BC: Association for the Advancement of Computing in Education, 2007: 2134–42.

Adams, Vincanne. *Markets of Sorrow, Labors of Faith: New Orleans in the Wake of Katrina.* Durham, NC: Duke University Press, 2013.

Adler, Judith. "Origins of Sightseeing." *Annals of Tourism Research* 16, no. 1 (1989): 7–29.

Adler, Matthew, and Eric Posner. "Happiness Research and Cost-Benefit Analysis." *Journal of Legal Studies* 37, no. S2 (2008): S253–S292.

Agrawal, Arun. *Environmentality: Technologies of Government and the Making of Subjects.* Durham, NC: Duke University Press, 2005.

Alliance to Save Energy. *The History of Energy Efficiency.* Washington, DC: Alliance to Save Energy, 2013. https://www.ase.org/sites/ase.org/files/resources /Media%20browser/ee_commission_history_report_2-1-13.pdf.

Amazon Web Services. "BlocPower Decarbonizes Buildings by Deploying SaaS BlocMaps Using HPC to Manage Big Data on AWS." *HPCwire*, October 11, 2022. https://www.hpcwire.com/solution_content/aws/blocpower-decarbonizes -buildings-by-deploying-saas-blocmaps-using-hpc-to-manage-big-data-on -aws.

American Petroleum Institute. "Intro—2022 State of American Energy." YouTube, January 20, 2022. https://www.youtube.com/watch?v=noMfHjmkKcM.

Amrute, Sareeta. "What the Facebook Files Tell Us about Racial Capitalism." *Interactions* 29, no. 2 (March–April 2022): 59–61.

Anand, Nikhil. *Hydraulic City: Water and the Infrastructures of Citizenship in Mumbai.* Durham, NC: Duke University Press, 2017.

Angouri, Jo, and Francesca Bargiela-Chiappini. "'So What Problems Bother You and You Are Not Speeding Up Your Work?' Problem Solving Talk at Work." *Discourse and Communication* 5, no. 3 (August 2011): 209–29.

Apollo Alliance and Green for All. *Green-Collar Jobs in America's Cities: Building Pathways out of Poverty and Careers in the Clean Energy Economy.* Washington,

DC: Apollo Alliance, 2008. http://cdn.americanprogress.org/wp-content /uploads/issues/2008/03/pdf/green_collar_jobs.pdf

Appel, Hannah. *The Licit Life of Capitalism: US Oil in Equatorial Guinea.* Durham, NC: Duke University Press, 2019.

Arora, Payal. "The Leisure Factory: Production in the Digital Age." *LOGOS,* no. 3 (2015): 88–119.

Asafu-Adjaye, John, Linus Blomqvist, Stewart Brand, Barry Brook, Ruth DeFries, Erle Ellis, Christopher Foreman, et al. *An Ecomodernist Manifesto.* April 2015. http://www.ecomodernism.org/s/An-Ecomodernist-Manifesto.pdf.

Ballew, Matthew, Edward Maibach, John Kotcher, Parrish Bergquist, Seth Rosenthal, Jennifer Marlon, and Anthony Leiserowitz. "Which Racial/Ethnic Groups Care Most about Climate Change?" Yale Program on Climate Change Communication, April 16, 2020. https://climatecommunication.yale.edu /publications/race-and-climate-change.

Barad, Karen. *Meeting the Universe Halfway: Quantum Physics and the Entanglement of Matter and Meaning.* Durham, NC: Duke University Press, 2007.

BBC News. "China Solar Panel Factory Shut after Protests." September 19, 2011. http://www.bbc.com/news/world-asia-pacific-14968605.

Beck, Abaki. "Tribal Solar Projects Provide More Than Climate Solutions." *Yes! Magazine.* September 16, 2021. https://www.yesmagazine.org/environment /2021/09/16/native-solar-projects-climate-solutions.

Benjamin, Walter. "The Work of Art in the Age of Mechanical Reproduction." In *Illuminations: Essays and Reflections,* edited by Hannah Arendt, 217–52. New York: Schocken, 2007.

Berger, John. *About Looking.* New York: Vintage, 1992.

Berger, John. *Ways of Seeing.* New York: Penguin, 1990.

Berger, Martin. *Sight Unseen: Whiteness and American Visual Culture.* Berkeley: University of California Press, 2005.

Berlant, Lauren. *Cruel Optimism.* Durham, NC: Duke University Press, 2011.

Berry, Wendell. *The Unsettling of America: Culture and Agriculture.* San Francisco: Sierra Club Books, 1977.

Bertrand, Marianne, and Sendhil Mullainathan. "Are Emily and Greg More Employable Than Lakisha and Jamal? A Field Experiment on Labor Market Discrimination." *American Economic Review* 94, no. 4 (September 2004): 991–1013.

Bishop, Matthew, and Michael Green. *Philanthrocapitalism: How the Rich Can Save the World.* New York: Bloomsbury, 2008.

BlocPower. "About Us." Accessed March 17, 2023. https://www.blocpower.io /about-us.

BlocPower. "The BlocPower Smart Cities Platform Connects Investors to Green and Healthy Building Projects in Local Communities." Archived May 28, 2017. https://web.archive.org/web/20170328020732/http://www.blocpower .io/#updates.

BlocPower. "Discover Projects: Invest in Your Community and the Planet."

Archived June 19, 2018. https://web.archive.org/web/20180619195757
/https://marketplace.blocpower.io/discover.

Bloomquist, Jennifer. "The Minstrel Legacy: African American English and the Historical Construction of 'Black' Identities in Entertainment." *Journal of African American Studies* 19, no. 4 (December 2015): 410–25.

BlueGreen Alliance. "Manufacturing to Build Back Better." Video, October 27, 2021. https://www.bluegreenalliance.org/site/for-america-by-america/our -ads-calling-for-congress-to-build-back-better-for-america-by-america.

Boerum Hill Association. "Schools YES—Towers NO in Boerum Hill." Care2 Petitions, 2017. https://www.thepetitionsite.com/671/717/537/schools-yes-towers -no-in-boerum-hill.

Boggs, Kyle. "The Material-Discursive Spaces of Outdoor Recreation: Rhetorical Exclusion and Settler Colonialism at the Arizona Snowbowl Ski Resort." *Journal for the Study of Religion, Nature and Culture* 11, no. 2 (2017): 175–96.

Bond, David. *Negative Ecologies: Fossil Fuels and the Discovery of the Environment.* Berkeley: University of California Press, 2022.

Bonnell, Victoria. *Iconography of Power: Soviet Political Posters under Lenin and Stalin.* Berkeley: University of California Press, 1998.

Borunda, Alejandra. "How 'Nature Deprived' Neighborhoods Impact the Health of People of Color." *National Geographic*, July 29, 2020. https://www.national-geographic.com/science/article/how-nature-deprived-neighborhoods -impact-health-people-of-color.

Bourdieu, Pierre. *Outline of a Theory of Practice.* Cambridge: Cambridge University Press, 2013.

Bourdieu, Pierre. "The Social Space and the Genesis of Groups." *Theory and Society* 14, no. 6 (November 1985): 723–44.

Bowen, Frances. *After Greenwashing: Symbolic Corporate Environmentalism and Society.* Cambridge: Cambridge University Press, 2014.

Boyer, Dominic. "Anthropology Electric." *Cultural Anthropology* 30, no. 4 (2015): 531–39. https://doi.org/10.14506/ca30.4.02.

Boyer, Dominic. *Energopolitics: Wind and Power in the Anthropocene.* Durham, NC: Duke University Press, 2019.

Boyer, Dominic. *The Life Informatic: Newsmaking in the Digital Era.* Ithaca, NY: Cornell University Press, 2013.

Bright Power. "Via Verde." Accessed June 12, 2023. https://www.brightpower.com /projects/via-verde.

Brown, Jayna. "Body." In *Keywords for African American Studies*, edited by Erica Edwards, Rodrick Ferguson, and Jeffrey Ogbar. 29–32. New York: New York University Press, 2018.

Brown, Wendy. *Undoing the Demos: Neoliberalism's Stealth Revolution.* New York: Zone, 2015.

Brown, Wendy. "What Exactly Is Neoliberalism?" Interview by Timothy Shenk. *Dissent*, April 2, 2015. https://www.dissentmagazine.org/blog/booked-3-what -exactly-is-neoliberalism-wendy-brown-undoing-the-demos.

Browne, Simone. *Dark Matters: On the Surveillance of Blackness*. Durham, NC: Duke University Press, 2015.

Bruns, Eric, Jim Rast, Christa Peterson, Janet Walker, and Jone Bosworth. "Spreadsheets, Service Providers, and the Statehouse: Using Data and the Wraparound Process to Reform." *American Journal of Community Psychology* 38, nos. 3–4 (December 2006): 201–12.

Bullard, Robert. *Dumping in Dixie: Race, Class and Environmental Quality*. Boulder, CO: Westview, 2000.

Byrne, JKevin. "Open Data Visualized from Police Spreadsheets: A Case of Where Force, Race, and Place Collided." *SSRN*, January 29, 2021. https://ssrn.com/abstract=3775925.

Çalışkan, Koray, and Michel Callon. "Economization, Part 1: Shifting Attention from the Economy towards Processes of Economization." *Economy and Society* 38, no. 3 (2009): 369–98.

Campbell, David. "Geopolitics and Visuality: Sighting the Darfur Conflict." *Political Geography* 26, no. 4 (May 2007): 357–82.

Campt, Tina. *Listening to Images*. Durham, NC: Duke University Press, 2017.

Canella, Gino. "Social Movement Documentary Practices: Digital Storytelling, Social Media and Organizing." *Digital Creativity* 28, no. 1 (2017): 24–37.

Carroll, William K. "Fossil Capitalism, Climate Capitalism, Energy Democracy: The Struggle for Hegemony in an Era of Climate Crisis." *Socialist Studies / Études Socialistes* 14, no. 1 (2020): 1–26.

Carse, Ashley. *Beyond the Big Ditch: Politics, Ecology, and Infrastructure at the Panama Canal*. Cambridge, MA: MIT Press, 2014.

Carville, Justin. "Photography, Tourism and Natural History." In *Irish Tourism: Image, Culture and Identity*, edited by Michael Cronin and Barbara O'Connor, 215–38. Bristol: Channel View, 2013.

Cecelski, Elizabeth. *Re-thinking Gender and Energy: Old and New Directions*. ENERGIA/EASE Discussion Paper, Energy, Environment and Development (EED), 2004.

Cheng, Anne Anlin. "Shine: On Race, Glamour, and the Modern." *PMLA* 126, no. 4 (October 2011): 1022–41.

Choose Energy. "About Us." Accessed December 2, 2022. https://www.chooseenergy.com/about-us.

Christensen, Clayton, Michael Raynor, and Rory McDonald. 2015. "What Is Disruptive Innovation?" *Harvard Business Review*, December 2015.

Christin, Angèle, and Yingdan Lu. "The Influencer Pay Gap: Platform Labor Meets Racial Capitalism." *New Media and Society*, first published online April 29, 2023. https://doi.org/10.1177/14614448231164995.

Clemmer, Steve. "Federal Renewable Energy Tax Credits: Creating American Jobs and Investment in State and Local Economies." Testimony before the Subcommittee on Energy, Committee on Energy and Commerce, US House of Representatives, March 29, 2017.

Coleman, Kevin, and Daniel James. *Capitalism and the Camera: Essays on Photography and Extraction*. New York: Verso, 2021.

Collier, Stephen. *Post-Soviet Social: Neoliberalism, Social Modernity, Biopolitics*. Princeton, NJ: Princeton University Press, 2011.

Cronon, William. *Uncommon Ground: Rethinking the Human Place in Nature*. New York: W. W. Norton, 1995.

Cross, Jamie. "The Solar Good: Energy Ethics in Poor Markets." *Journal of the Royal Anthropological Institute* 25, no. S1 (April 2019): 47–66.

Daggett, Cara. "Petro-Masculinity: Fossil Fuels and Authoritarian Desire." *Millennium* 47, no. 1 (2018): 25–44.

Dash, Glen. "Occam's Egyptian Razor: The Equinox and the Alignment of the Pyramids." *Journal of Ancient Egyptian Architecture* 2 (2017): 1–8.

Davis, Douglas. "The Work of Art in the Age of Digital Reproduction." *Leonardo* 28, no. 5 (1995): 381–86.

Debord, Guy. *Society of the Spectacle*. Detroit: Black and Red, 2002.

Degani, Michael. *The City Electric: Infrastructure and Ingenuity in Postsocialist Tanzania*. Durham, NC: Duke University Press, 2022.

Deger, Jennifer. "Shimmer." In *The International Encyclopedia of Anthropology*, edited by Hillary Callan, 1–3. Oxford: Wiley, 2018.

Deleuze, Gilles. *Foucault*. Minneapolis: University of Minnesota Press, 1988.

Deleuze, Gilles. *Francis Bacon: The Logic of Sensation*. Minneapolis: University of Minnesota Press, 1981.

De Onís, Catalina. *Energy Islands: Metaphors of Power, Extractivism, and Justice in Puerto Rico*. Berkeley: University of California Press, 2021.

Doss, Erika. "Toward an Iconography of American Labor: Work, Workers, and the Work Ethic in American Art, 1930–1945." *Design Issues* 13, no. 1 (Spring 1997): 53–66.

DSA Ecosocialists. "DSA's Green New Deal Principles." February 28, 2019. https://ecosocialists.dsausa.org/dsas-green-new-deal-principles.

Du Bois, W. E. B. *The Philadelphia Negro: A Social Study*. Philadelphia: University of Pennsylvania Press, 1995.

Du Bois, W. E. B. *The Souls of Black Folk*. Boston: Bedford, 1997.

Ducarme, Frédéric, Gloria Luque, and Franck Courchamp. "What Are 'Charismatic Species' for Conservation Biologists?" *BioSciences Master Reviews* 10 (July 2013): 1–8.

Easterly, William. *The Elusive Quest for Growth: Economists' Adventures and Misadventures in the Tropics*. Cambridge: MIT Press, 2002.

Elyachar, Julia. *Markets of Dispossession: NGOs, Economic Development, and the State in Cairo*. Durham, NC: Duke University Press, 2005.

Energy Information Administration. "EIA State Electricity Profile, New York." 2018. https://www.eia.gov/state/analysis.php?sid=NY.

Evans, Robert. "Talking about Money: Public Participation and Expert Knowledge in the Euro Referendum." *British Journal of Sociology* 55, no. 1 (March 2004): 35–53.

Fairchild, Denise, and Al Weinrub, eds. *Energy Democracy: Advancing Equity in Clean Energy Solutions.* Washington, DC: Island, 2017.

Fancher, Patricia. "Composing Artificial Intelligence: Performing Whiteness and Masculinity." *Present Tense: Journal of Rhetoric and Society* 6, no. 1 (2016): 1–18.

Fanon, Frantz. *Black Skin, White Masks.* New York: Grove, 2008.

Farrell, John. *Feed-In Tariffs in America: Driving the Economy with Renewable Energy Policy That Works.* Minneapolis: New Rules Project, 2009.

Farrell, Justin. *Billionaire Wilderness: The Ultra-wealthy and the Remaking of the American West.* Princeton, NJ: Princeton University Press, 2020.

Faucher, Kane. *Social Capital Online: Alienation and Accumulation.* London: University of Westminster Press, 2018.

Feldman, Kiera. "Why Private Waste Management Is One of the Nation's Most Hazardous Jobs." *PBS News Hour,* January 4, 2018. https://www.pbs.org /newshour/nation/why-private-waste-management-is-one-of-the-nations -most-hazardous-jobs.

Ferguson, James. *The Anti-Politics Machine: Development, Depoliticization, and Bureaucratic Power in Lesotho.* Minneapolis: University of Minnesota Press, 1994.

Ferrer, Alexander. "It's Time to Ditch the Stakeholder Discourse." *Progressive City,* 2020. https://www.progressivecity.net/single-post/it-s-time-to-ditch -the-stakeholder-discourse.

Finney, Carolyn. *Black Faces, White Spaces: Reimagining the Relationship of African Americans to the Great Outdoors.* Chapel Hill: University of North Carolina Press, 2014.

Fleetwood, Nicole. *Troubling Vision: Performance, Visuality, and Blackness.* Chicago: University of Chicago Press, 2010.

Folch, Christine. *Hydropolitics: The Itaipu Dam, Sovereignty, and the Engineering of Modern South America.* Princeton, NJ: Princeton University Press, 2019.

Foreman, Gabrielle P., and Labanya Mookerjee. "Computing in the Dark: Spread-sheets, Data Collection and DH's Racist Inheritance." In *Always Already Computational: Collections as Data,* edited by Thomas Padilla et al., 108–9. Washington, DC: Institute of Museum and Library Services, 2019.

Fortun, Kim. "From Bhopal to the Informating of Environmentalism: Risk Communication in Historical Perspective." *Osiris* 2, no. 19 (2004): 283–96.

Foucault, Michel. *The Birth of Biopolitics: Lectures at the Collège de France, 1978–1979.* New York: Palgrave Macmillan, 2008.

Fraser, Nancy. "From Redistribution to Recognition? Dilemmas of Justice in a Post-socialist Age." *New Left Review,* no. I/212 (July/August 1995): 68–93.

Friedman, Robert. "The Perversity of Diversity." *Tampa Bay Times,* September 27, 2005. https://www.tampabay.com/archive/2000/08/17/perversity-of -diversity.

Frost, Stephen, and Raafi-Karim Alidina. *Building an Inclusive Organization: Leveraging the Power of a Diverse Workforce.* New York: Kogan Page, 2019.

Fullagar, Kate. *The Savage Visit: New World People and Popular Imperial Culture in Britain, 1710–1795*. Berkeley: University of California Press, 2012.

Gabriel, Yiannis. "Against the Tyranny of PowerPoint: Technology-in-Use and Technology Abuse." *Organization Studies* 29, no. 2 (February 2008): 255–76.

Gahin, Fikry S., and Susan A. Chesteen. "Executives Contemplate the Call of the Wild." *Risk Management* 35, no. 7 (July 1988): 44–51.

Ganti, Tejaswini. "Neoliberalism." *Annual Review of Anthropology* 43 (2014): 89–104.

Gershon, Ilana. *Down and Out in the New Economy: How People Find (or Don't Find) Work Today*. Chicago: University of Chicago Press, 2017.

Gilmore, Ruth Wilson. "Abolition Geography and the Problem of Innocence." In *Futures of Black Radicalism*, edited by Gaye Johnson and Alex Lubin. New York: Verso, 2017.

Giridharadas, Anand. *Winners Take All: The Elite Charade of Changing the World*. New York: Knopf, 2018.

Glassdoor. "How Much Does a Solar Installer Make in New York City, NY?" March 7, 2023. https://www.glassdoor.com/Salaries/new-york-city-solar-installer -salary-SRCH_IL.0,13_IM615_KO14,29.htm.

Gottlieb, Robert. *Forcing the Spring: The Transformation of the American Environmental Movement*. Washington, DC: Island, 1993.

Graeber, David. *Bullshit Jobs: A Theory*. New York: Simon and Schuster, 2018.

Graeber, David. *Debt: The First 5,000 Years*. New York: Melville House, 2011.

Green, Michael. "(A)Woke Workplaces." *Wisconsin Law Review* 2023, no. 3 (2023): 811–72.

Groom, Nichola. "Prison Labor Helps U.S. Solar Company Manufacture at Home." *Reuters*, June 10, 2015. http://www.reuters.com/article/solar-prison-suniva -idUSL1N0YP17Y20150610.

Gross, Terry. "How 'Modern-Day Slavery' in the Congo Powers the Rechargeable Battery Economy." *Fresh Air*, NPR, February 1, 2023. https://www.npr.org /sections/goatsandsoda/2023/02/01/1152893248/red-cobalt-congo-drc -mining-siddharth-kara.

Gurevitch, Leon. "Google Warming: Google Earth as Eco-Machinima." *Convergence: The International Journal of Research into New Media Technologies* 20, no. 1 (February 2014): 85–107.

Guskin, Emily, and Brady Dennis. "Majority of Americans Now Say Climate Change Makes Hurricanes More Intense." *Washington Post*, September 28, 2017.

Gusterson, Hugh. *Nuclear Rites: A Weapons Laboratory at the End of the Cold War*. Berkeley: University of California Press, 1996.

Halawa, Mateusz, and Marta Olcoń-Kubicka. "Digital Householding: Calculating and Moralizing Domestic Life through Homemade Spreadsheets." *Journal of Cultural Economy* 11, no. 6 (2018): 514–34.

Haraway, Donna. "Situated Knowledges: The Science Question in Feminism and the Privilege of Partial Perspective." *Feminist Studies* 14, no. 3 (Autumn 1988): 575–99.

Harris, Lee. "Workers on Solar's Front Lines." *American Prospect*, December 7, 2022. https://prospect.org/labor/workers-on-solars-front-lines.

Hartman, Saidiya. *Scenes of Subjection: Terror, Slavery, and Self-Making in Nineteenth-Century America*. Oxford: Oxford University Press, 1997.

Hartman, Saidiya. *Wayward Lives, Beautiful Experiments: Intimate Histories of Riotous Black Girls, Troublesome Women, and Queer Radicals*. New York: W. W. Norton, 2019.

Harvey, Penelope. "Cementing Relations: The Materiality of Roads and Public Spaces in Provincial Peru." *Social Anthropology* 54, no. 2 (Summer 2010): 28–46.

Hawkin, Paul, Amory Lovins, and Hunter Lovins. *Natural Capitalism: Creating the Next Industrial Revolution*. Washington, DC: US Green Building Council, 2000.

Heidegger, Martin. *Being and Time*. New York: State University of New York Press, 1996.

Hernández, Diana, and Stephen Bird. "Energy Burden and the Need for Integrated Low-Income Housing and Energy Policy." *Poverty and Public Policy* 2, no. 4 (2010): 5–25.

Hertwich, Edgar, Thomas Gibon, Evert A. Bouman, Anders Arvesen, Sangwon Suh, Garvin A. Heath, Joseph D. Bergesen, Andrea Ramirez, Mabel I. Vega, and Lei Shi. "Integrated Life-Cycle Assessment of Electricity-Supply Scenarios Confirms Global Environmental Benefit of Low-Carbon Technologies." *Proceedings of the National Academy of Sciences of the United States of America* 112, no. 20 (May 2015): 6277–82.

Ho, Karen. *Liquidated: An Ethnography of Wall Street*. Durham, NC: Duke University Press, 2008.

Howe, Cymene. *Ecologics: Wind and Power in the Anthropocene*. Durham, NC: Duke University Press, 2019.

Hudgins, Anastasia, and Amanda Poole. "Framing Fracking: Private Property, Common Resources, and Regimes of Governance." *Journal of Political Ecology* 21, no. 1 (2014): 304–19.

Hulme, Mike. *Why We Disagree about Climate Change: Understanding Controversy, Inaction and Opportunity*. Cambridge: Cambridge University Press, 2009.

Ikerd, John. *Sustainable Capitalism: A Matter of Common Sense*. Bloomfield, CT: Kumarian, 2005.

INCITE! Women of Color Against Violence, eds. *The Revolution Will Not Be Funded: Beyond the Non-Profit Industrial Complex*. Cambridge, MA: South End, 2007.

Ingold, Tim. *The Perception of the Environment: Essays on Livelihood, Dwelling and Skill*. New York: Routledge, 2000.

Ingraham, Chris, and Joshua Reeves. "New Media, New Panics." *Critical Studies in Media Communications* 33, no. 5 (2016): 455–67.

Institute for Policy Integrity. "Environmental Value of Distributed Energy Resources in New York State." July 11, 2018. https://policyintegrity.org/documents/NY_DPS_E_Value_Presentation_071118.pdf.

International Bottled Water Association. "Clear, Consistent, and Increased Demand for Bottled Water." May 18, 2021. https://bottledwater.org/nr /clear-consistent-and-increased-demand-for-bottled-water-as-a-healthy -alternative-to-other-packaged-drinks.

International Climate Justice Network. "Bali Principles of Climate Justice." Corp-Watch, August 28, 2002. https://www.corpwatch.org/article/bali-principles -climate-justice.

Jackson, Ronald. *Scripting the Black Masculine Body: Identity, Discourse, and Racial Politics in Popular Media.* Albany: State University of New York Press, 2006.

Jackson, Zakiyyah. *Becoming Human: Matter and Meaning in an Antiblack World.* New York: New York University Press, 2020.

Jones, Van. *The Green Collar Economy: How One Solution Can Fix Our Two Biggest Problems.* New York: HarperOne, 2008.

Kanoi, Lav, Vanessa Koh, Al Lim, Shoko Yamada, and Michael R. Dove. "'What Is Infrastructure? What Does It Do?': Anthropological Perspectives on the Workings of Infrastructure(s)." *Environmental Research: Infrastructure and Sustainability* 2, no. 1 (2022).

Kantrowitz, Jonathan. "American Beauty and Bounty: The Judith G. and Steaven K. Jones Collection of Nineteenth-Century Painting." *Art History News*, February 16, 2019. http://arthistorynewsreport.blogspot.com/2019/02/american -beauty-and-bounty-judith-g-and.html.

Kavanagh, Donncha. "Ocularcentrism and Its Others: A Framework for Metatheoretical Analysis." *Organization Studies* 25, no. 3 (March 2004): 445–64.

Kharrazi, Ali. "Resilience." In *Encyclopedia of Ecology,* edited by Brian Fath, 414–18. Amsterdam, Netherlands: Elsevier, 2019.

King, Tiffany. *The Black Shoals: Offshore Formations of Black and Native Studies.* Durham, NC: Duke University Press, 2019.

Klein, Naomi. *This Changes Everything: Capitalism versus the Climate.* New York: Simon and Schuster, 2014.

Larkin, Brian. "The Politics and Poetics of Infrastructure." *Annual Review of Anthropology* 42 (2013): 327–43.

Larkin, Brian. *Signal and Noise: Media, Infrastructure, and Urban Culture in Nigeria.* Durham, NC: Duke University Press, 2008.

Leach, William. *Land of Desire: Merchants, Power and the Rise of New American Culture.* New York: Vintage, 1994.

Leland, John. "Why an East Harlem Street Is 31 Degrees Hotter Than Central Park West." *New York Times*, August 20, 2021. https://www.nytimes.com/2021 /08/20/nyregion/climate-inequality-nyc.html.

Lennon, Myles. "Decolonizing Energy: Black Lives Matter and Technoscientific Expertise amid Solar Transitions." *Energy Research and Social Science* 30 (August 2017): 18–27.

Lennon, Myles. "Energy Transitions in a Time of Intersecting Precarities: From Reductive Environmentalism to Antiracist Praxis." *Energy Research and Social Science* 73 (March 2021): 101930.

Lennon, Myles. "Postcarbon Amnesia: Toward a Recognition of Racial Grief in Renewable Energy Futures." *Science, Technology, and Human Values* 45, no. 5 (September 2020): 934–62.

Lennon, Myles. "The Problem with Solutions: Apolitical Optimism in the Sustainable Energy Industry." *Current Anthropology*, forthcoming.

Lerner, Steve. *Sacrifice Zones: The Front Lines of Toxic Chemical Exposure in the United States.* Cambridge, MA: MIT Press, 2012.

Lewis, Verlan. "The Problem of Donald Trump and the Static Spectrum Fallacy." *Party Politics* 27, no. 4 (July 2021): 605–18.

Li, Tania. *The Will to Improve: Governmentality, Development, and the Practice of Politics.* Durham, NC: Duke University Press, 2007.

Lipsitz, George. "What Is This Black in the Black Radical Tradition?" In *Futures of Black Radicalism*, edited by Gaye Johnson and Alex Lubin, 108–19. New York: Verso, 2017.

Lohmann, Larry. "Carbon Trading, Climate Justice and the Production of Ignorance: Ten Examples." *Development* 51 (September 2008): 359–65.

Lohmann, Larry. *Energy Alternatives: Surveying the Territory.* UK: Corner House, 2013.

Luke, Nikki, and Nik Heynen. "Community Solar as Energy Reparations: Abolishing Petro-Racial Capitalism in New Orleans." *American Quarterly* 72, no. 3 (September 2020): 603–25.

MacDougall, David. "The Visual in Anthropology." In *Rethinking Visual Anthropology*, edited by Marcus Banks and Howard Morphy, 276–95. New Haven, CT: Yale University Press, 1997.

Macintosh, Norman, Teri Shearer, Daniel Thornton, and Michael Welker. "Accounting as Simulacrum and Hyperreality: Perspectives on Income and Capital." *Accounting, Organizations and Society* 25, no. 1 (January 2000): 13–50.

Marcuse, Herbert. *One Dimensional Man: Studies in the Ideology of Advanced Industrial Society.* Boston: Beacon, 1964.

Marx, Karl. *Capital.* Vol. 1. London: Penguin Classics, 1990.

Marx, Karl. *Capital.* Vol. 3. London: Penguin Classics, 1992.

Marx, Karl. *Economic and Philosophic Manuscripts of 1844.* Mineola, NY: Dover, 2012.

Mason, Arthur, and Maria Stoilkova. "Corporeality of Consultant Expertise in Arctic Natural Gas Development." *Journal of Northern Studies* 6, no. 2 (2012): 83–96.

Massumi, Brian. "The Autonomy of Affect." *Cultural Critique*, no. 31 (Autumn 1995): 83–110.

Matthews, John. *Greening of Capitalism: How Asia Is Driving the Next Great Transformation.* Stanford, CA: Stanford University Press, 2014.

Mayer, Richard. "Problem Solving." In *The Oxford Handbook of Cognitive Psychology*, edited by D. Reisberg, 769–78. Oxford: Oxford University Press, 2013.

Mayhew, Helen, Tamim Saleh, and Simon Williams. "Making Data Analytics Work for You—Instead of the Other Way Around." *McKinsey Quarterly*, Octo-

ber 7, 2016. https://www.mckinsey.com/capabilities/mckinsey-digital/our
-insights/making-data-analytics-work-for-you-instead-of-the-other-way
-around.

McCright, Aaron, and Riley Dunlap. "The Politicization of Climate Change and
Polarization in the American Public's Views of Global Warming, 2001–2010."
Sociological Quarterly 52, no. 2 (2011): 155–94.

McKittrick, Katherine. *Demonic Grounds: Black Women and the Cartographies of
Struggle*. Minneapolis: University of Minnesota Press, 2006.

McKittrick, Katherine. "Mathematics Black Life." *Black Scholar* 44, no. 2 (Summer
2014): 16–28.

Melamed, Jodi. "Racial Capitalism." *Critical Ethnic Studies* 1, no. 1 (Spring 2015):
76–85.

Meskell, Lynn. "Objects in the Mirror Appear Closer Than They Are." In *Materiality*, edited by Daniel Miller, 51–71. Durham, NC: Duke University Press,
2005.

Mills, C. Wright. *White Collar: The American Middle Class*. Oxford: Oxford University Press, 2002.

Mitchell, Katharyne, and Matthew Sparke. "The New Washington Consensus:
Millennial Philanthropy and the Making of Global Market Subjects." *Antipode*
48, no. 3 (June 2016): 724–49.

Mitchell, Timothy. *Carbon Democracy: Political Power in the Age of Oil*. New York:
Verso, 2011.

Movement Generation. *From Banks and Tanks to Cooperation and Caring: A Strategic Framework for a Just Transition*. Berkeley, CA: Movement Generation, 2016.
https://movementgeneration.org/wp-content/uploads/2016/11/JT_booklet
_English_SPREADs_web.pdf.

Mulgaonkar, Priya. "Testimony to New York State Department of Environmental
Conservation." New York City Environmental Justice Alliance, June 9, 2016.
https://www.nyc-eja.org/wp-content/uploads/2016/08/dec_part360_hearing
testimony_060816_pm.pdf.

Mulvaney, Dustin. "Solar's Green Dilemma." *IEEE Spectrum* 51, no. 9 (September
2014): 30–33.

Murphy, Laura, and Nyrola Elimä. *In Broad Daylight: Uyghur Forced Labour and
Global Solar Supply Chains*. Sheffield, UK: Sheffield Hallam University Helena
Kennedy Centre for International Justice, 2021.

Murphy, Michelle. *The Economization of Life*. Durham, NC: Duke University Press,
2017.

Murphy, Michelle. *Seizing the Means of Reproduction: Entanglements of Feminism,
Health, and Technoscience*. Durham, NC: Duke University Press, 2012.

Murtugudde, Raghu. "10 Reasons Why Climate Change Is a Wicked Problem."
Wire, December 11, 2019. https://thewire.in/environment/climate-change
-wicked-problem.

Nader, Laura. "The Harder Path—Shifting Gears." In *The Energy Reader*, edited by
Laura Nader, 517–34. Oxford: Wiley-Blackwell, 2010.

Nader, Laura. "The Politics of Energy: Toward a Bottom-Up Approach." In *The Energy Reader*, edited by Laura Nader, 313–17. Oxford: Wiley-Blackwell, 2010.

Nader, Laura, and Norman Milleron. "Dimensions of the 'People Problem' in Energy Research and 'the' Factual Basis of Dispersed Energy Futures." *Energy* 4, no. 5 (October 1979): 953–67.

Nagel, Thomas. *The View from Nowhere*. Oxford: Oxford University Press, 1986.

NASEO. *Diversity in the U.S. Energy Workforce*. Arlington, VA: NASEO, 2021.

National Renewable Energy Laboratory. "Community Solar." Accessed October 23, 2024. https://www.nrel.gov/state-local-tribal/community-solar.html.

Navarro, Mireya. "Public Housing in New York Reaches a Fiscal Crisis." *New York Times*, August 11, 2014.

New York City Department of Health and Mental Hygiene. *Brooklyn Community District 16: Brownsville*. New York: New York City Department of Health and Mental Hygiene, 2015.

Ngai, Sianne. *Ugly Feelings*. Cambridge, MA: Harvard University Press, 2005.

Nixon, Rob. *Slow Violence and the Environmentalism of the Poor*. Cambridge, MA: Harvard University Press, 2013.

Nugent, Daniel, and Benjamin Sovacool. "Assessing the Lifecycle Greenhouse Gas Emissions from Solar PV and Wind Energy: A Critical Meta-survey." *Energy Policy* 65 (February 2014): 229–44.

Nye, David. *American Technological Sublime*. Cambridge, MA: MIT Press, 1994.

Olafsson, Anton, Maia Møller, Thomas Mattijssen, Natalie Gulsrud, Bas Breman, and Arjen Buijs. "Social Media and Experiences of Nature: Towards a Plurality of Senses of Place." In *Changing Senses of Place: Navigating Global Challenges*, edited by Christopher M. Raymond, Lynne C. Manzo, Daniel R. Williams, Andrés Di Masso, and Timo von Wirth, 271–84. Cambridge: Cambridge University Press, 2021.

Organic Consumers Association. "'As World Teeters on Brink of Climate Catastrophe,' 600+ Groups Demand Congress Back Visionary Green New Deal." January 10, 2019. https://www.organicconsumers.org/news/world-teeters-brink-climate-catastrophe-600-groups-demand-congress-back-visionary-green-new.

Ottinger, Gwen, and Benjamin Cohen. *Technoscience and Environmental Justice: Expert Cultures in a Grassroots Movement*. Cambridge, MA: MIT Press, 2011.

Özden-Schilling, Canay. *The Current Economy: Electricity Markets and Techno-Economics*. Stanford, CA: Stanford University Press, 2021.

Özden-Schilling, Canay. "Economy Electric." *Cultural Anthropology* 30, no. 4 (2015): 578–88.

Pager, Devah. "The Mark of a Criminal Record." *American Journal of Sociology* 108, no. 5 (March 2003): 937–75.

Peck, Jamie, and Nik Theodore. "Still Neoliberalism?" *South Atlantic Quarterly* 118, no. 2 (April 2019): 245–65.

Pellow, David, and Lisa Park. *The Silicon Valley of Dreams: Environmental Injustice*,

Immigrant Workers, and the High-Tech Global Economy. New York: New York University Press, 2002.

Peters, John, John Elliot, and Stephen Cullenberg. "Economic Transition as a Crisis of Vision: Classical versus Neoclassical Theories of Equilibrium." *Eastern Economic Journal* 28, no. 2 (Spring 2002): 217–40.

Porter, Theodore. *Trust in Numbers: The Pursuit of Objectivity in Science and Public Life*. Princeton, NJ: Princeton University Press, 1995.

Positive Energy Solar. "Go Solar and Save the Planet—It's Easy with No Out of Pocket Costs!" October 8, 2018. https://www.positiveenergysolar.com/blog /2018/october/go-solar-save-the-planet-its-easy-with-no-out-of.

Povinelli, Elizabeth. *The Cunning of Recognition: Indigenous Alterities and the Making of Australian Multiculturalism*. Durham, NC: Duke University Press, 2002.

Povinelli, Elizabeth. "Defining Security in Late Liberalism." In *Times of Security: Ethnographies of Fear, Protest and the Future*, edited by Martin Holbraad and Morton Axel Pedersen, 28–32. New York: Routledge, 2013.

Povinelli, Elizabeth. "The Social Projects of Late Liberalism." *Dialogues in Human Geography* 3, no. 2 (July 2013): 236–39.

Pratt, Mary Louise. *Imperial Eyes: Travel Writing and Transculturation*. New York: Routledge, 1992.

Pulido, Laura. "Geographies of Race and Ethnicity II: Environmental Racism, Racial Capitalism and State-sanctioned Violence." *Progress in Human Geography* 41, no. 4 (August 2017): 524–33.

Pulido, Laura. "Rethinking Environmental Racism: White Privilege and Urban Development in Southern California." *Annals of the Association of American Geographers* 90, no. 1 (2000): 12–40.

Queens Gazette. "Power of the People: Queens Leads in Solar Energy." October 21, 2020. https://www.qgazette.com/articles/power-of-the-people-queens-leads -in-solar-energy.

Realtor.com. Listing for 555 Tenth Ave Unit 20G, Manhattan, NY 10018. Accessed October 23, 2024. https://www.realtor.com/realestateandhomes-detail/555 -10th-Ave-Apt-20G_New-York_NY_10018_M49015-13429.

Reitman, Meredith. "Uncovering the White Place: Whitewashing at Work." *Social and Cultural Geography* 7, no. 2 (April 2006): 267–82.

Rignall, Karen. "Solar Power, State Power, and the Politics of Energy Transition in Pre-Saharan Morocco." *Environment and Planning A: Economy and Space* 48, no. 3 (March 2016): 540–57.

Roane, J. T. "Towards Usable Histories of the Black Commons." *Black Perspectives*, February 28, 2017. https://www.aaihs.org/towards-usable-histories-of-the -black-commons.

Roberts, Dorothy. *Killing the Black Body: Race, Reproduction, and the Meaning of Liberty*. New York: Penguin Random House, 1998.

Robinson, Cedric. *Black Marxism: The Making of the Black Radical Tradition*. Chapel Hill: University of North Carolina Press, 2000.

Robles-Anderson, Erica, and Patrik Svensson. "'One Damn Slide after Another':

PowerPoint at Every Occasion for Speech." *Computational Culture*, no. 5 (January 2016). http://computationalculture.net/one-damn-slide-after-another-powerpoint-at-every-occasion-for-speech.

Rocky Mountain Institute. "Carbon War Room." Accessed October 23, 2024. https://rmi.org/carbon-war-room.

Romele, Alberto, and Dario Rodighiero. "Digital Habitus or Personalization without Personality." *Journal of Philosophical Studies* 13, no. 37 (July 2020): 98–126.

Safina, Carl. *Song for the Blue Ocean: Encounters along the World's Coasts and beneath the Seas*. New York, Henry Holt, 1998.

Sato, Courtney. "Settler Colonial Projections: The Visual Politics of the Interwar Pan-Pacific Movement." *Verge: Studies in Global Asias* 8, no. 2 (Fall 2022): 201–32.

SAV. *The War on Carbon: The End of the Beginning?* Surrey, UK: SAV Systems, 2019. Archived September 16, 2021, at Archive.org. https://web.archive.org/web/20210916034603/https://www.sav-systems.com/wp-content/uploads/2020/01/The-War-on-Carbon-Web.pdf.

Schreurs, Bert, Hetty Van Emmerik, Nele De Cuyper, Guy Notelaers, and Hans De Witte. "Job Demands-Resources and Early Retirement Intention: Differences between Blue- and White-Collar Workers." *Economic and Industrial Democracy* 32, no. 1 (February 2011): 47–68.

Schweickart, David. "Is Sustainable Capitalism Possible?" *Procedia—Social and Behavioral Sciences* 2, no. 5 (2010): 6739–52.

Scott, James C. *Seeing like a State*. New Haven, CT: Yale University Press, 1998.

Seltzer, Mark. *Bodies and Machines*. New York: Routledge, 1992.

Shakur, Tupac. *The Rose That Grew from Concrete*. New York: MTV Books, 2002.

Shange, Savannah. *Progressive Dystopia: Abolition, Antiblackness, and Schooling in San Francisco*. Durham, NC: Duke University Press, 2019.

Sharpe, Christina. *In the Wake: On Blackness and Being*. Durham, NC: Duke University Press, 2016.

Shepard, Peggy, and Cecil Corbin-Mark. "Climate Justice." *Environmental Justice* 2, no. 4 (December 2009): 163–66.

Shortsleeve, Cassie. "Adventure Trips Are the New Business Dinners." *Condé Nast Traveler*, April 3, 2018. https://www.cntraveler.com/story/adventure-trips-are-the-new-business-dinners.

Silicon Valley Clean Energy. *2020 Integrated Resource Plan*. September 1, 2020. https://www.svcleanenergy.org/wp-content/uploads/2020/02/SVCE-IRP-Narrative-Report-1.pdf.

Singh, Madanjeet. *The Sun: Symbol of Power and Life*. New York: Harry N. Abrams, 1993.

Sivaramakrishnan, K., and Ismael Vaccaro. "Introduction. Postindustrial Natures: Hyper-mobility and Place-Attachments." *Social Anthropology* 14, no. 3 (October 2006): 301–17.

Sluyter, Andrew. "Colonialism and Landscape in the Americas: Material/

Conceptual Transformations and Continuing Consequences." *Annals of the Association of American Geographers* 91, no. 2 (June 2001): 410–28.

Solar Foundation. *2017 U.S. Solar Industry Diversity Study*. Washington, DC: Solar Foundation, 2017.

Solmetric. "SunEye-210 Shade Tool." Accessed October 23, 2024.https://www.solmetric.com/buy210.html.

Sontag, Susan. *On Photography*. New York: Picador, 2001.

Spencer, Rochelle. "CSR for Sustainable Development and Poverty Reduction? Critical Perspectives from the Anthropology of Development." In *Disciplining the Undisciplined? Perspectives from Business, Society and Politics on Responsible Citizenship, Corporate Social Responsibility and Sustainability*, edited by Martin Brueckner, Rochelle Spencer, and Megan Paull, 73–87. New York: Springer, 2018.

Starosielski, Nicole. *The Undersea Network*. Durham, NC: Duke University Press, 2015.

Stein, Jill. "Green New Deal." April 22, 2016. https://www.jill2016.com/green_new_deal. Archived November 29, 2019, at Archive.org. https://web.archive.org/web/20161129023148/http://www.jill2016.com/green_new_deal.

Stewart, Nikita, et al. "Underground Lives: The Sunless World of Immigrants in Queens." *New York Times*, October 23, 2019.

Stone, C. H. "A Neglected Topic in Chemistry Teaching." *School Science and Mathematics* 35, no. 8 (1935): 795–98.

Strathern, Marilyn. *Relations: An Anthropological Account*. Durham, NC: Duke University Press, 2020.

Strauss, Sarah, Stephanie Rupp, and Thomas Love. *Cultures of Energy: Power, Practices, Technologies*. London: Routledge, 2016.

Strings, Sabrina. *Fearing the Black Body: The Racial Origins of Fat Phobia*. New York: New York University Press, 2019.

Stuelke, Patricia. *The Ruse of Repair: US Neoliberal Empire and the Turn from Critique*. Durham, NC: Duke University Press, 2021.

Sturken, Marita, and Lisa Cartwright. *Practices of Looking: An Introduction to Visual Culture*. Oxford: Oxford University Press, 2001.

Taylor, Kate. "Progressive Caucus Seeks to Influence Race for New York Council Speaker." *New York Times*, November 19, 2013.

Thill, David. "Chicago Nonprofits Train South Side Youth as Energy Efficiency 'Ambassadors.'" *Energy News Network*, October 1, 2019. https://energynews.us/2019/10/01/chicago-nonprofits-train-south-side-youth-as-energy-efficiency-ambassadors.

Thomas, Deborah. *Political Life in the Wake of the Plantation: Sovereignty, Witnessing, Repair*. Durham, NC: Duke University Press, 2019.

Thompson, Michael. "Among the Energy Tribes: A Cultural Framework for the Analysis and Design of Energy Policy." *Policy Sciences* 17, no. 3 (September 1984): 321–39.

Thoreau, Henry David. *Walden*. San Diego: ICON Group International, 2005.

Tillery, Denise. "The Plain Style in the Seventeenth Century: Gender and the History of Scientific Discourse." *Journal of Technical Writing and Communication* 35, no. 3 (July 2005): 273–89.

Tishman Environment and Design Center. "Environmental Justice and Philanthropy: Challenges and Opportunities for Alignment Gulf South and Midwest Case Studies." New York: Tishman Environment and Design Center, 2020.

Tsing, Anna. *The Mushroom at the End of the World: On the Possibility of Life in Capitalist Ruins*. Princeton, NJ: Princeton University Press, 2015.

Turow, Joseph, and Lokman Tsui, eds. *The Hyperlinked Society: Questioning Connections in the Digital Age*. Ann Arbor: University of Michigan Press, 2008.

Tuttle, Brendan. "A Trip to the Zoo: Colonial Sightseeing and Spectacle in Sudan (1901–1933)." *Journal of Tourism History* 11, no. 3 (2019): 217–42.

US Department of Energy. *The Solarize Guidebook: A Community Guide to Collective Purchasing of Residential PV Systems*. Washington, DC: US Department of Energy, 2014.

Vaughn, Sarah. *Engineering Vulnerability: In Pursuit of Climate Adaptation*. Durham, NC: Duke University Press, 2021.

Wacquant, Loic. "Crafting the Neoliberal State: Workfare, Prisonfare, and Social Insecurity." *Sociological Forum* 25, no.2 (June 2010): 197–220.

Wall, Tyler, and Travis Linnemann. "Staring Down the State: Police Power, Visual Economies, and the 'War on Cameras.'" *Crime Media Culture* 10, no. 2 (August 2014): 133–49.

Wang, Seaver, and Juzel Lloyd. *Sins of a Solar Empire: An Industry Imperative to Address Unethical Solar Photovoltaic Manufacturing in Xinjiang*. Berkeley, CA: Breakthrough Institute, 2021.

Wanger, Leonard, James Ferwerda, and Donald Greenberg. "Perceiving Spatial Relationships in Computer-Generated Images." *IEEE Computer Graphics and Applications* 12, no. 3 (May 1992): 44–58.

Weber, Max. *The Protestant Work Ethic and the Spirit of Capitalism*. Mineola, NY: Dover, 2003.

Welker, Marina. *Enacting the Corporation: An American Mining Firm in Post-authoritarian Indonesia*. Berkeley: University of California Press, 2014.

White, Damian, and Chris Wilbert. *Technonatures: Environments, Technologies, Spaces, and Places in the Twenty-First Century*. Waterloo, IA: Wilfrid Laurier University Press, 2010.

Whittington, Jerome. *Anthropogenic Rivers: The Production of Uncertainty in Lao Hydropower*. Ithaca, NY: Cornell University Press, 2018.

Whyte, Kyle. "Settler Colonialism, Ecology, and Environmental Injustice." *Environment and Society* 9, no. 1 (September 2018): 125–44.

Widick, Richard. "Flesh and the Free Market (On Taking Bourdieu to the Options Exchange)." *Theory and Society* 32 (December 2003): 679–723.

Wilderson, Frank. *Afropessimism*. New York: Liveright, 2020.

Williams, Miriam. "Care-Full Justice in the City." *Antipode* 9, no. 3 (2017): 821–39.

Wilmott, Clancy. "Surface." Theorizing the Contemporary, *Fieldsights*, October 24, 2017. https://culanth.org/fieldsights/surface.

Windham, Lane. *Knocking on Labor's Door: Union Organizing in the 1970s and the Roots of a New Economic Divide*. Chapel Hill: University of North Carolina Press, 2017.

Winther, Tanja, and Harold Wilhite. "Tentacles of Modernity: Why Electricity Needs Anthropology." *Cultural Anthropology* 30, no. 4 (2015): 569–77.

Wollerton, Megan. "US Solar Has to Be More Affordable If We're Going to Address Climate Change." *CNET*, June 21, 2022. Archived July 2, 2022, at Archive.org. https://web.archive.org/web/20220702132426/https://www.cnet.com/home/energy-and-utilities/us-residential-solar-has-to-be-more-affordable-if-were-going-to-address-climate-change.

Wolske, Kimberly, Kenneth Gillingham, and Wesley Shultz. "Peer Influence on Household Energy Behaviours." *Nature Energy* 5 (January 2020): 202–12.

Wynter, Sylvia. "Ethno or Socio Poetics." *Alcheringa/Ethnopoetics* 2, no. 2 (1976): 78–94.

Wynter, Sylvia. "Unsettling the Coloniality of Being/Power/Truth/Freedom." *CR: New Centennial Review* 3, no. 3 (Fall 2003): 257–337.

Yang, Ethan. "Market Governance and Polycentrism." American Institute for Economic Research, February 19, 2021. https://www.aier.org/article/market-governance-and-polycentrism.

Yusoff, Kathryn. *A Billion Black Anthropocenes or None*. Minneapolis: University of Minnesota Press, 2018.

Zapata, Laura, and Gavin McCormick. "How Tech Can Boost Emissions 'Bang for the Buck' from Corporate Climate Investments." *Canary Media*, May 17, 2021. https://www.canarymedia.com/articles/corporate-procurement/how-tech-can-boost-the-emissions-bang-for-solar-buck-from-corporate-climate-promises.

ing, 33, 40, 160, 168, 171; paradigm of, 13, 36; physical labor, 218, 223; purity of outdoors, 39, 53, 55, 66; racial capitalism, 14, 43, 45, 74, 76, 81–82; "recognition justice" framework, 209; redressing, 80, 116, 120, 123, 137, 211; relating energy to, 273; reproducing, 24, 32, 138, 188, 193, 216–18; of settler colonialism, 259; shrine to, 265; slavery, 44; of solar production, 183; spatialized, 113, 264; structures of, 271; systems of capital, 26; of technological innovations, 222; transforming, 277; two axes of difference, 54; working-class individualism, 258

diffuse spatiality: energy efficiency and renewable energy (EERE), 86, 93–95, 98, 112, 116, 119–20; social equality, 100; solar energy, 4, 83–85, 107, 111–12, 142, 145; solar infrastructure, 80; sustainable energy, 81–82

dispossession, 36, 51, 215, 216, 259

domesticate sublime, 57

Dominic, 204–6

Don, 24–25, 27

double consciousness, 278–79

Du Bois, W. E. B., 109, 216, 278, 279

Easterly, William, 227

ecomodernism: assertion, 268; dream, 275; human/nature split, 217–18; intelligence, 219, 247; neoliberal technofetishism, 215; SunLight, 218–19, 220, 223, 227; visions, 214

ecosocialist visuals, 254

Elusive Quest for Growth, The, 227

energy efficiency and renewable energy (EERE): bureaucrats, 174; companies, 131; corporations, 83, 84–85; data, 129; decentralization, 96, 99, 106; diffuse spatiality, 86, 93–94, 95, 112, 116, 119–20; equicratic work, 113; expertise, 133, 233, 237–38; infrastructures, 202, 229, 232; markets, 97–98, 100–103; organizations, 201; programs, 117; projects, 135; services, 123; SunLight, 228; uptake, 121; work, 132, 215

energy infrastructure: advertisements, 283n7; consumers, 97–98; controlling, 139; decentralization, 101; development of, 145, 148; diffuse spatiality, 142; financing, 157; governance, 135, 144; images of, 273; materiality of, 31; modern, 220; modularity, 143; racial capitalism, 27; relations that constitute, 36; renewable, 182; transforming, 175; visibility, 5–6

environment: benefits, 46; built, 1, 5, 6, 20, 34, 53; conservation, 216–17; corporeal relations with, 220; desecration, 43; destruction, 44–45; differential value, 264; environmentalist ideals, 69; environmental shifts, 51; impacts, 92, 121, 184, 237, 255–57, 268; injustices, 12–13, 48, 66, 77, 79, 141; naturalist ethos, 49; philanthropies, 153–54; policy, 108; problems, 189–90, 258; protection, 31, 162; racism, 28; stewardship, 136–37, 167, 214; sunny, 61; technological, 261; transforming, 80; urban, 39, 42, 71

environmental injustices: affective texture of, 149; areas burdened by, 207–8; concrete, 77, 79; corporeal terrain, 210; encounters with, 209; environmental justice (EJ) campaigns, 272; equicratic order indifferent to, 277; everyday, 192; experiencing, 141; Instagram post, 13; labor exploitation, 12; necropolitics of, 48; New Yorkers, 66; persistent, 199; problems, 202

environmentalism, 19, 36, 54–55, 73, 181–82, 249

environmental justice (EJ): activism, 19, 35, 111, 141, 148, 192; campaigns, 272; communities, 170, 179; ecocentric ethos absent in, 36; "EJ value," 207, 208, 209–10; Environmental Quality for All (EQUAL), 2, 141, 148, 183–84; ethos, 190, 293n37; modularity, 35, 141; movements, 109, 145–47, 149, 152, 156–57, 168; organization, 17, 31–32, 100, 153, 188; organizing, 18, 140, 201; politics, 20, 107–8, 193; principles, 204; problems, 183; protests, 112; solar infrastructure, 140, 153; solar-powered, 41

uralistic, 49–50; philanthrocapitalism, 200; quantitative dimensions of, 232; racial capitalism, 4, 11–14, 74; recycling, 184; spatiality of, 109; sun and, 62; sustainable, 3; visualizing, 231; visual renderings of, 33

Instagram: images, 9–10, 14, 156; microgrids, 13, 23; page, 7–8; posts, 14, 19–21, 24, 26, 33

intersectionality: activists, 79; affect, 240; alignment, 242; character, 93, 239, 257; contexts, 258; critiques, 85, 111; difference, 28; divisions, 32; ethos, 91–92; framework, 245; hierarchies, 9; identity, 243; inequalities, 86; injustices, 195; intersectionally situated, 246, 256; intersectional profile, 94; labor politics, 36; oppression, 130; politics, 31, 215; positionality, 89, 105; power, 6, 8, 26, 117, 145, 149, 204; sensibilities, 93, 101; violence, 259; visions, 181

Jackson, 266–67
Janice, 161–62, 163–65, 166–68, 170–72, 175, 179
Joe, 248
Johanna, 70–71, 72–73
John, 83–85, 86–88, 124–26, 127–31
Juan, 252–53

Karim, 251–52
Kendall, 256
Kevin, 255
King, Martin Luther, 151

labor: affective, 109–11; blue-collar, 7, 8, 16, 128, 214, 216; capitalist, 229; capitalized, 270; classed, 262; control over, 44; corporeality of, 214; dispute, 113; divisions of, 103–4, 106; dynamics, 82; emotional, 228; energy transition, 215; exploitation, 12, 42, 134, 182–83; governance, 144; grievances, 256; hands-on, 3; haptic, 90–92, 249; human, 60, 61; infratructural, 11; intersectionality, 36; leisure and, 259–60; markets, 38, 89, 128, 265; movements, 145; organizing, 257; philanthrocapitalist, 198; politics, 8, 36, 103, 105, 254;

practices, 43, 134–36; racial capitalism, 42–46, 238, 247, 258; restrictions, 150; social justice, 118; solar energy, 45; sovereignty, 266; spatiality of, 93; spatialized stratifications, 243; standards, 169; transforming waste, 267; unions, 9, 100; valuing, 87, 90; virtual, 77; white-collar, 225, 230, 253

late liberal aesthetic, 38–39
late liberal fantasy, 40, 276
late liberal governance, 83, 168
late liberal hypocrisy, 131
late liberal ideal, 37
late liberal inclusivity, 275
late liberalism, 81, 159, 189, 202, 209, 239
late liberal order, 33
late liberal politics, 86
late liberal renderings, 40
late liberal screenwork: animating white-collar care, 18; apprehension, 216; Black body, 150; brands as inclusive, 26; collaborative economization, 171; commodities, 14; corrective to, 215; destabilizing normative distinctions, 32; differential value, 82; diffuse spatiality, 112; equicratic spectacle, 141; material properties and, 34; paradigmatic, 136; philanthrocapitalism, 156, 203; political consciousness, 16; possessed by, 21; quantifiability and, 176, 200; racial capitalism, 10, 188, 210, 277; sun and, 41; transfomation of energy infrastructure, 175; transforming energy governance, 27; transition and return, 272; visual economy, 129
Lohmann, Larry, 216

Manny, 261–65, 267–68, 269, 279
Marcuse, Herbert, 227, 229
Marcus Garvey Apartments (MGA): images of, 3; microgrids, 1–2, 5–6, 15–21, 22–28, 29; panels, 2; solar panels, 2
Mark, 15–16
Martha: activism, 229; intensity, 115–16; labor, 228; problem solving, 120–22; SunLight, 84–85, 113, 118, 130
Martin, 130
Marx, Karl, 34, 41, 60, 76, 79, 224–26
Matt, 231–33, 235, 236, 238, 260–62, 265

iency, 255; selling, 224; social context, 248, 251; solar projects, 104; solar technology, 9, 211–12, 237; wildlife restoration, 181

Pineda, Ceci, 269

Povinelli, Elizabeth, 14

quantifiability: activism, 150–51, 184, 201; electrical, 10, 32, 34–35, 160, 163–66, 175; environmental justice (EJ), 153; of renewable energy, 203; of solar electricity, 140–41; solar energy, 176, 188, 199–200, 208, 210

quantification, 31, 232

quantitative visuals, 185

racial capitalism: alienation of, 40; anthropocentric hierarchies, 271; Black people, 136; built on returns of, 233; colonial forms of difference, 22; commodities, 42, 63, 69, 73, 76; counteracting, 84; critiques of, 102, 106, 137, 161; digital imagery and, 278; differential value, 14, 43, 45, 74, 76, 81–82; discontent with, 94; dispossession, 215–16; divisions of, 123; ecological imprint, 77; economic exploitation, 211; energy infrastructure, 27; environmental burdens, 209; equicratic ideology, 97; extractive relations, 74, 131, 210, 215, 268; fallout from, 188; formation of, 95–96; forms of, 150; hegemony of, 112; human relations, 50; inequalities, 82; infrastructure, 4, 11–14, 74; intersectionality, 32; labor, 42–46, 238, 247, 258; life beyond, 21, 34; logics of, 19, 255; market analysis, 167; media and, 275; multispecies relations, 69; neoliberalism, 80, 83, 111, 135, 203; New Yorkers, 9, 14; ontological divisions, 39; opposing, 81, 85, 108, 158, 160; oppressive conditions, 207; policy interventions, 37; powering, 27, 36, 216; putting nature outside of, 55; ravages of, 23, 47, 119, 133; settler colonialism, 41; shadow, 265–66, 272, 279, 282; solar energy, 8–13, 47, 65, 69; solar inextricably enmeshed, 182–83; spreadsheet reinforces, 168; structural inequalities, 35;

structures of, 33, 80, 88, 128, 208, 272, 277; struggles against, 229; sun and, 61, 62, 68; values, 246, 270; violence, 52, 194

racism, 54

Randall, 50–51, 53–55, 72

Raymond, 252

reduction: burdens on grid, 173; consumption, 83, 97, 98, 100, 116; emissions, 29, 49, 76, 98, 153–54, 181; energy bills, 100, 102, 118, 133, 191; humans' physical imprint, 217; means of, 184–86, 190, 192–93, 210, 237, 268; price, 167, 189

relational spatiality, 92

Renee, 47–48

Reynolds, Diamond, 129

Rick, 253

Roane, J. T., 270

Robinson, Cedric, 10, 151

Ron, 239–40, 241–43, 244–47

Roxanne, 25–27, 28, 29

Safina, Carl, 227

Samir, 143–48, 154–56, 157, 168–69, 191–92

Scheer, Herman, 227, 229

Sean: beliefs, 63; on Black employee, 239; diffuse spatiality, 112; on intersectional inequality, 86–89, 91–93; labor, 90; on nature, 55–57, 58–59, 60; politics, 94–97, 99, 101–3, 105–7, 111; on SunLight, 116, 219–23, 260

Selena: anticapitalist consciousness, 200; brainstorming, 196; data, 29, 176, 177–78, 204; "energy democracy," 30; energy management screenwork, 33; environmental justice (EJ) legislation, 21–22; infrastructure accessibility, 203; Instagram post, 19–20; late liberalism, 202; philanthrocapitalism, 201; presentation, 197–99; screenwork for justice, 206; tour, 21–23

settler colonialism, 41, 44, 51–52, 259, 270

Shane, 249

Shange, Savannah, 128

slavery, 12, 43–45, 263

solar commodities, 99

solar data, 29–32

Solar Economy, The, 227

www.ingramcontent.com/pod-product-compliance
Lightning Source LLC
Chambersburg PA
CBHW071731270326
41928CB00013B/2629